CISM International Centre for Mechanical Sciences

Courses and Lectures

Volume 589

For more than 40 years the book series edited by CISM, "International Centre for Mechanical Sciences: Courses and Lectures", has presented groundbreaking developments in mechanics and computational engineering methods. It covers such fields as solid and fluid mechanics, mechanics of materials, micro- and nanomechanics, biomechanics, and mechatronics. The papers are written by international authorities in the field. The books are at graduate level but may include some introductory material.

More information about this series at http://www.springer.com/series/76

Andreas Müller · Dimiter Zlatanov
Editors

Singular Configurations of Mechanisms and Manipulators

Springer

Editors
Andreas Müller
Institute of Robotics
Johannes Kepler University Linz (JKU)
Linz, Austria

Dimiter Zlatanov
Department of Mechanical Engineering
University of Genoa
Genoa, Italy

ISSN 0254-1971 ISSN 2309-3706 (electronic)
CISM International Centre for Mechanical Sciences
ISBN 978-3-030-05218-8 ISBN 978-3-030-05219-5 (eBook)
https://doi.org/10.1007/978-3-030-05219-5

Library of Congress Control Number: 2018963281

This Springer imprint is published by the registered company Springer Nature Switzerland AG
The registered company address is: Gewerbestrasse 11, 6330 Cham, Switzerland

Preface

Singularity analysis is a central topic of mechanism and robot kinematics. It provides an insight of major practical and theoretical importance for the design, control, and application of robot manipulators. In special configurations, known as singularities, the kinetostatic properties of a mechanism undergo sudden and dramatic change. Hence there is an enormous practical value of a careful study and thorough understanding of the phenomenon for the design and use of manipulators. The key role played by kinematic singularities in mechanism theory is analogous to, and in fact a consequence of, the critical importance of singularities in algebraic geometry and in the theory of differentiable mappings.

This book provides an introduction to the broad topic of singularities and a summary of key methods, milestone results, and the prevailing main problems in singularity analysis. It is a compilation of contributions presented at the summer schools on Singularities of Mechanisms and Robotic Manipulators held at the Centre International des Sciences Mécanique (CISM) in Udine, Italy, on September 22–26, 2014, and at the Johannes Kepler University Linz, Austria, on September 18–22, 2017. The five chapters summarize the key result obtained by the contributing authors over the last decades. The book thus serves as a reference and inspiration for anyone pursuing research in this field.

The opening chapter, by Dimiter Zlatanov, sets up the robot singularity problem in a general form using simple conceptual definitions based on linear algebra. There are two general kinds of singular phenomena in robotic kinematics. The first are about the failure of constraint and are related to the geometry of the configuration space and the structure of the constraint equations of the kinematic chain. The second involve failure of control in the use of the mechanism as an input–output device and are dependent on the choice of input and output motions. Closer examination of the interplay of control and constraint degeneracies yields finer classifications of the possible singularities. The chapter emphasizes the dangers of nave generalizations of approaches to singularity that work fine for special simple manipulators. These problems are illustrated by multiple examples where unexpected phenomena occur in seemingly simple and predictable situations.

Chapter "Singularity Set Computation: A Hands-On Session with the CUIK Suite," by Oriol Bohigas, retains the general outlook of Chapter "Generalized Singularity Analysis," but is concerned with the practical and exhaustive identification of all singularity sets of a given mechanism. A general method for singularity identification based on advanced numerical techniques is proposed. The approach, which can find all singularities of arbitrary mechanisms with common algebraic joints (such as sliders, hinges, and spherical pairs), is illustrated with simple step-by-step examples. The algorithms were implemented in the open-source software CUIK, briefly introduced in the chapter.

Chapter "Cuspidal Robots" by Philippe Wenger focuses on singularities of the input–output control of serial robots used for positioning tasks (also called regional manipulators). The aim to avoid any singularity whatsoever during operation of a robotic manipulator has been a main motivation for research. A milestone result is that there is a class of serial manipulators that can change from one inverse kinematics solution to another without having to pass through singularities. Since this is related to the existence of cusp singularities, such manipulators are called cuspidal. The main features of cuspidal manipulators are described. The chapter discusses how to identify and classify them, with an extended focus on orthogonal 3R serial manipulators. Also briefly discussed are cuspidal 6R manipulators and the recent research on parallel manipulators. Singularities separate the workspace of a serial robot (according to the number of inverse kinematic solutions) and the configuration space of a mechanism. It is thus highly instructive to visualize this which is also discussed.

The mathematical methods that can be used to analyze a mechanism are inextricably related to the mathematical formulation of the kinematics. Algebraic formulations, in particular, allow for application of methods from computational algebraic geometry. This is discussed in Chapter "Mechanism Constraints and Singularities—The Algebraic Formulation," by Manfred L. Husty and Dominic R. Walter. Study parameters (or dual quaternions) are used for modeling the kinematics. It is shown how algebraic descriptions can be used to obtain a global description of mobility and singularities. Three approaches to deriving constraint equations are presented. The first involves using geometry-specific information, the second is the elimination method, and the third is the linear implicitization algorithm. The latter is described in detail since it can be applied even without deep kinematic and geometric insights into the properties of the kinematic chain. As application of these methods, a complete analysis of the 3-UPU parallel manipulator proposed by Tsai is presented.

From a local point of view, an in-depth understanding of the many phenomena accompanying a kinematic singularity, and even identifying critical points as actually being singular, necessitates a higher-order local analysis. The latter reveals possible finite singular and regular finite motions. This is addressed for linkages in Chapter "Local Investigation of Mobility and Singularities of Linkages" by Andreas Müller. As prerequisite, the phenomenology of kinematic singularities is discussed by means of several examples. The relevant terminology is summarized and a comprehensive stringent definition of singularities is given, which clarifies several misconceptions found in the literature. Aiming at an exhaustive analysis, the concept of a kinematic tangent cone is applied to possible smooth finite motions as well as smooth motions with certain rank. This also allows for identification of overconstrained and shaky linkages. For mobility determination, a higher-order approximation of the configuration space is used. All these investigations are supported by closed-form and recursive relations in terms of screw coordinates, which allow for easy implementation in symbolic or numerical software. Various examples are used to explain the described kinematic phenomena and to show the computational methods.

Linz, Austria/Shanghai, China — Andreas Müller
Genoa, Italy — Dimiter Zlatanov

Contents

Generalized Singularity Analysis

Dimiter Zlatanov

Abstract In this chapter, we consider kinematic singularity in a mechanism as a deviance or abnormality in its instantaneous kinematics. A configuration is said to be regular if in it there is a proper relationship between the input, output, and passive velocities of the mechanism. Namely, both the forward and the inverse instantaneous kinematics problems must be well-posed, allowing full control of the system. Otherwise, we have a degenerate or "singular" case. The obvious division of velocity parameters into input, output, and other (passive) speeds gives rise to a natural classification of the possible degeneracies. The taxonomy is described by six overlapping singularity types partitioning the singularity set into as many as 15 classes. The various possibilities are examined and interpreted with the help of simple examples. Special attention is given to a few cases which best illustrate the inadequacy of using an exclusively input–output relationship as a formulation of the kinematic constraints. The fact that standard serial manipulators have no passive (unactuated) joints, and therefore have a well-defined output map, suggests a false generalization, which has led analysts and designers of more complex robots into error in the past and may do so again unless the lack of generality is well understood.

1 Introduction

In current popular science, the words "robots" and "singularity" are linked in a fantastical scenario of futuristic doom and gloom leading to potential human extinction, or so many prominent figures would have us believe. It is doubtful that among the professionals who build and study mechanical robots (and who know how difficult it is to make them do the simplest things), the prospect of the machines taking over is taken seriously at all (visionary hype for grant-application purposes excluded). Yet, "singularity" in real robotics can be just as chilling a concept as its scifi counterpart. The mechanical phenomenon, to which the more mundane term refers, may not be as

D. Zlatanov (✉)
Università di Genova, Genoa, Italy
e-mail: zlatanov@dimec.unige.it

A. Müller and D. Zlatanov (eds.), *Singular Configurations of Mechanisms and Manipulators*, CISM International Centre for Mechanical Sciences 589,
https://doi.org/10.1007/978-3-030-05219-5_1

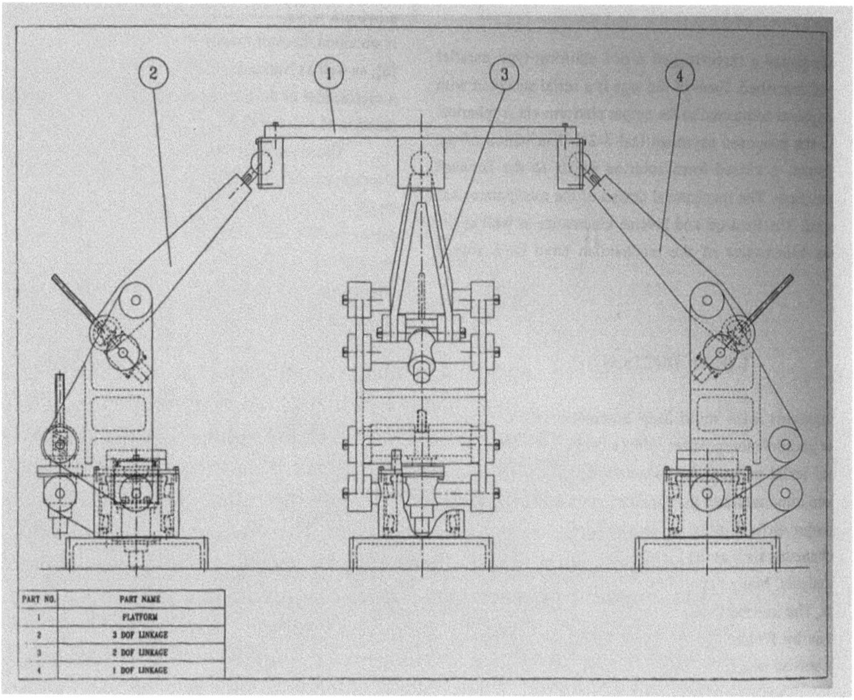

Fig. 1 A tripod six-dof parallel manipulator with a 3-2-1 distribution of the motors among the legs

apocalyptic but neither is it a speculative fantasy—it is completely real, ubiquitous, and troublesome.

Roboticists do not always agree on what they call (kinematic) singularity, but they are united in their desire to avoid it. Typically, it is a configuration where the mechanism cannot do what it is supposed to or at least we cannot "make it behave" by the usual means. This leads to somewhat parochial definitions of singularity skewed toward what occurs most often in the relatively simple popular robot architectures. As a consequence, although everyone agrees that robot singularity is a serious matter, many are under the misimpression that it is a simple one. Probably, the easiest way to disabuse ourselves from such views is to consider a nonstandard example.

Suppose we are given the task to analyze the six-degree-of-freedom (dof) parallel manipulator shown in Figs. 1 and 2 and its related architectures (Zlatanov et al. 1992). It is a platform with three legs, each a common PUMA-like kinematic chain where the wrist is replaced by a spherical joint. Unusually, the six rotary actuators are distributed unevenly among the legs: In leg A, all three R joints are active; in leg B, there are two actuators with the base revolute passive; and in leg C, only the third (elbow) joint is active.

This particular asymmetry leads to a simple closed-form formulation of both the forward and inverse kinematics avoiding the unpleasantness of either serial or parallel

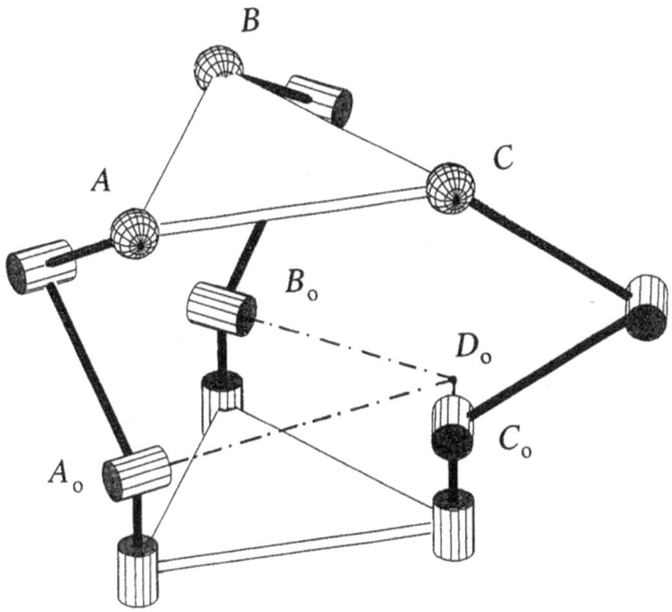

Fig. 2 A six-dof parallel manipulator in a singular configuration

robot position equations. The singularity analysis, however, quickly reveals that this type of mechanism can exhibit rather strange and complex kinematic behavior. Indeed, we discover phenomena not reported in virtually any of the innumerable papers where existing or proposed robots are analyzed. Consider the configuration in Fig. 3. A mechanism in this state will have the following properties:

(i) The kinematic chain has seven rather than six (instantaneous) dof.
(ii) The platform has four rather than six dof.
(iii) One actuated joint can move even if the platform is kept fixed.
(iv) Of the six actuated joints, only three are independent. (Choosing three input speeds determines the values of the others.)
(v) Even if the platform *and* all actuators are motionless, the kinematic chain retains two (purely passive) dof.

Clearly, each of these phenomena is not to be expected in the normal functioning of a mechanism, and its presence must be regarded as a form of irregularity. Yet, some of these states are virtually never mentioned in the robotics literature, e.g., (iv–v), while others, e.g., (i) and (ii), appear contradictory. Any thorough study of robot singularities must address and explain states like these, but they are ignored by standard treatments.

A typical defensive reaction when faced with such an example is to dismiss it as rare and exotic. This overlooks three, in my view very important, points: First, to disregard a singularity because it is exotic is a contradiction in terms—these phenomena are rare and unusual by definition. Second, irregularities of this kind occur in the simplest mechanisms such as four-bar linkages and slider cranks. Surely, any theory of a mechanism property which fails for four-bar leaves something to be desired. Third, robotics involves the creative design of novel complex devices. As products of human invention, these systems are "nongeneric" by their very nature. (A "generic" mechanism would be one produced by a random process.) Thus, new design solutions regularly exhibit one or another "exotic" phenomenon, which is characteristic of only a measure zero subset of some more general "design space."

Lacking adequate theoretical understanding of such "degenerate cases" designers will occasionally find themselves dumbfounded by the behavior of their brainchildren. As is appropriate for a course book on deviance and abnormality (all be it in robotic mechanisms), we tell some of these cautionary tales for the young (and not so young) who conceive and cradle the future of the field. Hopefully, familiarity with the ideas we present will help avoid some unpleasant future surprises and will allow the right questions to be asked when new mysteries arise.

2 The Standard View of Singularity

2.1 The Simplest Case: Serial Chain Singularity

It is safe to say that when a student or practitioner of robotics first encounters kinematic singularity, it is in a scenario as shown in Figs. 3 and 4. A planar serial 2R chain, with (actuated) joint angles (θ_1, θ_2), is used to control the position of a point P in the plane. In every configuration of the mechanism, the velocity kinematics can be given by the vector equation

$$\vec{v} = \dot{\theta}_1 \vec{u}_1 + \dot{\theta}_2 \vec{u}_2 \tag{1}$$

where $\vec{v}$, with coordinates $(\dot{x}, \dot{y})$ in some frame, is the velocity of P, while $\vec{u}_i$ is the (time-normalized) velocity that P has when *only* joint i is rotating with speed $\dot{\theta}_i = 1$, i.e., $\vec{u}_1 = \vec{k} \times \overrightarrow{OP}$, $\vec{u}_2 = \vec{k} \times \overrightarrow{AP}$. (Whenever needed, two vectors in the Oxy plane are treated as three-dimensional with z coordinate 0. The unit coordinate vectors are $\vec{i}$, $\vec{j}$, and $\vec{k}$.) This velocity kinematics equation is usually derived by differentiating some trigonometric expressions, but it is more enlightening to think of it as a direct consequence of the general principle that relative velocities (here, those due to motions in the joints) are additive.

In matrix form, (1) is written

$$\vec{v} = \boldsymbol{J}\dot{\boldsymbol{\theta}} \tag{2}$$

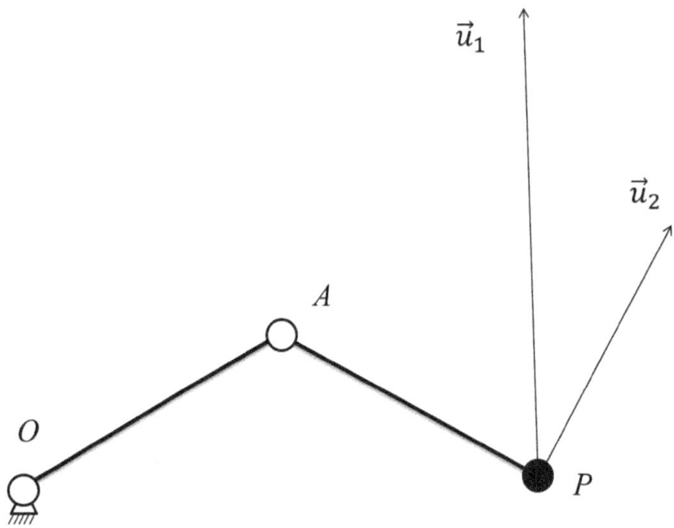

Fig. 3 A two-dof parallel manipulator. The possible velocities of P are linear combinations of the velocities due to the joints

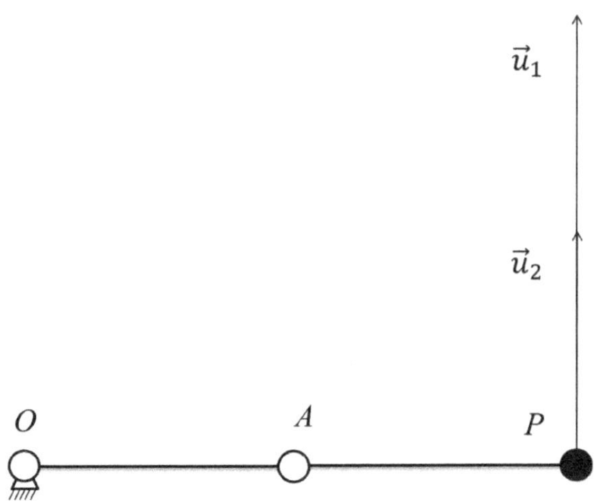

Fig. 4 A two-dof serial manipulator in a singular configuration

where $\dot{\boldsymbol{\theta}} = (\dot{\theta}_1, \dot{\theta}_2)^T$ and the 2×2 matrix $\boldsymbol{J}$ with columns $\vec{u}_1$ and $\vec{u}_2$ is called the manipulator Jacobian.

In a regular configuration, the vectors $\vec{u}_1$ and $\vec{u}_2$ are independent, the matrix $\boldsymbol{J}$ has full rank, and (2) defines a bijective linear map (an isomorphism) between the two spaces of the input and output instantaneous motion vectors, $\dot{\boldsymbol{\theta}}$ and $\vec{v}$.

In a singular configuration, $\vec{u}_1$ and $\vec{u}_2$ are linearly dependent, and the rank of $\boldsymbol{J}$ drops from two to one (not to zero because $\vec{u}_2$ can never vanish). Thus, the mapping from the input to the output velocities is neither injective nor surjective: It has a

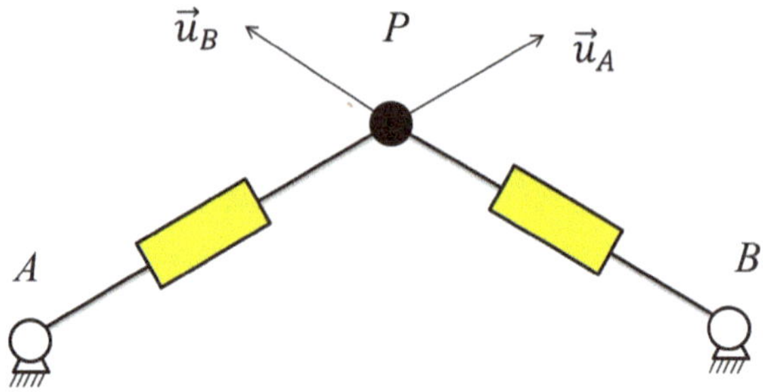

Fig. 5 A two-dof parallel manipulator. The speeds in the actuated prismatic joints are the projections of the velocity of the output point, P, on the slider directions, $\vec{u}_A$ and $\vec{u}_B$

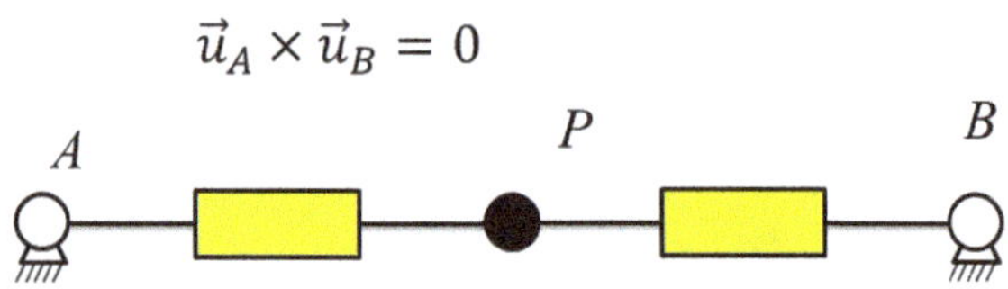

Fig. 6 A two-dof parallel manipulator in a singular configuration

nonzero (here, one-dimensional) kernel, ker $\boldsymbol{J} \neq 0$, and its image, im $\boldsymbol{J}$, is a proper linear subset of the codomain (the output space $\mathcal{O}$, here composed of all velocity vectors in the plane). In kinematic terms, this means that there is a one-space of input instantaneous motions for each of which point P has zero velocity and that the velocity vectors that P can actually have are restricted to a line rather than a plane. We refer to these two phenomena, ker $\boldsymbol{J} \neq \mathbf{0}$ and im $\boldsymbol{J} \subsetneq \mathcal{O}$, as redundant input (RI) and impossible output (IO), respectively. (For convenience, $\mathbf{0}$ is used to denote either the zero vector or a zero subspace, $\{\mathbf{0}\}$, in a vector space of any dimension.) .

Clearly, $\vec{u}_1$ and $\vec{u}_2$ are dependent, and the serial chain (configuration) is singular, if and only if the pin joint centers are aligned with point P.

2.2 *The Serial–Parallel Duality*

The next type of singularity situation one typically considers is of a parallel manipulator. Possibly, the simplest example is in Figs. 5 and 6. A (type of five-bar) single-loop RPRPR linkage has three revolute joints and two nonadjacent sliders. The prismatic joints are actuated and used to control the center of one pin joint, P, with respect to the (base) link between the other two, fixed in points A and B.

As in the serial chain case, one finds a relationship between the input and output velocity parameters in a given configuration. It is always true that the speed of the motion in the slider adjacent to A, $\dot{d}_A$, is given by the orthogonal projection of the velocity of P, $\vec{v}$, on the direction of the slider, $\vec{u}_A$, $|\vec{u}_A| = 1$, and similarly with B:

$$\dot{d}_A = \vec{u}_A \cdot \vec{v} \tag{3}$$

$$\dot{d}_B = \vec{u}_B \cdot \vec{v} \tag{4}$$

In matrix form,

$$\dot{\boldsymbol{d}} = \boldsymbol{K}\vec{v} \tag{5}$$

where $\dot{\boldsymbol{d}} = (\dot{d}_A, \dot{d}_B)^T$ and the matrix $\boldsymbol{K}$, the (inverse) parallel manipulator Jacobian, has as its rows the vectors $\vec{u}_A$ and $\vec{u}_B$.

Everything is as in the serial chain case, but input and output have switched places. In a regular configuration, $\boldsymbol{K}$ defines a bijection between the velocities in the plane and the space of input speeds, $\mathcal{I}$. In a singularity, $\vec{u}_A$ and $\vec{u}_B$ are linearly dependent, and the rank of $\boldsymbol{K}$ drops to one (never zero as these are unit vectors). We have $\ker \boldsymbol{K} \neq 0$ meaning that point P can move instantaneously with the actuators fixed, and $\operatorname{im} \boldsymbol{K} \subsetneq \mathcal{I}$, indicating that the input velocities cannot be chosen independently but only from a (one-dimensional) subset. We will call these phenomena redundant output (RO) and impossible output (II), respectively.

Clearly, $\boldsymbol{K}$ is singular only when the base joints are aligned with P.

2.3 A Generalization: The Input–Output Equation

The duality between the two simple cases considered above seems complete and obvious. It is a point of view with some utility which still dominates the analysis of the common fully parallel manipulators, such as the Stewart-Gough platform (Merlet 1989). However, as we shall see, the picture it paints is incomplete, when all singularities need to be considered. An obvious reason is that a parallel kinematic chain is not some inversion of a serial one but a merging of several such chains (the so-called legs or limbs) with coinciding terminal links. If any of these leg serial chains is singular, the end effector (likely) loses mobility, there is no longer a output-to-input mapping, and Eq. (5) is no longer valid.

Gosselin and Angeles (1990) introduced an approach which seemed to account for such "serial-like" singularities of parallel manipulators. Moreover, the authors believed that the scheme was applicable to arbitrary kinematic chains with the same number, n, of dof, inputs, and outputs.

Rather than assume there is a forward or inverse input–output mapping, the starting point is an equation relating the input and output velocities. Typically, one starts with an expression relating the input, $\boldsymbol{q}$, and output, $\boldsymbol{p}$, position parameters, $\boldsymbol{f}(\boldsymbol{q}, \boldsymbol{p}) = \boldsymbol{0}$. This is differentiated to obtain the velocity input–output equation

$$\boldsymbol{L}_{\mathcal{I}}\boldsymbol{v}_{\mathcal{I}} + \boldsymbol{L}_{\mathcal{O}}\boldsymbol{v}_{\mathcal{O}} = \boldsymbol{0} \tag{6}$$

where $\boldsymbol{v}_{\mathcal{I}}$ and $\boldsymbol{v}_{\mathcal{O}}$ are the n-dimensional input and output velocity vectors, while $\boldsymbol{L}_{\mathcal{I}}$ and $\boldsymbol{L}_{\mathcal{O}}$ are $n \times n$ matrices dependent on the configuration (but not on the velocities).

(It is usually assumed that $\boldsymbol{v}_{\mathcal{I}} = \dot{\boldsymbol{q}}$ and $\boldsymbol{v}_{\mathcal{O}} = \dot{\boldsymbol{p}}$, but this is not necessarily so: The parameterizations used for displacements and velocities may be quite different; e.g., the components of the angular velocity of a body are not the derivatives of the Euler angles. Even the number of variables can be different: Redundant parameterizations are sometimes used for nonlinear spaces, like the rotations.)

Equation (6) yields (2) and (5) as special cases (when $\boldsymbol{L}_{\mathcal{O}}$ or $\boldsymbol{L}_{\mathcal{I}}$ is minus the identity). So, the natural conclusion must be that that, for a general chain, the velocity kinematics can degenerate in two ways: Either $\boldsymbol{L}_{\mathcal{I}}$ or $\boldsymbol{L}_{\mathcal{O}}$ is rank deficient and then, the device has a singularity which is like that of a serial or a parallel manipulator.

(There is also the logical possibility that both matrices in (6) are singular. In Gosselin and Angeles (1990), it is claimed that this is a coincidence which never happens in generic cases, but only for special geometries. Obviously, this is a simple error, describing correctly only one-dof mechanisms. For higher values of n, generically there will be a nonempty $(n-2)$-dimensional set of singularities belonging to both types. See, for example, (Bonev et al. 2003).)

The above formulation of the singularity problem remains one of the most influential in the literature. It is important to understand its critical limitations.

2.4 The Counterexamples

Natural and elegant as serial–parallel dualities may be, any general singularity framework based on them quickly collapses under closer examination. As it turns out, very simple kinematic chains readily provide counterexamples.

Probably, one of the easiest to understand is presented in Fig. 7. Consider the one-dof linkage constraining the motions of three sliders using three hinges and two bars. The linear speeds in the sliders (and of the hinge centers) are v_A, v_B, and v_C. Points A and B trace the same line, which is perpendicular to the linear path of C. Suppose that the input and output motions are in sliders A and B, respectively. It is easy to prove that in *any* configuration the following scalar velocity equation is true.

$$y_A v_A = y_B v_B \tag{7}$$

where y_A and y_B are the oriented distances from the line traced by C to A and B, respectively. (To prove it without differentiation, add the planar rigid constraints $\vec{v}_A \cdot \overrightarrow{AC} = \vec{v}_C \cdot \overrightarrow{AC}$ and $\vec{v}_B \cdot \overrightarrow{CB} = \vec{v}_C \cdot \overrightarrow{CB}$ to eliminate v_C.)

When A, B, and C are vertically aligned, Eq. (7) is also correct and neither y_A nor y_B is 0. If we base our singularity analysis on an input–output equation, we have to conclude that the configuration is nonsingular. It is true that (7) provides a linear isomorphism between the input and output velocities. However, a simple inspection of the linkage reveals that the configuration, shown on the right-hand side of Fig. 7, is a caricature rather than an example of kinematic regularity. The input and output velocities are "co-determined" only in the sense that they both must necessarily be equal to zero, which is hardly regular in a one-dof input–output device. Moreover, the linear motion of C here is completely independent from either input or output and can be arbitrary. In other words, both end effector and actuator are blocked, while

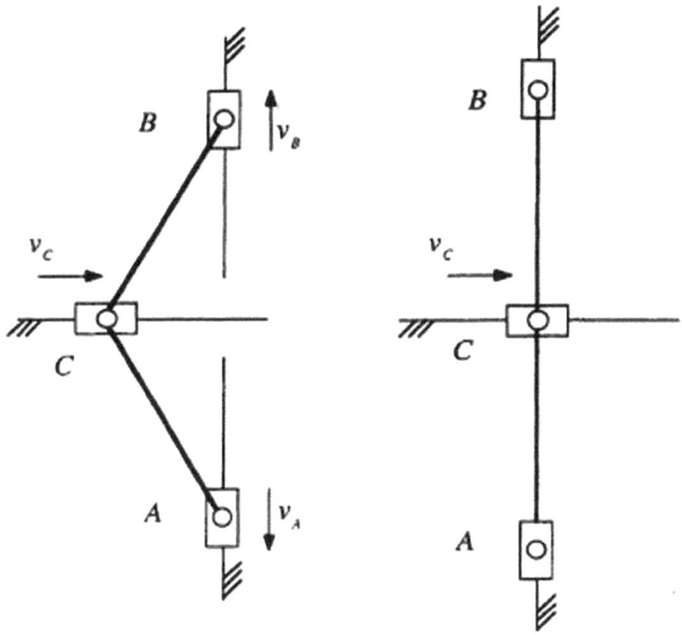

Fig. 7 A one-dof mechanism in a general (left) and a special configuration (right). The input and output are the motions of points A and B

the mechanism has (passive) motions that cannot be controlled. This last singular phenomenon is called redundant passive mobility (RPM).

This example, analyzed in detail in Chap. 2, makes it clear that hopes to use something like Eq. (6) for comprehensive singularity analysis of arbitrary chains must be abandoned. Could it be that this formulation remains valid at least for fully parallel manipulators?

The answer is negative, as can be discovered by taking a closer look at the elementary parallel manipulator in Fig. 5, already (somewhat deceptively) discussed above. Consider its configurations in Fig. 8, where the point P coincides with joint center A. This is easy to realize physically as is maybe clearer in Fig. 9, where an equivalent mechanism is presented in a regular configuration and in another with $P = A$.

With two nonadjacent revolute axes coinciding, the links inbetween can rotate together, like a rigid body, about the common pivot, yielding a single infinity of configurations with $P = A$, each of which exhibits RPM, Fig. 8.

Apart from passive self-motion, all these configurations also manifest IO, as the velocity of P must be equal to that in slider A. Furthermore, as the motion of slider A determines the velocity of P, this in turn univocally decides the speed $\dot{d}_B$ (because leg B cannot be singular at the same time as A, assuming $A \neq B$). Therefore, the inputs, $\dot{d}_A$ and $\dot{d}_B$, are not independent; i.e., there is impossible input (II).

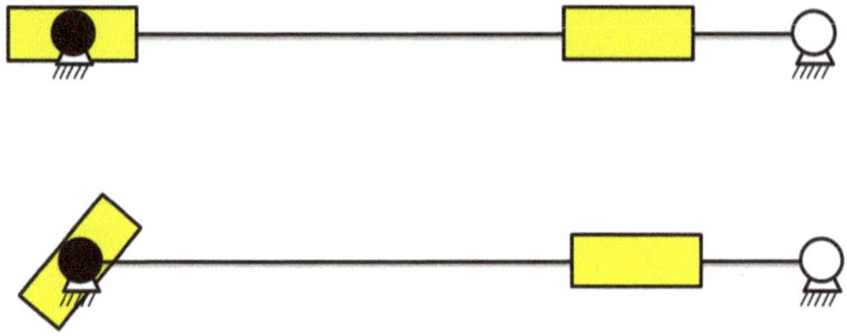

Fig. 8 A two-dof parallel mechanism in a configuration with redundant passive mobility

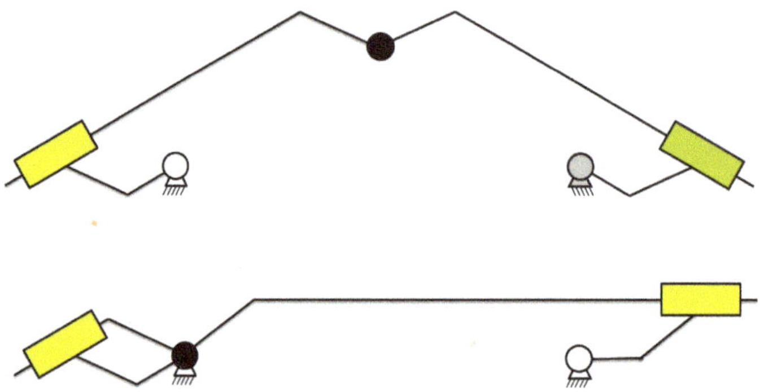

Fig. 9 A different physical realization of the two-dof parallel mechanism allowing a better view of the redundant passive mobility

Thus, there can be no doubt that the configurations with $P = A$ are all singular. Yet, the Eq. (5), which remains true in all of them, degenerates only if the directions of the sliders coincide (as in the top part of Fig. 8). Even in this case, (5) does not correctly describe the velocity kinematics. Indeed, although it does yield that $\dot{d}_A = -\dot{d}_B$ (describing the II phenomenon), (5) is satisfied by *any* vector $\vec{\boldsymbol{v}}$, not only by ones parallel to the sliders. Thus, the occurrence of IO remains hidden in (5).

These examples appear puzzling because the reasoning in Sect. 2.2, as well as in Gosselin and Angeles (1990), is simple and convincing. Where is the error?

The fundamental problem is the assumption that the constraints can be fully formulated using only input and output parameters. Closed-loop chains have many unactuated joints and the number of joint freedoms exceeds the dof of the mechanism. If one writes the constraints imposed on the motion (finite or instantaneous) of the joints by the presence of loops, one has to use all joint parameters. In general, there is no reason to expect that equivalent constraint equations can be written in terms of only the input and output coordinates.

The fact that an input–output equation like (6) is properly derived and proven to be correct for any configuration does not imply that it is a valid description of the velocity kinematics. It only means that it is a *necessary* condition for the velocities to be feasible in the given configuration. To use it for analysis, we need to verify that there are no other constraints, i.e., that it is also a *sufficient* condition for the velocities to be possible. Paraphrasing the famous Gershwin song, the problem with the input–output formulation is not that "it ain't necessarily so" but that it "ain't sufficient to be so." In other words, Eq. (6) may be telling only the truth, but not all the truth.

In the following sections, we will provide more general definitions and formulations of the singularity problem.

3 Singularity via Differentiable Maps

3.1 Serial Chains

In serial robots, kinematics is described by the so-called kinematic map, $\boldsymbol{f}_{\mathcal{O}} : \mathcal{Q} \rightarrow \mathcal{O}$, the input–output transformation yielding the pose of the end effector as a function of the joint parameters.

The domain of $\boldsymbol{f}_{\mathcal{O}}$, $\mathcal{Q}$, is joint space, a smooth manifold, defined as the Cartesian product of the configuration space manifolds of the individual joints. Thus, the joint space of a hinge is a circle, and for a kinematic chain composed of N revolute joints, $\mathcal{Q}$ is the N-torus, T^N. More generally, we can assume that the individual joint space can be any manifold.

The codomain of the kinematic map is the output space, $\mathcal{O}$, a manifold containing the workspace, the set of poses of the end effector, as a codimension zero subset. Often, $\mathcal{O}$ is the Euclidian group itself, but it may also be either a lower-dimensional subset or a factor space of $SE(3)$. Examples of the former is when the end effector is restricted to planar motions or has a fixed point; the latter case is when we are only interested in the position of a point or a line in the end effector.

We say that $\boldsymbol{q} \in \mathcal{Q}$ is a nonsingular (regular) or singular configuration when it is a regular or a critical point of $\boldsymbol{f}_{\mathcal{O}}$ in $\mathcal{Q}$, respectively, i.e., when the tangent (Jacobian) map has full rank. For a nonredundant serial robot, the joint and workspace have the same dimension and the configuration is nonsingular if and only $\boldsymbol{f}_{\mathcal{O}}$ is locally a diffeomorphism.

3.2 Chains with Closed Loops

The generalization to kinematic chains with an arbitrary tree topology is easy; however, for chains with closed loops, the kinematics is significantly more complex.

Constraint Equations First, not all elements of $\mathcal{Q}$ are feasible configurations. The presence of loops means that the displacements of different joints are not independent. The configurations must satisfy the constraint equations. This can be expressed by a constraint map $\boldsymbol{F} : \mathcal{Q} \rightarrow \mathcal{E}$ defining the *configuration space*, the set of *feasible* configurations as the inverse image, $\mathcal{D} = \boldsymbol{F}^{-1}(\boldsymbol{e}) = \{\boldsymbol{q} \in \mathcal{Q} \mid \boldsymbol{F}(\boldsymbol{q}) = \boldsymbol{e}\}$ of a fixed point $\boldsymbol{e} \in \mathcal{E}$. For example, $\boldsymbol{F}$ can be obtained by writing the product of exponentials formula for every independent loop in the kinematic chain. Then, $\mathcal{E} = SE(3)^l$, where l is the number of independent loops and $\boldsymbol{e} = (id_{SE(3)}, \ldots, id_{SE(3)})$. More details on generating constraint equations in this way are given in Chap. 5.[1]

In practice, in order to solve equations, the spaces $\mathcal{Q}$ and $\mathcal{C}$ are parameterized and the constraint equations, $\boldsymbol{F}(\boldsymbol{q}) = \boldsymbol{e}$, are written in the chosen coordinates in a (globally or at least locally) equivalent form $\boldsymbol{f}(\boldsymbol{x}) = \boldsymbol{0}$, where $\boldsymbol{f}$ is an analytic mapping between real Cartesian coordinate spaces. For many practical mechanisms, the constraints are formulated as algebraic equations, allowing the use of powerful methods for their solution, as is done in Chaps. 2 and 4 of this book.

In mechanism kinematics, it is natural and common to define the configuration space as a subset of joint space. However, there are infinitely many other equivalent possibilities for the ambient space on which constraints are imposed, and when a specific kinematic chain is analyzed, one chooses whatever is most convenient. For example, the equations may be defined on the space of the absolute displacements of all or some of the rigid links, $SE(3)^k$, for a chain with $M \geqslant k$ moving bodies. Alternatively, it is often simpler (especially for planar linkages) to use as free variables the positions of different material or fictitious points. Examples of different formulations can be found in Chaps. 2 and 4. A simple one is given by the three sliders in Fig. 7: The state of the kinematic chain is completely determined by the position of A, B, and C along the axes on which they are constrained to move, and so $\mathcal{Q}$ can be chosen as the Cartesian three-space with these three coordinates, (y_A, y_B, x_C).

Input and Output Maps: The second difference between chains with and without closed loops is in the nature of the input joint motion. If a chain with tree topology is to be controlled using joint actuators, then all joints must be actuated, because the motion in each is independent from the others. This is not so in the presence of closed loops. In a nonredundant mechanism, only part of the joints is actuated or active, and the others are referred to as passive.

A mechanism is a kinematic chain used to transmit motion. This is understood in the sense that a mechanism is intended to be an input–output device. Mathematically, this is modeled by introducing an input and an output space.

The definition of the output space, $\mathcal{O}$, in the general case is not different in principle from the serial chain kinematics discussed above. When the mechanism is referred to as a manipulator, an output space is typically a subset or a factor space of $SE(3)$. However, applications may require different output motions. In all cases, there is an output manifold $\mathcal{O}$ and a smooth (in fact analytic) map to it from one of the ambient spaces where the configuration space is defined, e.g., $\boldsymbol{F}_{\mathcal{O}} : \mathcal{Q} \to \mathcal{O}$. The restriction of this map to the configuration space is called the output map, $\boldsymbol{f}_{\mathcal{O}} : \mathcal{D} \to \mathcal{O}$.

The input motion is modeled in a similar way. There is a manifold, $\mathcal{I}$, called the input space, the motions in which are used to control the mechanism. A map from the ambient manifold specifies the state of the input given the state of the whole

[1]In this book, constraints are always holonomic and scleronomic, i.e., can be formulated in terms of some coordinates without needing the coordinate derivatives or time to be explicit variables.

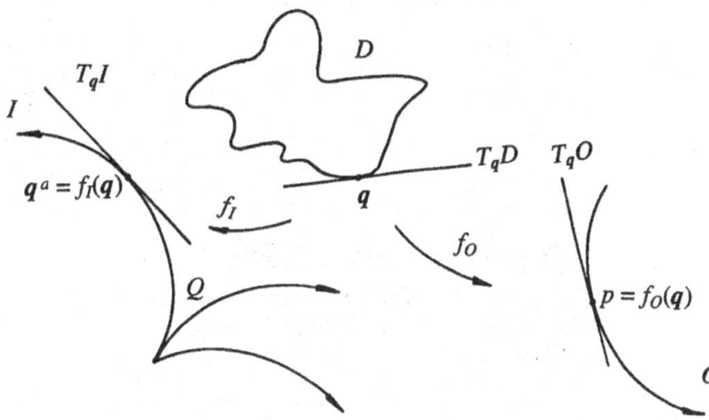

Fig. 10 Input and output in a mechanism

mechanism, $\boldsymbol{F}_{\mathcal{I}} : \mathcal{Q} \rightarrow \mathcal{I}$. The restriction of this map to the configuration space is called the input map, $\boldsymbol{f}_{\mathcal{I}} : \mathcal{D} \rightarrow \mathcal{I}$.

It is important to note and understand that the input space is the target, not than the source, of the input map. Each motion of the kinematic chain defines unique input and output motions in $\mathcal{I}$ and $\mathcal{O}$, via the input and the output maps. The inverse it not necessarily true.

The kinematics of a mechanism is represented schematically in Fig. 10. The configuration space, $\mathcal{D}$, shown as a curve, is the domain of the input and output maps. The input space is shown as a submanifold of joint space, as is often the case in practice.

3.3 *Regular and Singular Configurations*

In a regular state, a nonredundant mechanism allows to prescribe any motion in the output space by choosing a motion in the input space. The theory and the analysis of mechanism singularities study the cases where this is not possible. In particular, degenerate or singular configurations are usually defined as those where either the input, or the output map cannot be inverted. When this happens, one can distinguish two most general cases:

(A) The lack of invertibility is due to some degeneracy of the local properties of the *kinematic chain*; no choice of input or output would make the resulting mechanism function regularly at this configuration. The singular phenomenon can be identified by analyzing the constraint equations. In the neighborhood of a regular configuration, the configuration space is a smooth manifold. (As will be shown in Chap. 5, the reverse is not true: If the C-space is, locally or globally, a manifold, a configuration may still be singular if the constraint map loses rank.)

(B) The constraints of the kinematic chain do not degenerate. The lack of invertibility is specific to the input or output map.

To summarize the distinction, we can say that (A) and (B) represent loss of constraint and of control, respectively.

In Chaps. 4 and 5, the main interest is in (A), i.e., degenerate points of the kinematic constraints, which are singular regardless of the choice of input or output. In Chap. 3, only case (B) is considered: The focus is on serial chains where $\mathcal{Q} = \mathcal{D} = \mathcal{I}$ is always a manifold and there are no constraints. In this and the following chapters, both kinds of singular events are addressed.

We can now summarize the view of mechanism singularity that emerges from the model of mechanism kinematics outlined in this section. We are given a nonredundant mechanism with n inputs and outputs, and n-dimensional configuration space. A configuration $\boldsymbol{q} \in \mathcal{D}$ is regular if (a) it is a regular point of the constraint map, $\boldsymbol{F}$; and (b) it is a regular point of both the input and output maps, $\boldsymbol{f}_{\mathcal{I}}$ and $\boldsymbol{f}_{\mathcal{O}}$.

Note that given (a), a neighborhood of $\boldsymbol{q}$ is an n-manifold, and each of $\boldsymbol{f}_{\mathcal{I}}$ and $\boldsymbol{f}_{\mathcal{O}}$ is a mapping between n-manifolds.

A configuration is singular if, for whatever reason, it is not regular.

4 Singularity via Instantaneous Kinematics

In this section, we will redefine singularity in purely instantaneous terms. Rather than the manifolds and differentiable maps introduced in the previous section, we use their linearizations: the tangent spaces and maps.

The main justification is the greater simplicity and clarity of a formulation based on linear algebra rather than differentiable geometry. From a practical point of view, linearization makes it easier to formulate and solve the relevant systems of equations, as Chap. 2 demonstrates.

There are also theoretical advantages in modeling directly the velocity kinematics. As mentioned before, for nonlinear spaces such as $SO(3)$, the preferred coordinates to parameterize the tangent space are not the derivatives of the parameters used on the manifold. It is then useful to obtain the infinitesimal formulations directly rather than through differentiation. Moreover, in many cases, positions on the manifold are calculated from the tangent vectors using (precise or approximate) integration and exponentiation.

Considerations like these also determine the growing importance of screw theoretical formalisms. The main focus of screw theory is the instantaneous motion of rigid bodies. The linear algebra of twists and wrenches plays a key role in the instantaneous analysis of mechanism singularity.

4.1 Instantaneous Motions

The instantaneous motions at a given feasible configuration, $\boldsymbol{q}$, of a kinematic chain are the solutions of the linearized constraint equations, i.e., the velocity constraints. This linearization can be obtained by differentiating the nonlinear constraint equations, but in practice, it is often more convenient to derive them directly. Thus, one can easily write velocity equations imposed by the presence of loops by equating to zero the sum of the relative instantaneous twists in the joints. In fact, the nonlinear equations can be obtained in terms of the joint twists using the product of exponential formula, see Chap. 5.

The instantaneous motions form a vector space, a subspace of the tangent space to the ambient manifold, $T_{\boldsymbol{q}}\mathcal{Q}$. The dimension of this subspace at configuration $\boldsymbol{q}$ is the instantaneous mobility, $n_{\boldsymbol{q}}$. We can think of instantaneous motions as velocities or as linearized displacements, sometimes called *flexes*, of the system.

If the kinematic chain is used as an input–output device, there are also instantaneous input and output motions, elements of the tangent spaces to $T_{f_I(\boldsymbol{q})}\mathcal{I}$ and $T_{f_O(\boldsymbol{q})}\mathcal{O}$, which we call I and O for simplicity.

Every potential instantaneous motion $d\boldsymbol{q} \in T_{\boldsymbol{q}}\mathcal{Q}$ (feasible or not) determines in a unique way both the input and the output instantaneous motions as images of the linearizations (differentials) of the maps $\boldsymbol{F}_I$ and $\boldsymbol{F}_O$ defined in the previous section. Therefore, without loss of generality, we can think of a velocity vector (or flex, or instantaneous motion) for the mechanism at $\boldsymbol{q}$ as the triple, of $\boldsymbol{v} = (\boldsymbol{v}_I, \boldsymbol{v}_O, \boldsymbol{v}_P)$, specifying the input, output, and passive velocities. The velocity vectors form a vector space, $V = I \times O \times P$. An element of P, a passive velocity vector $\boldsymbol{v}_P$, is some array of speeds or velocities which completely prescribes, together with $\boldsymbol{v}_I$ and $\boldsymbol{v}_O$ the instantaneous motion. One can take, for example, $P = T_{\boldsymbol{q}}\mathcal{Q}$, however, usually $P \subsetneq T_{\boldsymbol{q}}\mathcal{Q}$, for which $I \oplus O \oplus P = I \oplus O \oplus T_{\boldsymbol{q}}\mathcal{Q}$. Most typically, $T_{\boldsymbol{q}}\mathcal{Q} = I \oplus P$ and $V = O \times T_{\boldsymbol{q}}\mathcal{Q}$, when $\mathcal{Q}$ is joint space and $\boldsymbol{v}_I$ and $\boldsymbol{v}_P$ are arrays of joint speeds, in the actuated and passive joints, respectively.

The velocity constraints at $\boldsymbol{q}$ augmented to include the input and output velocities have the general matrix form

$$\boldsymbol{L}\boldsymbol{v} = \boldsymbol{L}_o \boldsymbol{v}_{\mathcal{O}} + \boldsymbol{L}_i \boldsymbol{v}_I + \boldsymbol{L}_p \boldsymbol{v}_P = \boldsymbol{0} \quad (8)$$

which we call the *velocity equation*. The solution space M_q is the (instantaneous) *motion space*, composed of the feasible motions in the configuration. As mentioned at the end of Sect. 2, it is important that the set of constraints is complete; i.e., Eq. (8) is a necessary and sufficient condition for a flex, $\boldsymbol{v}$, to be possible at the configuration.

We can think of Eq. (8) as describing how the placement of the subspace M_q within V, including its orientation with respect to the input and output spaces, I and O, changes with $\boldsymbol{q}$. For practical purposes, it is useful to have the matrices of coefficients as explicit continuous functions of $\boldsymbol{q}$, at least in a neighborhood if not globally.

As the differential input and output in any configuration are a (linear) function of the joint velocities, the dimension of the motion space is equal to the instantaneous mobility n_q. The projection transforms, $\boldsymbol{\pi}_I$ and $\boldsymbol{\pi}_O$, which map V onto I and O, send any velocity vector into the array of its input and output coordinates, respectively. Of special importance, for the singularity properties of the mechanism, are the restrictions, $\boldsymbol{p}_I$ and $\boldsymbol{p}_O$, of these transforms to the motion space. They map every *feasible* motion vector into its input and output, respectively. Note that unlike $\boldsymbol{\pi}_I$ and $\boldsymbol{\pi}_O$, the maps $\boldsymbol{p}_I$ and $\boldsymbol{p}_O$ (and their ranks) are dependent on the configuration. For simplicity, this dependency is not denoted explicitly.

4.2 Instantaneous Definition of Singularity

The instantaneous kinematics analysis of an input–output device has to provide solutions to two problems:

The forward instantaneous kinematics problem (FIKP): where for a given configuration, $\boldsymbol{q}$, the instantaneous motion of the mechanism is determined when the input $\boldsymbol{v}_I$ is given and

The inverse instantaneous kinematics problem (IIKP): where for a given configuration $\boldsymbol{q}$, the instantaneous motion of the mechanism is determined when the output $\boldsymbol{v}_O$ is given.

The FIKP and the IIKP are equivalent to the problem of finding the inverse of the maps $\boldsymbol{p}_I$ and $\boldsymbol{p}_O$, respectively.

A configuration is *regular* or *nonsingular* when both these problems have unique solutions for any input or output, that is, when both projections $\boldsymbol{p}_I$ and $\boldsymbol{p}_O$ are one-to-one mappings of the motion space on the input and output, respectively. Otherwise, the configuration is *singular*.

Thus, singularity occurs whenever the instantaneous kinematics becomes indeterminate.

For the configuration to be regular, it is necessary that $n = n_q$. We will assume that the input and output are chosen so that for any configuration $n \leq n_q$. Usually, n is equal to the global degree of freedom or to the local one, if the chain has variable mobility.

For practical purposes, it is best if in Eq. (8) there are n fewer scalar equations than variables; i.e., the minimum co-rank is n, and so $n \leq n_q$. In most cases, like the examples in Chap. 2, it is easy to find such a matrix L as a continuous explicit function of $\boldsymbol{q}$. There are, however, exceptions, some of which will be alluded to in Chap. 5. For example, for the so-called overconstrained linkages, the natural way of writing (8) yields a greater number of equations, which are always dependent. It may then be difficult to reduce the system to a simpler one in a globally continuous way.

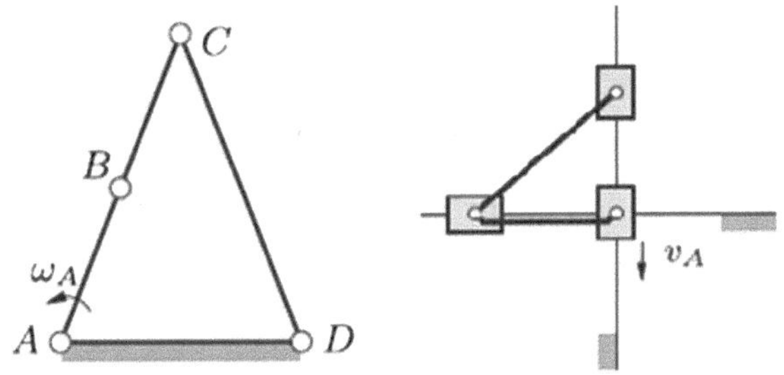

Fig. 11 Examples of RI configurations

4.3 Singularity Types and Classes

In this subsection, six types of singularities are defined in terms of the properties of $\boldsymbol{p}_I$ and $\boldsymbol{p}_O$. Each is characterized by the occurrence of a certain physical phenomenon resulting from the indeterminateness of either the forward or the inverse instantaneous kinematics, or both. Most of these singularity types were already encountered in the examples considered in Sect. 2.

Redundant Input: A configuration is of the redundant input (RI) type, if there is a feasible motion, $\boldsymbol{v}$, with zero output and nonzero input, $\boldsymbol{p}_O(\boldsymbol{v}) = \boldsymbol{0}$ and $\boldsymbol{p}_I(\boldsymbol{v}) \neq \boldsymbol{0}$, i.e.,

$$\ker \boldsymbol{p}_O - \ker \boldsymbol{p}_I \neq \varnothing \tag{9}$$

We illustrate all singularities using a four-bar double rocker, as well as with a linkage with three sliders as in Fig. 7, but choosing different length ratios of the two bars. The input joint of the four bars is at A; the output angular motion is (usually) that of hinge D. Input and output of the three sliders, as discussed before, are the velocities of points A and B, respectively.

The configurations in Fig. 11 exhibit redundant input because the input joint can move while the output motion is blocked.

Redundant Output: A configuration is of the redundant output (RO) type, if there is a feasible motion, $\boldsymbol{v}$, with zero input and nonzero output, $\boldsymbol{p}_I(\boldsymbol{v}) = \boldsymbol{0}$ and $\boldsymbol{p}_O(\boldsymbol{v}) \neq \boldsymbol{0}$, i.e.,

$$\ker \boldsymbol{p}_I - \ker \boldsymbol{p}_O \neq \varnothing \tag{10}$$

The configurations in Fig. 12 are of the RO-type because the output can be arbitrary while the input joint is transitorily blocked.

Impossible Output: A configuration is of the impossible output (IO) type, if there are output motions, $\boldsymbol{v}_O$, which cannot be achieved with any input, i.e.,

$$O - \operatorname{im} \boldsymbol{p}_O \neq \varnothing \tag{11}$$

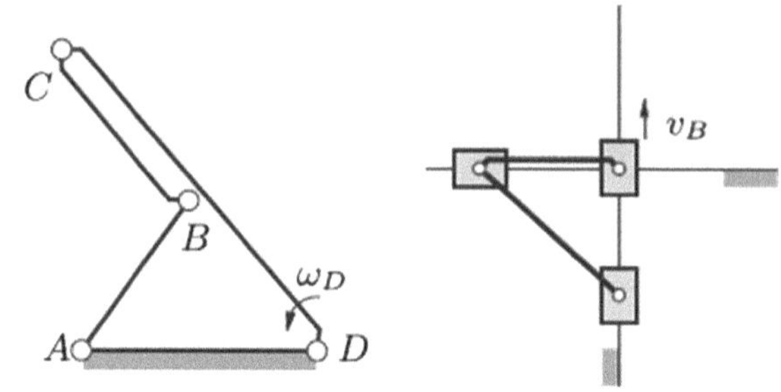

Fig. 12 Examples of RO configurations

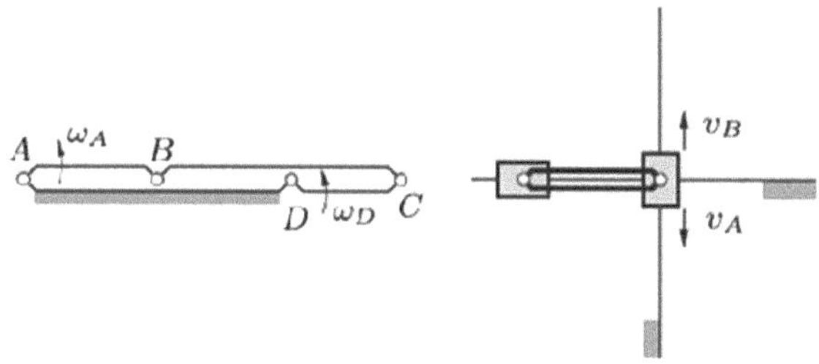

Fig. 13 Examples of IIM configurations

The configurations of Fig. 11 are also examples of IO singularities. Indeed in both of them, the output instantaneous motion must be zero, and any nonzero output is impossible.

Impossible Input: A configuration is of the impossible input (II) type, if there are input motions, $\boldsymbol{v}_I$, which cannot be feasible for any output, i.e.,

$$I - \mathrm{im}\, \boldsymbol{p}_I \neq \varnothing \tag{12}$$

The configurations of Fig. 12 are also examples of II singularities. Indeed in both of them, the input instantaneous motion must be zero, any nonzero input is impossible.

Increased Instantaneous Mobility A configuration is of the increased instantaneous mobility (IIM) type, if there are more than n instantaneous dof, i.e.,

$$n < n_q \tag{13}$$

Figure 13 shows two examples of IIM-type singularities. On the left, the flattened four bars have two instantaneous degrees of freedom and blocking any one joint does not stop the motion. Similarly, on the right, the velocities of sliders A and B are independent and arbitrary.

It can be noted that each of the configurations in Fig. 13 also exhibits both RI and RO, but neither IO nor II. This shows that RI and IO (or similarly RO and II) are not two faces of the same coin: Sometimes they co-occur; sometimes they do not.

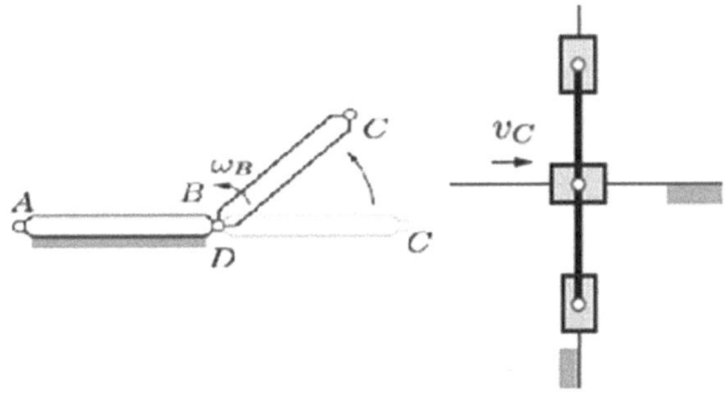

Fig. 14 Examples of RPM configurations

Redundant Passive Motion: A configuration is of the redundant passive motion (RPM) type, if there is a feasible nonzero motion, $\boldsymbol{v}$, with zero output and zero input, $\boldsymbol{v}_I = \boldsymbol{v}_O = \mathbf{0}$ and $\boldsymbol{v}_P \neq \mathbf{0}$, i.e.,

$$\ker \boldsymbol{p}_O \cap \ker \boldsymbol{p}_I \neq \mathbf{0} \tag{14}$$

Two examples are shown in Fig. 14. On the right is the three-slider configuration already discussed in Sect. 2. On the left is a specially proportioned four-bar linkage, which allows two hinges to become coincident, $B = D$. The other two opposing joints are the input and output motion. The flexing of the linkage with pivot $B = D$ is a purely passive motion, an RPM. (There can be no RPM if the input and output angles are adjacent.)

Figure 14 also exemplifies the II and IO phenomena, which occur here unaccompanied by RO or RI. Indeed in both mechanisms, the input and output joints have necessarily zero speeds.

4.4 *Classes of Singularities*

The six types of singularities are overlapping, as the examples show. In fact, it can be proven that every singular configuration must belong to at least two types. To obtain a proper classification of a set, one needs to divide it into nonintersecting classes. The finest intersections of the six types generate such a classification. Two singularities are of the same class if they belong to exactly the same types.

Various rules can be established for the simultaneous occurrence of the types; see Zlatanov (1998), Zlatanov et al. (1994a, b). For example, it can be shown that any singularity of any mechanism must belong to at least one "R type" (RI, RO, or RPM) and one "I type" (II, IO, or IIM). This means that all possible classes are among the 49 cells of the table in Fig. 15.

More detailed rules allow a general classification result to be deduced: Any singularity of any mechanism must belong to one of the 21 classes with nonempty cells in the table. Moreover, examples can be found for each of these 21 cases.

	IO	II	IIM	IO *and* II	IO *and* IIM	II *and* IIM	IO *and* II *and* IIM
RI	Y						
RO		Y					
RPM			N	Y			Y
RI *and* RO			N	Y	Y	N	Y
RI *and* RPM				Y	Y		Y
RO *and* RPM				Y		N	Y
RI *and* RO *and* RPM			N	Y	Y	N	Y

Fig. 15 Classification of singularities

Naturally, if interest is limited to particular mechanisms, architectures, or families of architectures, fewer classes may be possible. Thus, the architecture used as an example in the introduction, Fig. 1, can exhibit 15 classes, denoted by "Y" in the table. This is also true for a large class of parallel mechanisms.

5 Common and Uncommon Singularities

As was mentioned in the Introduction to the chapter, it can be dangerous to dismiss certain singular phenomena as unusual and unlikely to occur in practice.

5.1 Generic and Nongeneric Singularity Types

Obviously, the conditions for a configuration to belong to some of the types listed in the preceding section are stricter than for others. For example, assuming that the velocity equation matrix has N rows and $N+n$ columns, the IIM condition involves the rank deficiency of the whole $N \times (N+n)$ matrix L, i.e., the singularity of all its square submatrices, while for IO or II to occur it is sufficient only that one particular submatrix has zero determinant. Similarly, comparing RPM and RI, the former requires the singularity of an $N \times N$ matrix, while one of its $N \times (N-n)$ submatrices L_p is regular; the latter occurs only when L_p is rank deficient.

From such estimates, it would appear that, among a sample of mechanism designs, some singularity types will occur with probability zero, while others will be generic. This is strictly true, however, only if the sample is random. As no one designs mechanisms at random, such considerations are of limited utility.

Probably, the best example of this is IIM-type singularities, sometimes called uncertainty configurations (Hunt 1978). The conditions are the strictest of all types, and it would seem obvious that a "normal" kinematic chain would not have such singularities. Almost all examples of such configurations discussed in the literature prior to 2001 were single-loop linkages with specially chosen bar lengths. However, at approximately that time, there was a surge of interest in constrained parallel manipulators. It turns out that within that category of kinematic chains IIM-type singularities are commonplace.

We discuss these so-called constraint singularities in Sect. 6. In the remainder of this section, we look at the "surprising" occurrence of RPM-type configurations.

5.2 Parallel Mechanisms with Extendable Legs

Consider again the mechanism in Fig. 5, among the simplest and seemingly unexceptional parallel manipulators. As we showed, it has RPM-type configurations, Fig. 8. In fact, RPM can occur only because two passive hinges, A and P, can coincide. For this to be possible, the RPR serial subchain between A and P must have zero offset, allowing the two hinge centers to merge by sliding the prismatic joint. In a general RPR with offset (DH parameter), a_1, the two hinge centers can be no closer than a_1. The difference is illustrated by Figs. 16 and 17. If the mechanism in Fig. 5 had its zero offset RPR legs replaced by generic ones, an RPM singularity could never occur. Instead, when one of the legs becomes singular, i.e., the distance between the hinges reaches the minimum, the parallel manipulator will have an (RI, IO) singularity: If point P is fixed, any feasible self-motion will involve movement in an actuated slider. In contrast, RI can never occur for the two-dof manipulator in Fig. 5: If points A are P fixed, there can be no motion in the slider.

Interestingly, virtually, all RPR chains used in planar manipulators, either in analysis or design, have zero offset. (Exceptionally, Bonev et al. (2003) consider arbitrary offsets.) As a result, RPM-capable mechanisms are ubiquitous, rather than nonexistent. Designers really make "generic" choices, aiming instead for symmetry, simplicity, and to meet nonkinematic mechanical requirements such as better force distribution and reduced friction.

Fig. 16 An RPR chain with zero offset

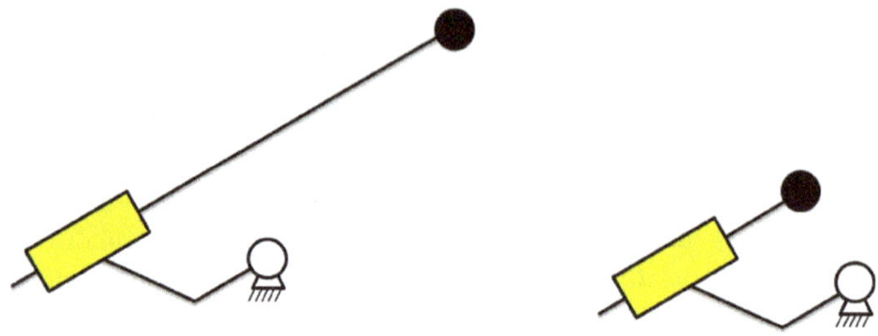

Fig. 17 An RPR chain with nonzero offset

The spatial analog of the RPR leg is the UPS (or SPS) chain with a single linear actuator used in many parallel manipulators such as the Stewart-Gough platforms. Almost all such legs in the literature or in practice have zero offset, while the "generic" case is never even mentioned. In a zero offset leg, the joint twists span a less-than-six-dimensional space only if all four or more hinge axes are concurrent or slightly more generally when the S-joint center is in the plain of the U-joint axes. If the terminal body is held fixed, all these configurations allow RPM and none permits motion in the slider. This means that with all legs of this type, the *only* serial-like singularities the parallel manipulator can have are of the RPM and none are of the RI type. (Note that, unlike in the planar case, a nonzero offset does not rule out RPM configurations; they still exist, as well as RI-type singularities.) The velocity equations that are used to model such mechanisms are similar to Eq. 5; they cannot detect RPM configurations.

Fortunately, in most realizations of spatial platform mechanisms with extendable legs, mechanical interference prevents coinciding joint centers. Not all parallel manipulator architectures are so protected by good luck.

5.3 *The Isotropic Singularity of the Orthoglide*

The Orthoglide is a translational parallel kinematic machine (PKM) designed and produced in France's IRCCyN, (Chablat and Wenger 2003), Fig. 18.

There are three identical PRPaR legs, Pa indicating a parallelogram joint. The final leg design, Fig. 19, is a modification from the original proposal shown in Fig. 20 (Chablat and Wenger 2003; Majou et al. 2007). There is a good reason for the change.

In the original leg design, Fig. 20, the direction of the P joint and the adjacent revolute axis are parallel, keeping the latter invariant. The three legs are placed in the base in such a way that these three fixed R axes intersect at right angles.

The two R joint axes in each leg are kept parallel by the Pa joint, and thus, each limb imposes a two-space of constraint couples on the tool. So, together, the legs allow the end effector to have translational twists only. Away from singularities, the

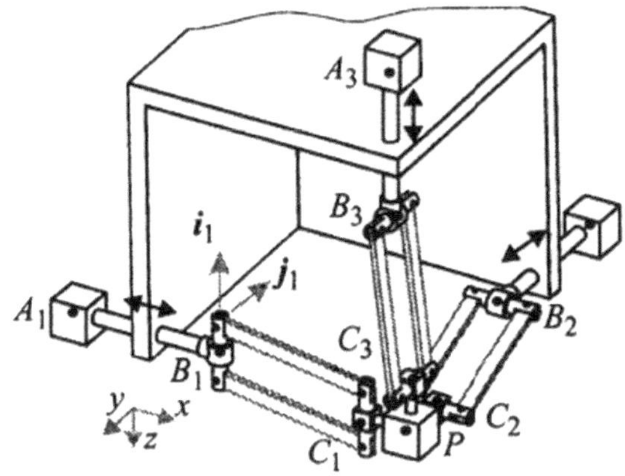

Fig. 18 The Orthoglide (Chablat and Wenger 2003)

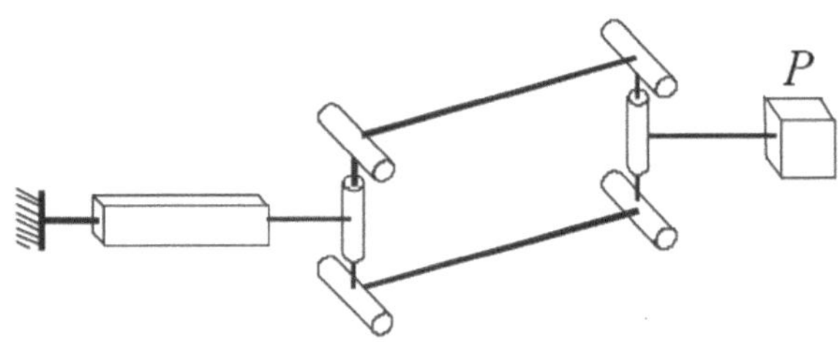

Fig. 19 The modified leg of the Orthoglide (Majou et al. 2007)

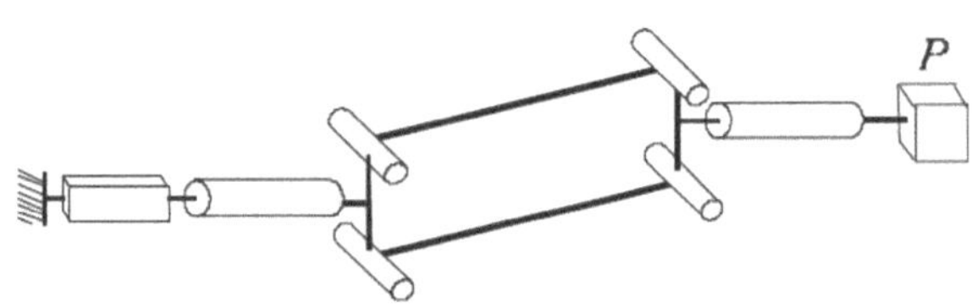

Fig. 20 The original leg of the Orthoglide (Majou et al. 2007)

end effector can have three translational dof. The motion is controlled by the linear actuators of the P joints.

The velocity and singularity analysis, if performed using an input output equation in the manner of Sect. 2.3, reveal no dangerous singularities (at least in the central part of the workspace). Moreover, in the most central configuration, when the axes of the six R joints are aligned on three mutually perpendicular lines intersecting in the tool center point, P (see Figs. 21 and 22), the input–output relationship becomes an identity: The speeds of the sliding actuators are simply the three orthogonal components of the velocity vector of the end effector. Such configurations are called "isotropic"; they are deemed particularly desirable, and operation of the robot in (or near) them is expected to be optimal. In particular, it is often said that an isotropic configuration is as "far away" as possible to a singularity (because the conditioning number of the relevant matrix is as far from zero as possible).[2]

[2]Isotropy and related concepts derived from the conditioning number of one or another Jacobian matrix have had a notoriously unfortunate track record of attempted applications in robotics. The

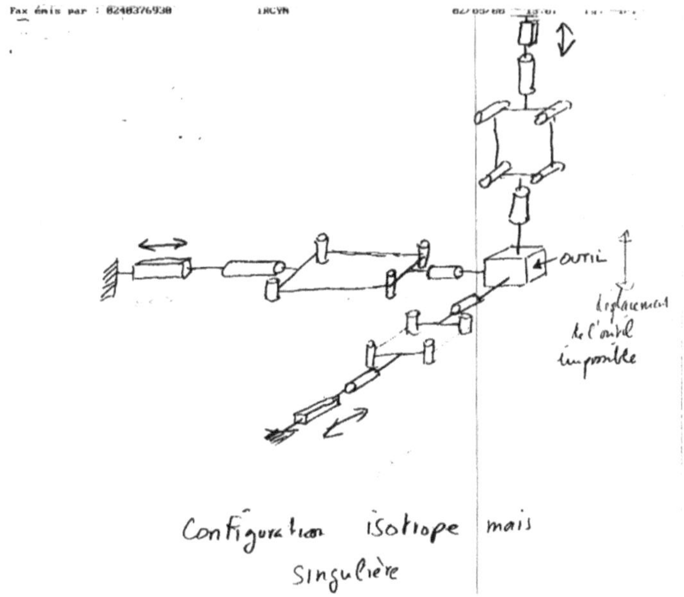

Fig. 21 The central "isotropic but singular" configuration of the original design (communication by P. Wenger 2000)

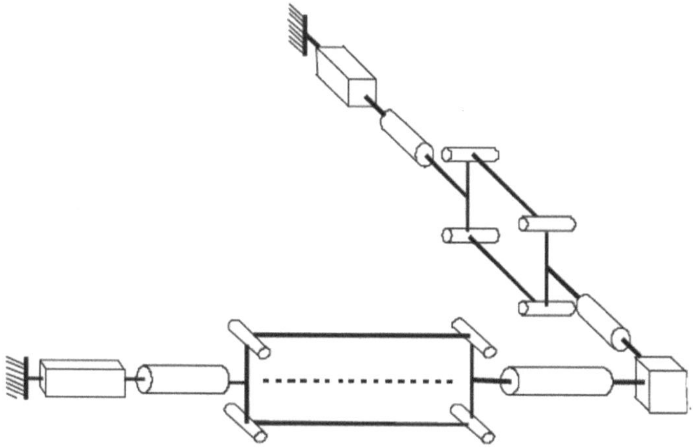

Fig. 22 Two legs in the central configuration (Majou et al. 2007)

In this particular case, the reality is far from optimal. If any prototype of the mechanism is configured with the tool in the central position, and motion is attempted, it is immediately discovered that the "isotropy" is of a very special kind. The good

worst errors, related to physical unit inconsistency, do not apply in this case. Mercifully, we do not need to dwell on this history here.

news is that, indeed, the input and output velocities must be identical. The bad news is that (almost certainly) they *all* have to be equal to zero. An "isotropic configuration" is supposed to be the "sweet spot" of the workspace, but in this case, it is its "dead center." The end effector is stuck in its place and locked and so are the actuators. The kinematic chain as a whole retains three dof: Each of the three parallelograms can spin as a rigid body, independently from the others, about the coinciding R joint axes of the leg.

The reason for the singularity is simple: Considered as a separate serial chain, each leg is in a singular configuration—only three of its four joint twists are independent. This results in a three-space of constraint wrenches on the end effector, including a force through the tool center perpendicular to the parallelogram plane.

When the constraints of the legs are combined, the total dimension of the wrench space is five or six, depending on whether the parallelogram plane normals are independent. If they are, e.g., when the leg parallelograms are in three orthogonal planes as in (the nonsingular state in) Fig. 18, there is a six-space of constraints and no tool motion is possible. The singularity class is RPM, IO, II.

If only two plane normals are dependent (all three cannot be parallel), there are five independent constraints and the end effector can only translate and only in one direction. Examples of the latter case are presented in Fig. 21 (personal communication by Philippe Wenger 2000) and Fig. 22 from Majou et al. (2007).

In the former, two parallelograms are in horizontal planes, the third in a vertical one. This results in a concurrent planar pencil of force constraints through the tool center and in a vertical plane. The end effector has only one dof: the translation freedom of the third parallelogram. There is also only one input freedom: The instantaneous motions in the horizontal actuators are coupled, while the vertical slider must be with speed zero.

In Fig. 22 (which shows only two of the legs), all three parallelogram planes are vertical. The tool can only translate in the vertical direction, and the horizontal actuators are motionless.

It should be noted that in either case, Fig. 21 or Fig. 22, the permitted input/output linear freedom is *in excess* of the three uncontrollable spinning leg motions. This means that the kinematic chain as a whole has four dof and an IIM is present. The singularity class is RPM, IIM, IO, II. In a sense, this is closely analogous to the type of configurations discussed in the following section, where the platform of parallel manipulator gains extra transitory dof.

To summarize, a central position of the tool implies a triple-RPM configuration. To what degree freedom is lost at the input and output depends on how the parallelogram planes have been oriented. Even when some input/output motions are possible, the state is unstable and the device will be completely locked by any uncontrollable spin of a leg about its axis.

6 Constraint Singularities of Parallel Manipulators

6.1 Another "Isotropic" Configuration

Shortly after the final presentation at the Second Workshop on Computational Kinematics, May 22, 2001, held at Seoul National University, there was a laboratory tour for the participants. One of the parallel mechanisms shown, Fig. 23, had been intended as a three-dof device for translating a machine tool without change of orientation. The manipulator stood there motionless with its three legs set at equal lengths. All looked perfect until Professor Frank Park, who led the tour, pushed the mobile platform and the mechanism seemed to collapse under its weight. Once deviated from its initial pose, the platform was no longer parallel to the base. At that time,

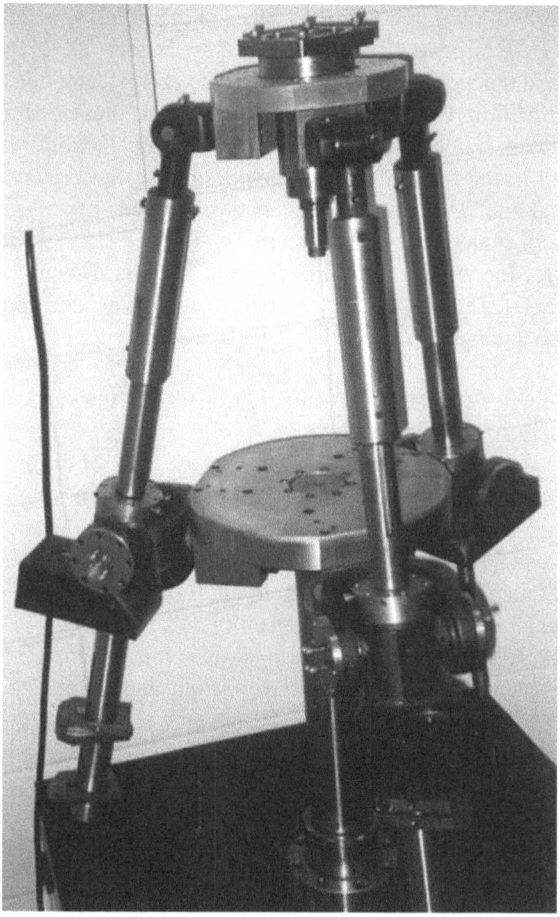

Fig. 23 A 3-UPU mechanism in a singular configuration (Seoul National University, 2001, courtesy of Prof. Frank Park)

Frank Park commented that this apparent singularity could not be explained by the kinematic analysis of the mechanism.

Indeed, this is another example where conventional analysis based on an input–output vector equation fails. The manipulator has a 3-UPU architecture with one linear actuator in each leg. An equation just like (5), but of dimension 3, can be written:

$$\dot{\boldsymbol{d}} = \boldsymbol{K}\vec{v} \tag{15}$$

where $\dot{\boldsymbol{d}}$ is the array of actuator speeds, $\vec{v}$ is the end effector linear velocity, and $\boldsymbol{K}$ is a matrix whose three rows are the directions of the P joints. There is nothing particularly singular about $\boldsymbol{K}$ in the shown configuration. In fact, we can imagine, without loss of generality, that the slider directions are mutually perpendicular making the configuration "isotropic" by definition and therefore presumably as far from being singular as one can be. Yet, singular it is, because locking the actuators does not prevent all motion.

Once again, the problem with Eq. (15) is not that it is not true, but that it is not sufficient to describe the motion capabilities of the mechanism.

6.2 *Constraints, Freedoms, and Constrained Singularities*

Consider a general parallel mechanism with legs labeled by the letters $L = A, B, \ldots$. The leg chain L has n_L joints. The relationship between the instantaneous motion of the platform, the output twist $\boldsymbol{\xi} = (\vec{\omega}^T, \vec{v}^T)^T$, and the joint velocities $\dot{\theta}_i^L$ is given by the twist equations of the serial subchains:

$$\boldsymbol{\xi} = \sum_{i=1}^{n_L} \dot{\theta}_i^L \boldsymbol{\xi}_i^L, \quad L = A, B, \ldots, \tag{16}$$

where the joint screws are denoted by $\boldsymbol{\xi}_i^L$.

Unlike Eqs. (15), (16) is a necessary and sufficient condition for the output twist, $\boldsymbol{\xi}$, and the joint velocities, $\dot{\theta}_i^P$, to be feasible. Therefore, the space of all possible platform twists, $\mathcal{T}$, i.e., the system of *platform freedoms*, is given by:

$$\mathcal{T} = \bigcap_{L=A,B,\ldots} \mathcal{T}_P, \tag{17}$$

where $\mathcal{T}_L$ is the output twist space of the subchain L,

$$\mathcal{T}_L = \mathrm{Span}(\boldsymbol{\xi}_1^L, \ldots, \boldsymbol{\xi}_{n_L}^L), \quad L = A, B \ldots. \tag{18}$$

For a twist (wrench) system $\mathcal{L}$, let $\mathcal{L}^{\perp}$ be its reciprocal wrench (twist) system. (When a wrench exerts no power on a twist, they, and their screws, are called reciprocal.)

The reciprocal wrenches in $\mathcal{W}_L = \mathcal{T}_L^{\perp}$ are the *leg constraints*. If applied to the platform, they can be resisted by the leg L with zero torque from the actuators. The *platform constraints* are given by $\mathcal{W} = \mathcal{T}^{\perp}$. Since, for any $\mathcal{L}$, $(\mathcal{L}^{\perp})^{\perp} = \mathcal{L}$, Eq. (17) implies

$$\mathcal{T}^{\perp} = \sum_{L=A,B,\ldots}^{D} \mathcal{T}_L^{\perp} . \tag{19}$$

For any configuration, $\dim \mathcal{W} + \dim \mathcal{T} = 6$.

In all nonsingular configurations, $\dim \mathcal{W} = 6 - n$, where n is the global (full-cycle) mobility of the mechanism. In a singular configuration, the constraint system of a leg, $\mathcal{W}_L$, can only increase in dimension. However, due to intersections of the leg constraint systems, it is possible for the platform constraint system, $\mathcal{W}$, to lose rank and its dimension to become lower than $6 - n$. If this occurs, we will say that the parallel manipulator is in a *constraint singularity*. It immediately follows that in such a configuration the freedom system of the platform, $\mathcal{T}$, will be of dimension higher than n, and therefore the instantaneous mobility of the mechanism is higher than its full-cycle mobility.

This is precisely what happens in the configuration in Fig. 23: Each leg imposes on the platform a constraint couple with a vertical moment. This means that that the constraint system is one-dimensional, rather than three-dimensional, and the instantaneous degree of freedom of the platform is five. Clearly, this is an IIM-type configuration, and no three actuators can immobilize the kinematic chain.

6.3 Singularities of 3-RPS Manipulators

Consider another mechanism, a 3-RPS parallel manipulator and two of its singular configurations.

Figure 24, where one R joint axis is in the plane of the platform, shows an ordinary RO-type singularity. The platform has one uncontrollable freedom, namely a rotation about the line BC, when the prismatic actuators are locked. In Fig. 25, we see another singular configuration. Here, the uncontrollable motion is again a rotation, with an axis passing through both point C and the intersection point of legs A and B. The second configuration is, in fact, a constraint singularity.

Despite the apparent similarity, the two singularities are quite different. In Fig. 24, the constraint wrenches (forces through the S joints and parallel to the R joints) are linearly independent and the platform has three dof. In Fig. 25, however, the three constraint wrenches all pass through point C and lie in the platform plane and hence form only a two systems. The platform gains a freedom and can instantaneously move with four dof. Indeed, it can rotate arbitrarily about point C and translate vertically.

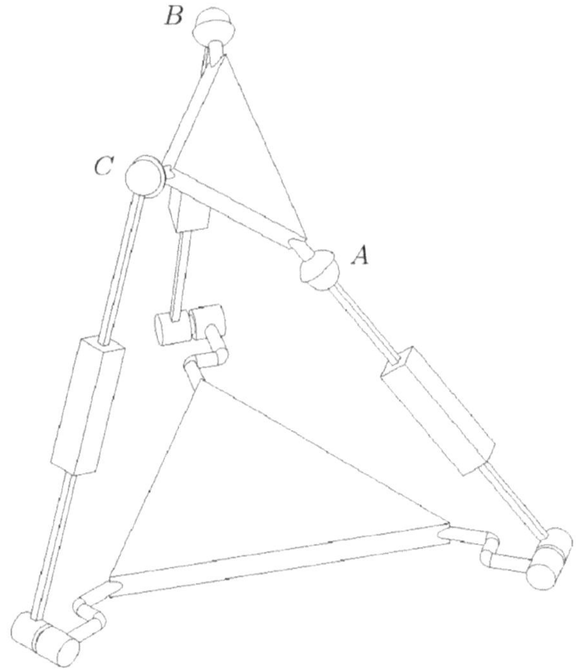

Fig. 24 An RO-type singularity of a 3-RPS mechanism

From a kinematic point of view, these are completely different phenomena. In one case, the loss of control is due to the (so to speak) unfortunate choice of the actuators. If, in Fig. 24, we actuate the base R joint rather than the P joint of leg B, and we would have full control of the platform. In Fig. 25, there is no choice of three actuators that will prevent the platform from moving instantaneously.

Let us identify all constraint singularities of the mechanism. Three nonparallel zero pitch screws are linearly dependent only if they span a planar pencil. Since the constraint forces are horizontal, it follows that the platform and base must be parallel. Moreover, it can be shown that the three force axes in the horizontal platform plane can intersect in one point, H, only if the platform is upside down as in Fig. 25. The feasible poses of the platform in every horizontal plane form a one-parameter set obtained by rolling the circle, c_m, drawn around the platform triangle, on the inside of a fixed circle, c_f, twice the radius of c_m and centered at G, right above the base center, Fig. 26.

The configuration in Fig. 25 is the special case when H coincides with a platform vertex. If this is not the case, the redundant motion with locked actuators will not be a pure rotation as in Fig. 25, but rather a screw motion, more precisely a linear combination of a rotation through H and the vertical translation.

Finally, we show another constraint singularity of a (different) 3-RPS mechanism, Fig. 27. The platform is in a much more unperturbed pose than in Fig. 25; however, the

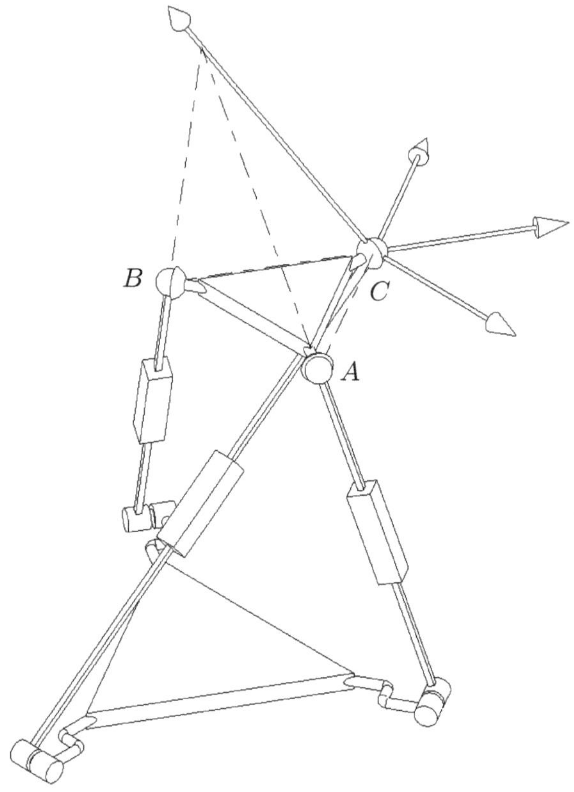

Fig. 25 A constraint singularity of a 3-RPS mechanism

three base R joint axes need to intersect in a point. The platform can instantaneously yaw (rotate about its normal axis) with locked actuators. This configuration was described in Huang et al. (1996).

6.4 The Simplest Constraint Singularity

To provide what must be the simplest example of a constraint singularity, let us consider the four-bar linkage in Fig. 28a. Assume that the link lengths are such that the mechanism can become flattened, Fig. 28b. In the shown configuration, the kinematic chain as a whole acquires an instantaneous mobility of two.

We assume that the output link is the coupler, a kind of "platform" of a one-dof planar "parallel manipulator" with two RR legs, one of which is passive. Now, in the singularity of Fig. 28b, the end effector gains a freedom and has two instantaneous dof. Indeed, it can rotate about any point on the line containing all joint centers and, of course, translate in the normal direction.

Each RR leg imposes one planar wrench constraint on the platform: a force through the two joint centers of the leg. Normally, these two lines are distinct and span two

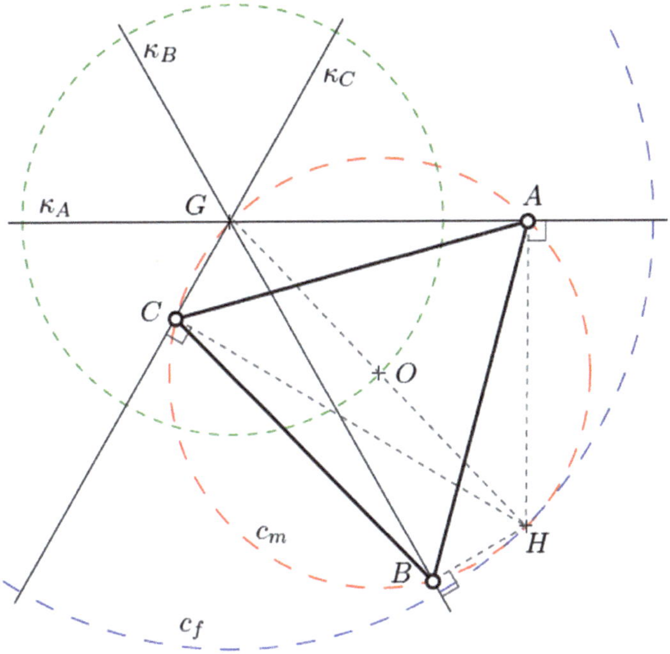

Fig. 26 Constraint singularities of a 3-RPS mechanism

systems. However, in Fig. 28b, the constraints become linearly dependent and span only a one-space.

For a one-dof constrained parallel manipulator (a four-bar), a constraint singularity requires specially proportioned link lengths. This, however, is not so if dimensions and leg number increase. Thus, for a constrained manipulator with three five-dof legs, constraint singularities occur generically, as is illustrated by the 3-UPU and 3-RPS examples above. This means that any designer of constrained parallel manipulator needs to be aware of the danger and be able to detect such configurations. Generally, this cannot be done by input–output models, as we have seen.

6.5 *Constraint Singularities as Branching Points*

As was suggested in Sect. 3, IIM configurations, and therefore the constraint singularities of parallel manipulators, are usually the singular points of the configuration space of the kinematic chain. These are critical points of self-intersection of different C-space branches (or sheets). In the present section, we look at singularities where the constraints on the end effector degenerate; this often implies that the intersecting branches will have different motion capability of the platform.

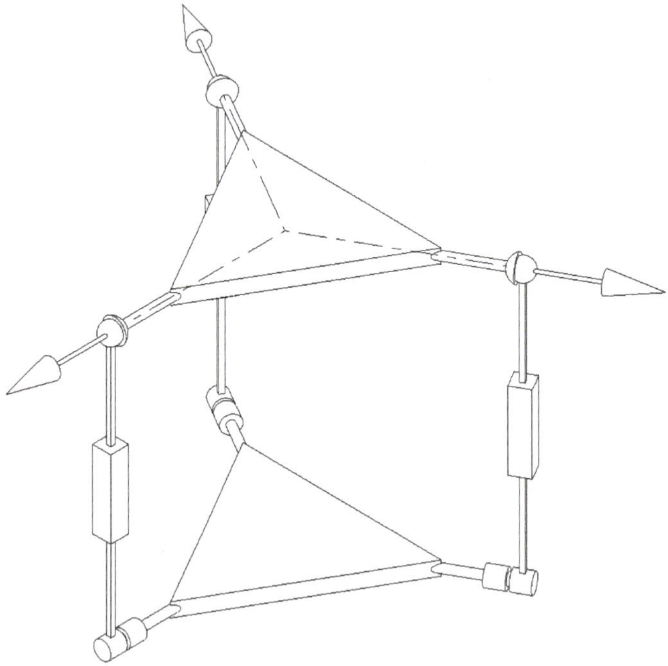

Fig. 27 A 3-RPS parallel mechanism in a constraint singularity

Fig. 28 A flattened four-bar mechanism in (a) a nonsingular configuration and (b) a constraint singularity

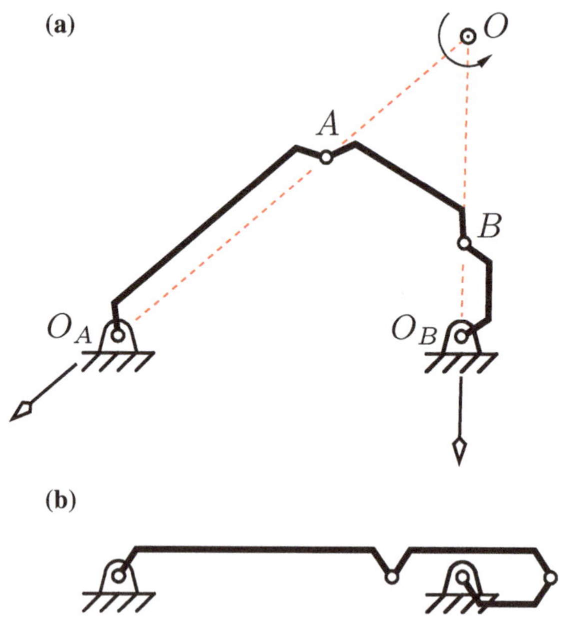

Transition between these different operation modes can completely transform the parallel mechanism, e.g., from one providing purely translational motion to the equivalent of a spherical wrist (Zlatanov et al. 2002b). This phenomenon is also addressed in Chap. 4 using as an example the 3-UPU discussed above. Here, we mention only three examples of such transitional configurations. The first is the 3-UPU configuration in the photograph above. The other two are, arguably, the simplest and the strangest examples of this phenomenon, and they are not considered in Chap. 4.

The 3-UPU: Two Shaft Couplings in One: The configurations like in Fig. 23 with various platform elevations form the one-dimensional intersection set of two three-dimensional sheets of the configuration space. In each of these, the platform has three dof, but with different motion pattern.

The two operation modes (as such subsets of the C-space are called) are illustrated in Fig. 29.

In the figure, we show a 3-URU rather than a 3-UPU. More generally, both can be seen as types of 3-RER, i.e., tripod devices whose each leg consists of two revolute joints constrained to perform relative planar motion (E is the three-dof planar joint). (This means that we ignore singularities and inverse kinematics multiplicities of the serial linkage realizing the planar pair.) The special features of the SNU 3-RERs are that the two revolutes within the RER leg stay in the same plane and that the joints in the base and platform form congruent planar and circularly symmetrical Y-shapes (Zlatanov et al. 2002a; Zlatanov and Gosselin 2004).

The configuration on the left in Fig. 29 is in translational mode, for which the SNU was built. The one on the right is clearly in a different mode as the platform is no longer horizontal.

It turns out that this second mode is characterized by the property that the platform, with the three revolute joints fixed in it, is the mirror image of the Y-shape in the base. The three-dimensional set of mirror symmetry planes defines the three-space of platform poses.

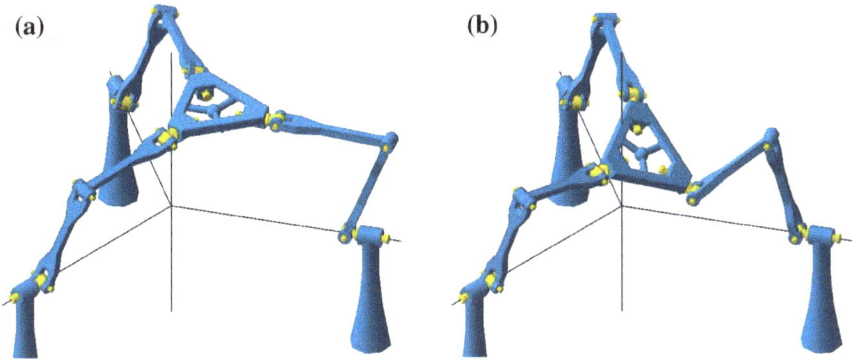

Fig. 29 The DYMO chain in (**a**) translation mode and (**b**) CVC mode

The mirror symmetry has a consequence of great practical utility: It is possible to fix two shafts to the base and platform, respectively, and use the kinematic chain as a transmission. The angular velocity on the shafts will be equal: a device known as a constant-velocity (or homokinetic) coupling (CVC).

So the configuration that mystified the attendees of CK 2001 is in fact a constraint singularity linking the translational and the CVC mode of the double-Y 3-RER kinematic chain.

Embarrassingly for kinematicians of the twenty-first century, this had already been established in the literature for nearly 30 years and for a more general class of 3-RERs. In 1972, Kenneth Hunt (Hunt 1973, 1978) was studying a 3-URU as a reflected CVC. His main interest was a chain slightly different from the one we presented: The base and platform R joints axes formed congruent triangles rather than Y-shapes. As the symmetry condition holds, CVC mode is also possible.

Hunt was amazed to discover the uncertainty configuration (a constraint singularity) in the central position, where the chain can enter translational mode—"a strange and fascinating phenomenon." He considered the use of the linkage as a coupling of parallel shafts (a translational PM) and to avoid the constraint singularity proposed to change the size of one of the triangles. Thus, by destroying the mirror symmetry between base and platform, he made reflected motion impossible and believed that the resulting coupling will no longer allow the shafts to angle.

It turns out, however, that when the base and platform are similar triangles, constraint singularities are not eliminated and only moved away from the center (and multiplied). Moreover, these special configurations still allow transition into a non-parallel state, and furthermore, the mechanism possesses several additional operation modes of which very little is known. More details including analysis of various modes 3-RERs with noncongruent base and platform are presented in (Zlatanov and Gosselin 2004).

The chain with triangular base and platform axes patterns is analyzed in Chap. 4 as well. This type of 3-UPU is known to many from the publications of Tsai in the late 1990s, although an 1992 patent exists (Appleberry 1992) and both post-date by decades the discussion of the kinematic chain by Hunt, who refers to even earlier practical CVC devices using similar linkages.

The Folding Rhombus: Next, the simplest case of mode changing: a four-bar linkage with equal lengths of the bars, Fig. 30.

When all four bars have different lengths, the C-space is homeomorphic to the figure eight; i.e., topologically, it may be described as two circles joined by a single common point (the flattened configuration). The C-space has two disjoint open connected nonsingular parts, and the mechanism can switch from one to the other (only) by passing through the IIM singularity. The two regions correspond to two different one-dof sets of poses of the coupler. However, in both regions, the allowed instantaneous coupler motion will be of the same type—an instantaneous rotation. (The screw system of the output freedoms remains the same.)

When all four bars are equal (Fig. 30), there are three constraint singularities, where the coupler has two instantaneous dof. The C-space of a four bars can be

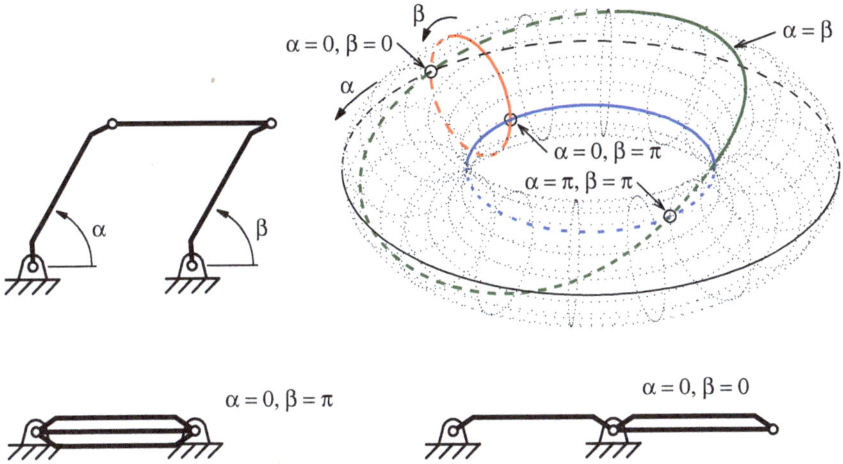

Fig. 30 C-space and singularities of the equal length four-bar mechanism

easily illustrated since it has a faithful projection on a torus, a configuration being defined in a unique way by (α, β), the base joint angles. Three smooth curves on the torus ($\alpha = 0$, $\beta = \pi$, and $\alpha = \beta$) form the C-space; their intersection points are the singularities. These three points divide the C-space into six nonsingular regions.

In Fig. 30, two distinct regions of the C-space may allow completely different coupler motions. While on one of the curves, $\alpha = \beta$, (i.e., in two of the C-space regions), the coupler remains parallel to the base, on each of the other two curves (four regions), the coupler rotates about a point fixed in the base. At two singularities, the mechanism has the options of entering two translational and two rotational regions, while at the third point, the four bars can choose among four rotational regions. We could say that the mechanism can be used both as a translational and as a rotational device and the transition between the two modes occurs via constraint singularities.

Stuck in Dead Center, Again: As our third example, we consider a state of the double-Y 3-RER which is not discussed in Chap. 4, but is closely related to some examples in the preceding sections above. It is illustrated by Fig. 31.

In each of the three legs, two revolute axes coincide. (This can happen because the base and platform joints form the same Y-shaped concurrent and coplanar pattern.) As a result, (i) part of the leg can spin about the common axis and (ii) only four leg joint twists are independent. With two constraints from each leg, if the legs are spun randomly about their axes, there will be six constraints and zero freedoms on the platform. However, there are special combinations of the three leg-spin angles for which the six constraints on the platform become dependent allowing it to escape from the lockup. In these cases, the mechanism as a whole will have four, five, or even six instantaneous dof. See (Zlatanov et al. 2002b) for more details.

It is interesting to note that this kind of "zero-dof working mode" is not detectable with methods (as in Chap. 4) which analyze parallel manipulators by parameterizing

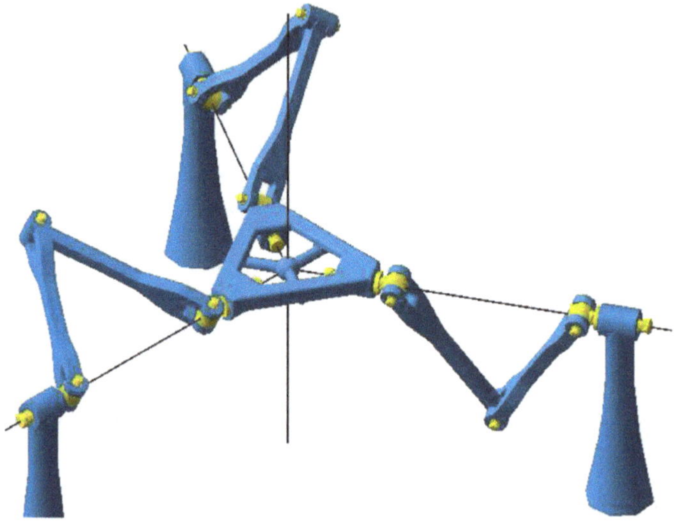

Fig. 31 A 3-URU chain in lockup mode

only the position and orientation of one body—the mobile platform. (Because the platform has the same pose for each of the infinitely many configurations obtained by the leg self-motions, and so a three-dimensional part of the C-space maps into a single point in the one-body parametrization.)

It turns out that not every IIM configuration is a self-intersection point of the configuration space. The planar curve given by the equation $y^3 + 2yx^2 - x^4 = 0$ defines an analytic manifold, but the point (0, 0) is singular. An analogous phenomenon has been recently discovered to occur with the configuration space of a single-loop linkage, see Chap. 5 and Müller and Zlatanov (2015). It is conceivable that the same applies to parallel manipulators.

7 Conclusions

In this chapter, we emphasized the perils of underestimating the capacity of mechanism to produce exceptional and exotic phenomena. Many conventional approaches to singularity analysis fail when faced with unforseen types of configurations, such as ones with purely passive self-motion (RPM) or constraint (IIM-type) singularities.

We also proposed a simple and more general approach, which allows to handle such cases. It is based on defining singularity in purely instantaneous terms as the solvability of both the forward and inverse kinematics problems.

It should be emphasized that in many cases it may be perfectly justified to use less general formulations, such as input–output equations. For example, if workspace or joints limits exclude the troublesome singularities where the simple methods fail.

However, one needs to be aware of the dangers in order to know how to decide when a more limited approach is harmless and when it can lead to disaster.

Another qualification needs to be added. The approach proposed in the chapter is also subject to a trade-off of simplicity and generality. Indeed, the definitions (and the related identification methods, see also Chap. 2) are relatively simple because they are formulated entirely in velocity terms and are based on the instantaneous kinematics relationships. However, this entails its own limitations.

For example, the interpretation of configurations as singular of various types depends on the choice of input and output. In some cases, a "bad choice," i.e., if the input–output device is "badly designed" it may be that certain degenerate properties, such as impossible input or output persist in whole branches of the configuration space, making it difficult to call them "singular."

Similar difficulties may be related with the choice of n, the number of inputs and outputs. It is supposed to be equal to the full-cycle dof of the kinematic chain, but this is not an instantaneous property, and its definition may be unclear for so-called kinematotropic mechanisms, where different parts of the configuration space have different dimensions.

References

W.T. Appleberry, Anti-rotation positioning mechanism (1992)

I.A. Bonev, D. Zlatanov, C.M. Gosselin, Singularity analysis of 3-DOF planar parallel mechanisms via screw theory. *J. Mech. Des.* **125**, 573 (2003)

D. Chablat, P. Wenger, Architecture optimization of a 3-DOF translational parallel mechanism for machining applications, the Orthoglide. IEEE Trans. Robot. **19**(3), 403–410 (2003)

C.M. Gosselin, J. Angeles, Singularity analysis of closed-loop kinematic chains. *IEEE Trans. Robot. Autom.* **6**(3), 281–290 (1990)

Z. Huang, W.S. Tao, Y.F. Fang, Study on the kinematic characteristics of 3 DOF in-parallel actuated platform mechanisms. *Mech. Mach. Theory* **31**(8), 999–1007 (1996)

K.H. Hunt, *Kinematic Geometry of Mechanisms* (Oxford University Press, 1978)

K.H. Hunt, Constant-velocity shaft couplings: a general theory. *Trans. ASME J. Eng. Ind.* **95B**, 455–464 (1973)

F. Majou, P. Wenger, D. Chablat, Design of a 3 axis parallel machine tool for high speed machining: the Orthoglide, in *Proceedings 4th International Conference Integrated Design and Manufacturing in Mechanical Engineering*, arXiv:0705.1271 [cs.RO], June 2007

J.-P. Merlet, Singular configurations of parallel manipulators and Grassmann geometry. *Int. J. Robot. Res.* **8**(5), 45–56 (1989)

A. Müller, D. Zlatanov, Kinematic singularities of mechanisms revisited, in *IMA Conference on Mathematics of Robotics (IMAR)* (2015). https://doi.org/10.19124/ima.2015.001.26

D. Zlatanov, *Generalized Singularity Analysis of Mechanisms*. Ph.D. thesis, University of Toronto (1998)

D. Zlatanov, M.Q. Dai, R.G. Fenton, B. Benhabib, Mechanical design and kinematic analysis of a three-legged six degree-of-freedom parallel manipulator, in *Robotics, Spatial Mechanisms, and Mechanical Systems, ASME*, vol. 45 (1992), pp. 529–536

D. Zlatanov, R.G. Fenton, B. Benhabib, Singularity analysis of mechanisms and robots via a motion-space model of the instantaneous kinematics, in *Proceedings of the IEEE International Conference on Robotics and Automation, ICRA (San Diego, USA)* (1994a), pp. 980–985

D. Zlatanov, R.G. Fenton, B. Benhabib, Singularity analysis of mechanisms and robots via a velocity-equation model of the instantaneous kinematics, in *Proceedings of the IEEE International Conference on Robotics and Automation, ICRA (San Diego, USA)* (1994b), pp. 986–991

D. Zlatanov, I.A. Bonev, C.M. Gosselin, Constraint singularities as C-Space singularities, in *Advances in Robot Kinematics: Theory and Applications*, ed. by J. Lenarcic, F. Thomas (Kluwer Academic Publishers, 2002a), pp. 183–192

D. Zlatanov, I.A. Bonev, C.M. Gosselin, Constraint singularities of parallel mechanisms, in *Proceedings of the IEEE International Conference on Robotics and Automation, ICRA (Washington D.C., USA)*, vol. 1 (2002b), pp. 496–502

D. Zlatanov, C.M. Gosselin, On the kinematic geometry of 3-RER parallel mechanisms, in *Proceedings of the 11th IFToMM World Congress in Mechanism and Machine Science, Tianjin, China* (2004)

Singularity Set Computation: A Hands-On Session with the CUIK Suite

Oriol Bohigas

Abstract This chapter provides an introduction to the analysis of the singularities of robot mechanisms using the CUIK suite software. The CUIK suite is an open-source toolbox for motion analysis of general closed-chain mechanisms, resulting from several years of research and development within the Kinematics and robot design group at the Institut de Robòtica i Informàtica Industrial. It is available under GPLv3 license from the CUIK Project home page. The intention is not to provide thorough definitions or developments, but to illustrate the basic concepts around singularity analysis through a short and simple presentation. The text assumes basic knowledge of the kinematics of robot mechanisms, such as the notions of configuration space, or the forward and inverse kinematics problems. An exhaustive description of the methods, algorithms, and the underlying mathematical concepts used by the CUIK suite can be found in Bohigas et al. (2016). For a better understanding, we do strongly recommend to install the CUIK suite and execute the examples provided while reading this chapter.

1 Introduction

The so-called singularities of a robot mechanism are special postures of the mechanism in which the kinetostatic behavior may degrade dramatically. Rigidity or dexterity losses arise, and there may appear unresolvable or uncontrollable end-effector forces, among other effects (Bohigas et al. 2016). Singularities generally pose problems for the normal operation of a robot, and thus, they should be analyzed before the actual construction of a prototype.

The CUIK suite provides a toolbox of methods for singularity computation and avoidance to this end. They are applicable to nonredundant mechanisms of large generality and allow a detailed study in the initial phases of the development of a mechanism. The computation of the singularity set provides comprehensive

---

O. Bohigas (✉)
Institut de Robòtica i Informàtica Industrial, CSIC-UPC, Barcelona, Spain
e-mail: bohigas@upc.edu; obohigas@iri.upc.edu

A. Müller and D. Zlatanov (eds.), *Singular Configurations of Mechanisms and Manipulators*, CISM International Centre for Mechanical Sciences 589,
https://doi.org/10.1007/978-3-030-05219-5_2

information on the local and global motion capabilities of a mechanism: Its projections onto the task and joint spaces determine the working regions in such spaces, may inform about the existence of different assembly configurations, and highlight areas where control or dexterity losses may arise. These projections also supply a fair view of the feasible movements of the mechanism, although they do not reveal all possible singularity-free motions. In order to also tackle this issue, the CUIK suite includes general path planning methods that can generate movements of the mechanism that avoid problematic singularities (Porta et al. 2014).

Overall, the CUIK suite can help eliminate barriers in design creativity, and it contributes to the general understanding on how the motions of complex multibody systems can be predicted, planned, and controlled in an efficient and reliable way.

This chapter concentrates on singularity computation, and the interested reader on singularity avoidance can refer to Bohigas et al. (2016).

1.1 Basic Concepts

Configuration space: Usually, all the feasible configurations of a mechanism can be characterized by a system of equations

$$\Phi(\mathbf{q}) = \mathbf{0}, \tag{1}$$

which expresses the assembly constraints between the different bodies of the robot, and $\mathbf{q}$ is the vector of configuration coordinates (see Sect. 2.2.1 of Bohigas et al. (2016)). The set of these possible configurations is known as the configuration space, or C-space, of the mechanism.

Velocity equation: For a given configuration $\mathbf{q}$ of the mechanism, it is possible to find a velocity equation

$$\mathbf{L}\,\mathbf{m} = \mathbf{0}, \tag{2}$$

which gives all possible motions of the manipulator at that specific configuration.

The velocity equation can be obtained in several ways, for example, by differentiating the assembly constraints, or using the tools of screw theory, or by other means (Zlatanov 1998). $\mathbf{L}$ is a matrix that depends on the configuration, and $\mathbf{m}$ is a velocity vector that globally characterizes the velocity state of the manipulator. In this vector, we can distinguish three different blocks. We have the output velocities that usually encode the velocity of a point, or the angular velocity or the twist of some body. The input velocities, which represent the velocities of the acutated joints, and the rest, which we call passive velocities:

$$\mathbf{m} = \begin{bmatrix} \mathbf{m}_u \\ \mathbf{m}_v \\ \mathbf{m}_p \end{bmatrix} \begin{matrix} \rightarrow output \\ \rightarrow input \\ \rightarrow passive \end{matrix}$$

Forward and inverse instantaneous kinematics problem: In general, the instantaneous kinematic analysis of a manipulator addresses two main problems:

- The *forward* instantaneous kinematics problem (FIKP): find $\mathbf{m}$ given the input velocity $\mathbf{m}_\upsilon$.
- The *inverse* instantaneous kinematics problem (IIKP): find $\mathbf{m}$ given the output velocity $\mathbf{m}_u$.

Note that in both cases, it is required to find *all* velocity components of $\mathbf{m}$, not just those referring to the output or input velocities. Following Zlatanov (1998), a configuration is said to be *nonsingular* when both the FIKP and the IIKP have unique solutions for any input or output velocity, and *singular* otherwise.

Forward and inverse singularities: Let $\mathbf{L}_y$, $\mathbf{L}_z$, and $\mathbf{L}_p$ be the submatrices of $\mathbf{L}$ obtained by removing the columns corresponding to the input, output, and both the input and output velocities, respectively. It is easy to see that the singular configurations are those in which either $\mathbf{L}_y$ or $\mathbf{L}_z$ is rank deficient (Bohigas et al. 2016).

Recall that if a matrix is rank deficient, its kernel has to be nonnull and, in particular, it must include a vector of unit norm. Thus, all singular configurations can be determined by solving the following two systems of equations:

$$\left.\begin{array}{l}\Phi(\mathbf{q})=\mathbf{0}\\ \mathbf{L}_y^{\mathsf{T}}\xi=\mathbf{0}\\ \|\xi\|^2=1\end{array}\right\},\quad \left.\begin{array}{l}\Phi(\mathbf{q})=\mathbf{0}\\ \mathbf{L}_z^{\mathsf{T}}\xi=\mathbf{0}\\ \|\xi\|^2=1\end{array}\right\}. \tag{3}$$

The first line of each system constrains $\mathbf{q}$ to be a feasible configuration of the mechanism, and the second and third lines enforce the existence of a nonzero vector in the kernel of the corresponding matrix. The solutions of the system on the left include all singularities where the FIKP is indeterminate, called *forward singularities*, while the solutions of the system on the right include all singularities where the IIKP is indeterminate, called *inverse singularities*.

All singularities of a mechanism can thus be obtained as the union of the solution sets of the systems of equations in Eq. (3).

1.2 An Illustrative Example

To illustrate the previous concepts on a very simple example, consider the three-slider mechanism in Fig. 1. Let (x_P, y_P) denote the coordinates of points $P = A$, B, or C relative to the reference frame OXY in the figure, and let L_1 and L_2 be the lengths of the connector links. Clearly, a configuration of the mechanism can be shortly described by the tuple $\mathbf{q} = (y_A, y_B, x_C)$, because $x_A = x_B = y_C = 0$ in any configuration. Since the distances from A to B and from B to C must be kept equal to L_1 and L_2, the assembly constraints for this mechanism can be written as

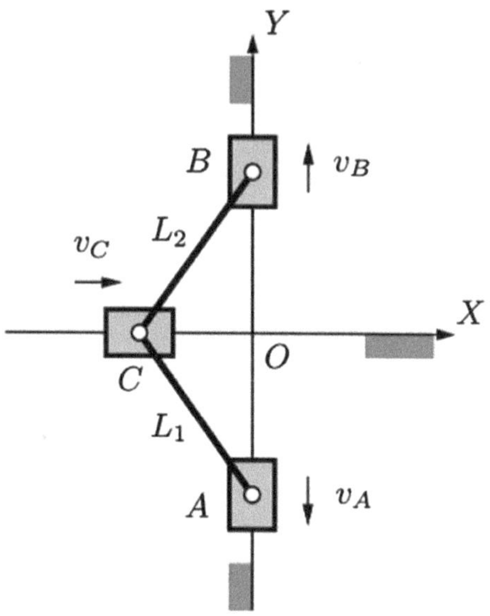

Fig. 1 A 1-DOF mechanism with three sliders. The prismatic joints at A and B are on a line perpendicular to the axis of the prismatic joint at C

$$\left.\begin{array}{l} {y_A}^2 + {x_C}^2 = {L_1}^2 \\ {y_B}^2 + {x_C}^2 = {L_2}^2 \end{array}\right\}, \tag{4}$$

which corresponds to Eq. (1) for this mechanism. It is easy to realize that in this case the C-space corresponds to the intersection of two cylinders in the (y_A, y_B, x_C) space, as illustrated in Fig. 2.

The velocity equation in Eq. (2) could now be obtained using the revolute- and prismatic joint screws (Zlatanov 1998), but a more compact expression can here be derived by differentiating Eq. (4) with respect to time. Taking v_A and v_B as the input and output velocities, the differentiation yields

$$\mathbf{L} \cdot \mathbf{m} = \begin{bmatrix} 0 & 2y_A & 2x_C \\ 2y_B & 0 & 2x_C \end{bmatrix} \cdot \begin{bmatrix} v_B \\ v_A \\ v_C \end{bmatrix} = \mathbf{0}.$$

From here, we rapidly obtain

$$\mathbf{L}_y = \begin{bmatrix} 0 & 2x_C \\ 2y_B & 2x_C \end{bmatrix},$$

$$\mathbf{L}_z = \begin{bmatrix} 2y_A & 2x_C \\ 0 & 2x_C \end{bmatrix},$$

$$\mathbf{L}_p = \begin{bmatrix} 2x_C \\ 2x_C \end{bmatrix},$$

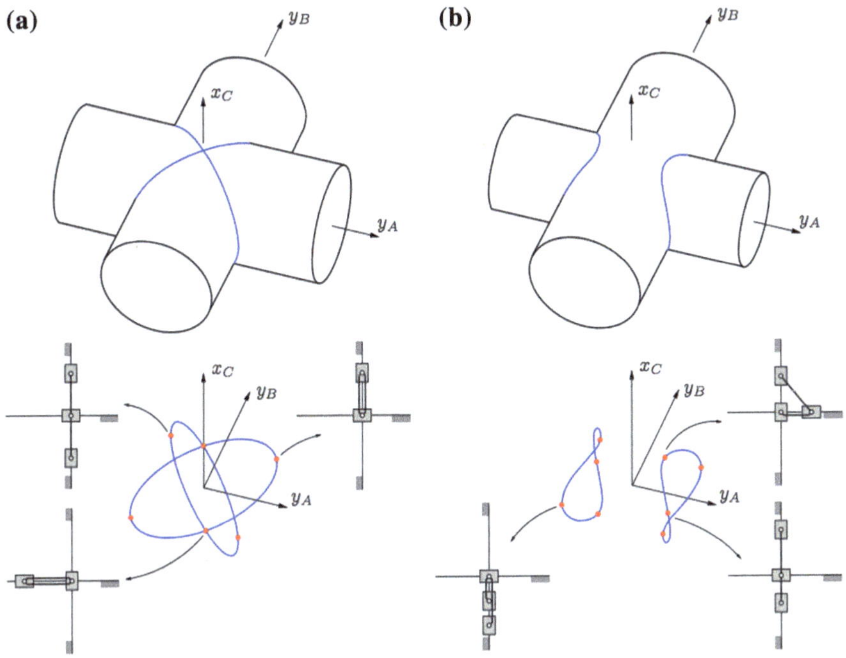

Fig. 2 C-space and singularities of the three-slider mechanism

which allow defining any of the systems in Eq. (3).

These systems can be solved analytically in this case. For example, if $L_1 = L_2 = 1$, the C-space has a single connected component composed of two ellipses intersecting on the x_C-axis (Fig. 2a), and the solutions of the systems reveal that the singularity set has six isolated configurations, marked in red in Fig. 2a, bottom, with the following values of $\mathbf{q}$:

$$(0, 0, 1),\ (0, 0, -1),$$
$$(1, 1, 0),\ (1, -1, 0),$$
$$(-1, -1, 0),\ (-1, 1, 0).$$

All of these configurations satisfy both systems in Eq. (3), so they are all forward and inverse singularities because both the FIKP and the IIKP are indeterminate on them. Therefore, the control of the input or the output velocity does not determine the overall motion of the mechanism. Note that the C-space self-intersects at two singularities and presents a bifurcation that allows to change the mode of operation from both sliders moving on the same side of the horizontal axis, $y_A y_B \geq 0$, to one slider moving on each side, $y_A y_B \leq 0$.

The topology of the C-space changes when $L_1 \neq L_2$, since it no longer presents any bifurcation. It is instead formed by two connected components (Fig. 2b). By solving Eq. (3) for $L_1 = 1$ and $L_2 = 0.8$, eight singularities are obtained (Fig. 2b, bottom):

$$\begin{array}{ll} (1, 0.8, 0), & (-1, -0.8, 0), \\ (-1, 0.8, 0), & (0.6, 0, -0.8), \\ (1, -0.8, 0), & (-0.6, 0, 0.8), \\ (0.6, 0, 0.8), & (-0.6, 0, -0.8). \end{array}$$

In this case, to change the operation mode from $y_A \geq 0$ to $y_A \leq 0$ the mechanism has to be disassembled.

2 Singularity Set Computation with the CUIK Suite

The previous example was simple enough so that the systems of equations describing the singularities could be solved by hand, but this is usually not the case. This section shows how to use the CUIK suite to obtain the singularities of the Dextar, a five-bar parallel robot designed to support teaching and research, with a more interesting singularity set (Campos et al. 2010).

Despite its simple architecture, Dextar exhibits forward and inverse singularities, which requires careful analysis and planning to keep it under control and to exploit its full motion capabilities.

The section will guide you in the process of computing Dextar's singularities: from the formulation of the assembly constraints and singularity set equations, to their numerical solution, to the visualization and interpretation of the results. Along the way, we shall explain the syntax of the involved CUIK files and commands.

2.1 Preparation

For a better understanding, we suggest you to read Sects. 2.1 and 2.2 of Bohigas et al. (2016) and, ideally, also Sects. 3.1–3.4. You should have the CUIK suite installed in your computer. Refer to the instructions in the CUIK Project home page if you have not installed it yet.

During this example, we shall build few files incrementally. We recommend you to follow the instructions below, but you can download the finished files as well if you wish at Tutorial Solutions.

Throughout, we will call CUIK's commands from a UNIX/Linux terminal window. Occasionally, we shall give computation times. Take them as rough estimations. The actual time will depend on your machine. Ours was an iMac 2010 with a 2.93 GHz Intel Core i7 processor and 12 Gb of RAM, running Mac OSX "Yosemite". Unless where noted, we used just one of the eight CPUs of the machine.

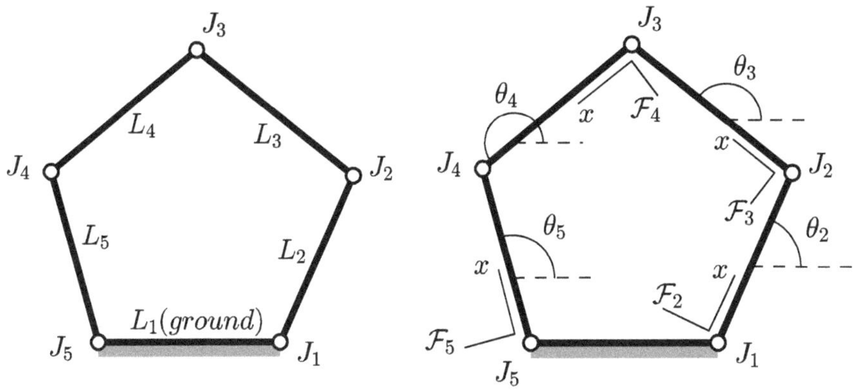

Fig. 3 Geometry of the Dextar robot. A reference frame F_i is attached to each link of the Dextar robot. Also, the orientation of each link is represented by angle θ_i

2.2 *Dextar's Geometry*

Dextar is a planar five-bar robot. It can be thought of as two-serial, three-revolute arms that share their distal joint. We start labeling its joints and links as $J_1, \ldots, J_5$ and $L_1, \ldots, L_5$, respectively (Fig. 3). We assume that l_i is the length of link L_i. The distance between J_1 and J_5 is d. We shall adopt the parameters of the Mecademic version of Dextar:

$$l_2 = l_3 = l_4 = l_5 = 90\,\text{mm}$$
$$d = 118\,\text{mm}$$

The actuated joints are J_1 and J_5. The end-effector is located in joint J_3, and only its x and y coordinates can be controlled.

For convenience, we assign a reference frame to each link L_i, indicated with F_i. We mark the x-axis of each frame with an "x". Frame F_1 is centered in J_5 and acts as the absolute frame. Frames $F_2, \ldots, F_5$ are centered and oriented as indicated in their respective links (Fig. 3).

We next see how to formulate Dextar's assembly constraints and the conditions characterizing its singularity locus. For simplicity, we shall derive the latter by taking the time derivative of Eq. (1). This will require us to include both the input and output coordinates in $\mathbf{q}$:

- The input coordinates will be θ_2 and θ_5.
- The output coordinates will be J_{3x} and J_{3y}.

2.3 Assembly Constraints

Equation (1) has to be formulated in a form that allows it to be converted into quadratic equation. There are basically two ways of achieving so: using joint-assembly constraints or using loop-closure constraints. We next explain both.

Using joint-assembly constraints: Under this method, we formulate Eq. (1) by gathering Eq. 3.1 in Bohigas et al. (2016):

$$\mathbf{r}_j + \mathbf{R}_j \mathbf{p}_i^{F_j} = \mathbf{r}_k + \mathbf{R}_k \mathbf{p}_i^{F_k}$$

for all joints J_i of the mechanism, which forces the assembly of links L_j and L_k at J_i (see Sect. 3.2.1 of Bohigas et al. (2016)), and:

- $\mathbf{r}_j$ is the absolute position vector of the origin frame F_j.
- $\mathbf{R}_j$ is the rotation matrix giving the orientation of F_j.
- $\mathbf{p}_i^{F_j}$ is the position vector of J_i in frame F_j, written in the vector basis of F_j.

Since this is a planar linkage, Eq. 3.2 in Bohigas et al. (2016) is not needed in this case. If we write the equation for each joint J_i, we obtain

$$\begin{bmatrix} d \\ 0 \end{bmatrix} = \begin{bmatrix} J_{1x} \\ J_{1y} \end{bmatrix} \tag{5}$$

$$\begin{bmatrix} J_{1x} \\ J_{1y} \end{bmatrix} + \mathbf{R}_2 \begin{bmatrix} l_2 \\ 0 \end{bmatrix} = \begin{bmatrix} J_{2x} \\ J_{2y} \end{bmatrix} + \mathbf{R}_3 \begin{bmatrix} 0 \\ 0 \end{bmatrix} \tag{6}$$

$$\begin{bmatrix} J_{2x} \\ J_{2y} \end{bmatrix} + \mathbf{R}_3 \begin{bmatrix} l_3 \\ 0 \end{bmatrix} = \begin{bmatrix} J_{3x} \\ J_{3y} \end{bmatrix} + \mathbf{R}_4 \begin{bmatrix} 0 \\ 0 \end{bmatrix} \tag{7}$$

$$\begin{bmatrix} J_{3x} \\ J_{3y} \end{bmatrix} + \mathbf{R}_4 \begin{bmatrix} l_4 \\ 0 \end{bmatrix} = \begin{bmatrix} J_{5x} \\ J_{5y} \end{bmatrix} + \mathbf{R}_5 \begin{bmatrix} l_5 \\ 0 \end{bmatrix} \tag{8}$$

$$\begin{bmatrix} J_{5x} \\ J_{5y} \end{bmatrix} = \begin{bmatrix} 0 \\ 0 \end{bmatrix} \tag{9}$$

which altogether form the system of Eq. (1). Note that, using these equations,

$$\mathbf{q} = (J_{3x}, J_{3y}, J_{1x}, J_{1y}, J_{2x}, J_{2y}, J_{5x}, J_{5y}, \theta_3, \theta_4, \theta_2, \theta_5),$$

where $\mathbf{v} = (\theta_2, \theta_5)$ and $\mathbf{u} = (J_{3x}, J_{3y})$, encompasses the input and output coordinates of the mechanism, respectively.

Since

$$\mathbf{R}_i = \begin{bmatrix} \cos\theta_i & -\sin\theta_i \\ \sin\theta_i & cos\theta_i \end{bmatrix},$$

Equations (5)–(9) are not quadratic and, thus, not directly solvable using CUIK. However, we can apply the change of variables

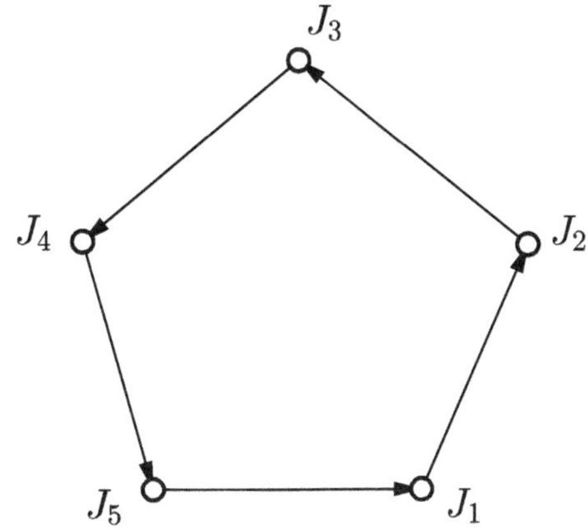

Fig. 4 Dextar robot has only one kinematic loop

$$c_i = \cos\theta_i$$
$$s_i = \sin\theta_i c$$

and include the circle equations

$$s_i^2 + c_i^2 = 1 \quad (10)$$

to readily obtain a quadratic system solvable by the CUIK suite.

This is a common technique in any system to be solved by CUIK. We first formulate the systems without worrying about the appearance of sines and cosines, and finally algebrize the systems with the mentioned change of variables.

In spatial mechanisms, we can directly formulate the systems in algebraic form following Sect. 3.2 in Bohigas et al. (2016).

Using loop-closure constraints: The previous method is simple, but sometimes it leads to a system $\Phi(\mathbf{q}) = \mathbf{0}$ with too many variables and equations. More compact systems of equations can often be obtained using loop-closure constraints and graph theory tools.

Dextar has only one kinematic loop. Forcing the assembly of this loop is equivalent to constraining the inter-joint vectors to form a closed polygon (Fig. 4). This condition can be expressed with the equations

$$d + l_2\cos\theta_2 + l_3\cos\theta_3 + l_4\cos\theta_4 - l_5\cos\theta_5 = 0 \quad (11)$$
$$l_2\sin\theta_2 + l_3\sin\theta_3 + l_4\sin\theta_4 - l_5\sin\theta_5 = 0 \quad (12)$$

which directly characterize the C-space of the robot.

Note that Eqs. (11) and (12) contain the input coordinates θ_2 and θ_5 as variables, but not the output coordinates J_{3x} and J_{3y}. To introduce the latter, we can simply add the equations

$$J_{3x} = d + l_2 \cos\theta_2 + l_3 \cos\theta_3 \tag{13}$$

$$J_{3y} = l_2 \sin\theta_2 + l_3 \sin\theta_3. \tag{14}$$

Then, Eq. (1) is the system formed by Eqs. (11)–(14), and the configuration vector of the system is

$$\mathbf{q} = (J_{3x}, J_{3y}, \theta_3, \theta_4, \theta_2, \theta_5),$$

again with $\mathbf{v} = (\theta_2, \theta_5)$ and $\mathbf{u} = (J_{3x}, J_{3y})$ encompassing the input and output coordinates of the mechanism, respectively.

Incidentally, note that Eqs. (11) and (12) can be obtained by adding Eqs. (5)–(9). This is true in a general mechanism. The loop-closure constraint of any kinematic loop can be obtained by adding the joint-assembly constraints of the loop joints. If our robot has multiple loops, Eq. (1) will include the loop-closure equations of a set of fundamental cycles of its connectivity graph. For details, see Porta et al. (2009), which also shows that the two formulations (via joint-assembly or loop-closure constraints) are equivalent.

Which formulation is better? Since Eqs. (11)–(14) involve less equations and variables than Eqs. (5)–(9), we shall prefer them to formulate the systems of the various singularity sets of Dextar.

Note however that, in other cases, there might be no gain in using loop-closure equations. For example, the general formulation of Bohigas et al. (2016) uses screw theory to formulate the velocity equation (Sect. 3.2.2) which requires the joints P_i to be available (see Table 3.1 of Bohigas et al. (2016) and the corresponding text). In such a case, it might be simpler to use joint-assembly constraints, which make all P_i points explicit and ready-to-use in the system.

2.4 *Inverse Singularity Conditions*

Inverse singularities occur when the mechanism is found to be shaky after locking its outputs. In Dextar, this happens when the left or right arms happen to be aligned. For example, if the left arm is aligned, we can feel the shakiness shown in Fig. 5.

Recall that such singularities are the configurations $\mathbf{q}$ that satisfy the right system in Eq. (3), where $\mathbf{L_z} = \Phi_\mathbf{z}(\mathbf{q})$ as shown in Sect. 2.2.1 of Bohigas et al. (2016). Since in our case $\Phi(\mathbf{q}) = \mathbf{0}$ is defined by Eqs. (11)–(14), and

$$\mathbf{z} = (\theta_3, \theta_4, \theta_2, \theta_5),$$

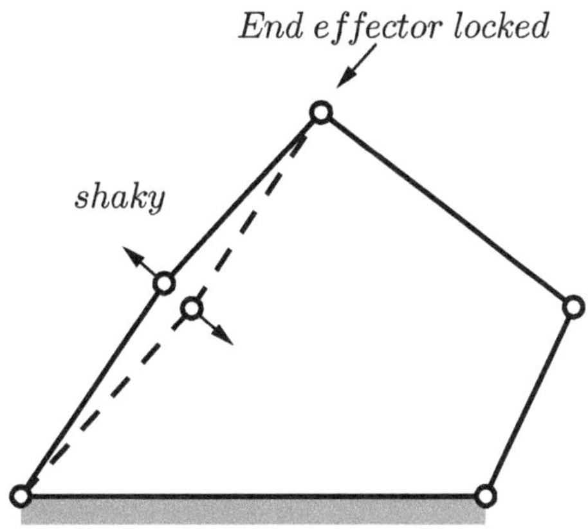

Fig. 5 An inverse singularity happens when the left (or right) arm of the Dextar is aligned

we have

$$\Phi_{\mathbf{z}}(\mathbf{q}) = \begin{bmatrix} -l_3 \sin\theta_3 & -l_4 \sin\theta_4 & -l_2 \sin\theta_2 & l_5 \sin\theta_5 \\ l_3 \cos\theta_3 & l_4 \cos\theta_4 & l_2 \cos\theta_2 & -l_5 \cos\theta_5 \\ -l_3 \sin\theta_3 & 0 & -l_2 \sin\theta_2 & 0 \\ l_3 \cos\theta_3 & 0 & l_2 \cos\theta_2 & 0 \end{bmatrix}. \tag{15}$$

The rows of $\Phi_{\mathbf{z}}(\mathbf{q})$ correspond to the partial derivatives of Eqs. (11)–(14) with respect to the $\mathbf{z}$ variables.

Again, the resulting system of equations is not quadratic, but it can be converted into a quadratic one by replacing the sines and cosines of θ_i by s_i and c_i, and adding the corresponding circle equations.

2.5 Forward Singularity Conditions

The forward singularities are the configurations in which the mechanism is shaky after locking the actuators. In Dextar, this happens when J_2, J_3, and J_4 happen to be aligned, as shown in Fig. 6.

These are the configurations that satisfy the left system in Eq. (3), with $\mathbf{L}_{\mathbf{y}} = \Phi_{\mathbf{y}}(\mathbf{q})$. In our case

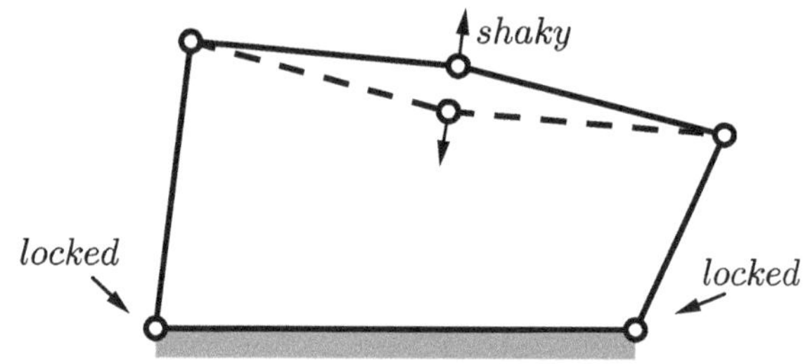

Fig. 6 A forward singularity happens when J_2, J_3, and J_4 are aligned

$$\Phi_{\mathbf{y}}(\mathbf{q}) = \begin{bmatrix} 0 & 0 & -l_3 \sin\theta_3 & -l_4 \sin\theta_4 \\ 0 & 0 & l_3 \cos\theta_3 & l_4 \cos\theta_4 \\ -1 & 0 & -l_3 \sin\theta_3 & 0 \\ 0 & -1 & l_3 \cos\theta_3 & 0 \end{bmatrix}, \tag{16}$$

where the rows correspond to the partial derivatives of Eqs. (11)–(14) with respect to the $\mathbf{y}$ variables

$$\mathbf{y} = (J_{3x}, J_{3y}, \theta_3, \theta_4).$$

2.6 Computation and Output Plots

We next see how to prepare the CUIK input files associated with the previous systems of equations and how to solve them using CUIK. We shall compute, in sequence, the following sets:

- The full C-space;
- The inverse singularity locus;
- The forward singularity locus.

These correspond to the sets indicated on the top of Fig. 7. Strictly speaking, the silhouettes of the C-space relative to the input and output spaces are called the input and output singularity loci. This is because these silhouettes are only defined for regular points of the C-space. However, Dextar does not have C-space singularities, also known as increased instantaneous mobility singularities (Zlatanov 1998). Thus, its inverse and forward singularity loci coincide with its output and input singularity sets, respectively.

Once computed, the three sets will be projected to the input and output spaces, producing what we call the input and output portraits of the mechanism (see Sect. 3.4 of Bohigas et al. (2016)). These portraits are also indicated in the bottom of Fig. 7.

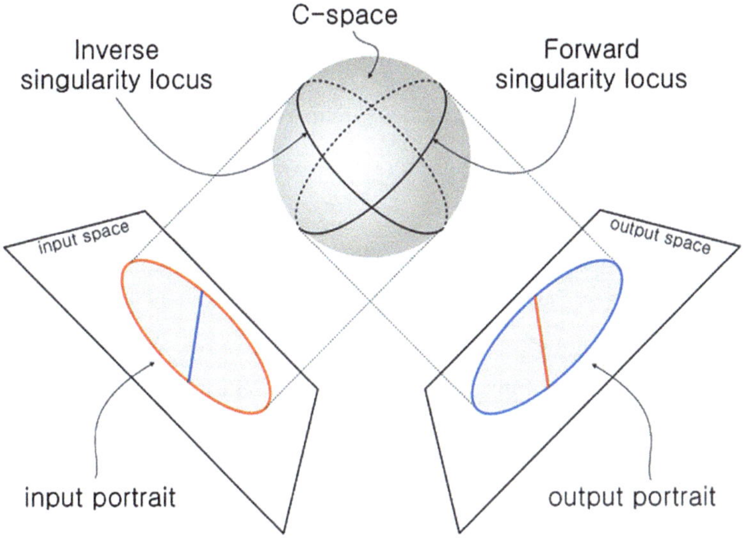

Fig. 7 Singularities seen as C-space silhouettes (in an abstract C-space)

Computing the configuration space: Open a new file with a text editor, and save it with the name `cspace.cuik`. Any cuik file has three sections which encompass, respectively, definitions of constants used in the equations, the list of equation variables, and the equations themselves:

```
[CONSTANTS]
...
[SYSTEM VARS]
...
[SYSTEM EQS]
...
```

In `cspace.cuik` type

```
[CONSTANTS]
l := 0.90
d := 1.18
l2 := 1
l3 := 1
l4 := 1
l5 := 1
```

These constants define the five link lengths. Although all moving links are of the same length in Dextar, we prefer to define the link lengths independently using "$l_i = value$" because this will allow us to test different geometric parameters.

Now let us define the variables in the system. Type in

```
[SYSTEM VARS]
J3x:[-l4-l4,l4+l5]
J3y:[-l4-l4,l4+l5]
```

The syntax is clear. A variable is created by specifying its name, followed by ":" and its feasibility interval. Since the position of joint J_3 is constrained to lie inside a circle of radius $l_4 + l_5$ centered in the origin of the absolute frame, J_{3x} and J_{3y} can be given the previous feasibility intervals.

We also define the sines and cosines of all angles $\theta_2, \ldots, \theta_5$, which take values in the range $[-1, 1]$:

```
ct2:[-1,1]
st2:[-1,1]
ct3:[-1,1]
st3:[-1,1]
ct4:[-1,1]
st4:[-1,1]
ct5:[-1,1]
st5:[-1,1]
```

Finally, we add Eqs. (11)–(14) and the circle equations relating the cti and sti variables:

```
[SYSTEM EQS]

% Loop equations
d + l2*ct2 + l3*ct3 + l4*ct4 - l5*ct5 = 0;
l2*st2 + l3*st3 + l4*st4 - l5*st5 = 0;

% Position of J3
J3x = d + l2*ct2 + l3*ct3;
J3y = l2*st2 + l3*st3;

% Circle equations due to angle algebraizations
ct2^2 + st2^2 = 1;
```

```
ct3^2 + st3^2 = 1;
ct4^2 + st4^2 = 1;
ct5^2 + st5^2 = 1;
```

Pay attention at the syntax employed to define sums, products, and the square of a variable, and note that all equations must be finished with a semicolon. Also observe that a percentage sign "%" can be used to insert a comment. It can be placed anywhere, and the result is that any text to the right of "%" will be neglected.

Now copy the following parameters into a new file `cspace.param`:

```
SIGMA = 0.2
EPSILON = 1e-6
RHO = 0.95
ERROR_SPLIT = TRUE
DUMMIFY = 1
LR2TM_Q =0.5
LR2TM_S = 0.1
MAX_NEWTON_ITERATIONS = 0
```

From the point of view of a user, the most important parameter is SIGMA. It defines the resolution at which the solution set will be computed. The smaller the sigma, the higher the resolution. SIGMA, in fact, is the maximum edge length allowed to any of the solution boxes returned.

Now we are ready to launch the first computation. Execute

```
cuik cspace
```

This will throw a bunch of numbers showing CUIK's progress on solving the equations. Their meaning is not much relevant to an end user.

Wait until the computation finishes (about 45 s in our case). Once it does, you will see that CUIK has created the file `cspace.sol`, which contains, in order:

- The numerical parameters used (SIGMA, RHO, and so on);
- The solved system of equations;
- A list of the solution boxes returned;
- Execution statistics.

To see the latter statistics, type

```
tail -30 cspace.cuik
```

We can inspect the `*.sol` file with a text editor if we wish, but a more useful choice is to obtain a plot of the solution boxes. Since there are ten variables defined in `cpsace.cuik`, the boxes returned will be ten dimensions, and we can only observe their projections onto three of the ten coordinates.

To obtain such projections, use cuikplot3d to convert `cspace.sol` to a file `cspace.gcl`:

```
cuikplot3d cspace 1 2 3 0 cpsace.gcl
```

The extension gcl stands for "Geomview Command Language", and the created `*.gcl` file contains all graphic commands needed to visualize the boxes using Geomview. In the previous command:

- `cspace` is the name of the input `*.sol` file.
- 1, 2, 3 are the dimensions onto which we project the solution boxes: 1 means the first variable that appears in `cspace.cuik`, 2 means the second variable, and so on (in our case, these three variables are J3x, J3y, and ct2).
- 0 indicates that no artificial enlarging of the box is to be applied. If you use a number different from zero here, then the output boxes will have their edge lengths equal to this number. This is useful to visualize boxes that are too tiny in comparison to other boxes.
- `cspace.gcl` is the desired output file.

Now we can readily view the boxes with Geomview. Type

```
geomview cspace.gcl
```

We can now obtain the views of the configuration space in Fig. 8, top. Nicer plots can be obtained by using a smaller SIGMA value. For example, using SIGMA = 0.05 (and waiting about 10 min), we get the refined plots in Fig. 8, bottom.

Observe that the first plot provides Dextar's workspace because it is the projection onto the J_{3x} and J_{3y} coordinates. The computation of the whole C-space, therefore, provides a simple way of obtaining workspace maps of any of the mechanism coordinates. In many cases, however, we shall prefer to obtain the workspace by finding a generalized singularity set, as explained in Chap. 4 of Bohigas et al. (2016). Such a procedure is advantageous as it yields the boundary and interior barriers of the workspace, and it also takes joint limits into account. It is often faster too because the boundaries and barriers are, generically, of lower dimension.

Computing the inverse singularity locus: Let us prepare the `*.cuik` file of the inverse singularity locus. To do so, we shall extend `cspace.cuik` with new equations forcing the rank deficiency of $\mathbf{\Phi_z(q)}$.

First copy `cspace.cuik` to `inv.cuik` and `cspace.param` to `inv.param`:

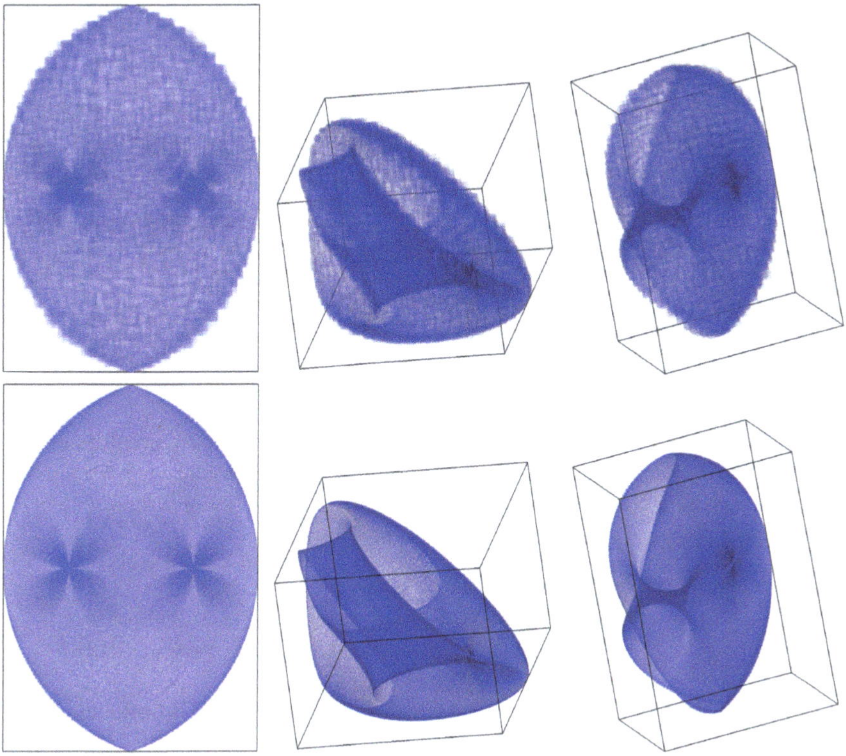

Fig. 8 Three different views of the C-space of the Dextar robot, using a different value for SIGMA (left and right)

```
cp cspace.cuik inv.cuik
cp cspace.param inv.param
```

and open `inv.cuik` with a text editor. Add the following variables to its `[SYSTEM VARS]` section:

```
[SYSTEM VARS]
...
xi1:[-1,1]
xi2:[-1,1]
xi3:[-1,1]
xi4:[-1,1]
```

These xi are the components of the ξ vector of Eq. (3). Now add the rank deficiency equations to the `[SYSTEM EQS]` section:

```
[SYSTEM EQS]
...
% Inverse singularity conditions
-l3*st3*xi1 - l4*st4*xi2 - l2*st2*xi3 + l5*st5*xi4 = 0;
 l3*ct3*xi1 + l4*ct4*xi2 + l2*ct2*xi3 - l5*ct5*xi4 = 0;
-l3*st3*xi1 - l2*st2*xi3 = 0;
 l3*ct3*xi1 + l2*ct2*xi3 = 0;

xi1^2 + xi2^2 + xi3^2 + xi4^2 = 1;
```

Note that when $\Phi_{\mathbf{z}}(\mathbf{q})$ is rank deficient, there will be at least two vectors ξ satisfying these equations. To avoid computing each singularity point twice, we can add the equation of a random half-plane through the origin, which will discard one of the points almost always. For example, let us add

```
0.5795*xi1 + 0.5683*xi2 + 0.3000*xi3 + 0.5013*xi4 >=0;
```

to the `[SYSTEM EQS]` section. Although adding this inequality is optional, it drastically reduces the computation time.

Now solve the equations

```
cuik inv
```

and plot the solutions in the (J3x, J3y, ct2) space:

```
cuikplot3d inv 1 2 3 0 inv.gcl
geomview inv.gcl
```

Using Geomview's controls, it is not difficult to obtain the plots of the inverse locus in Fig. 9, obtained for different values of SIGMA. Observe how, as expected, the plots delimit the workspace of Dextar in the (J3x, J3y) space.

The inverse locus is formed by two circular arcs and two isolated points. The arcs correspond to configurations in which the left or right arms are fully extended (Fig. 10, left), and the isolated points correspond to those in which the arms are folded back, with J_3 coinciding with J_2 or J_5 (Fig. 10, middle).

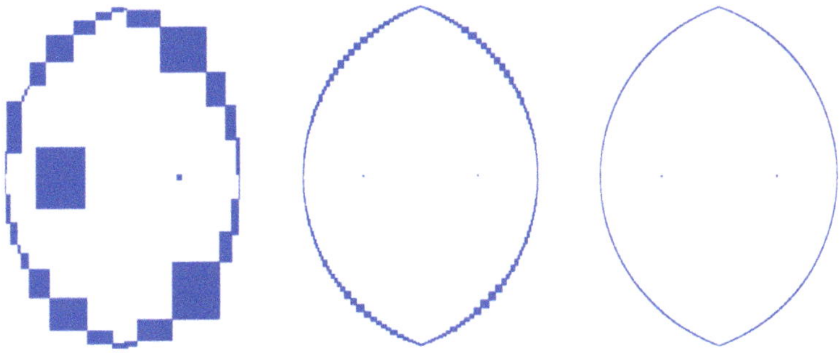

Fig. 9 Inverse singularity set of the Dextar robot computed at different SIGMA values. From left to right (in parentheses, the computation times needed in seconds): 1 (31), 0.1 (49), 0.01 (158)

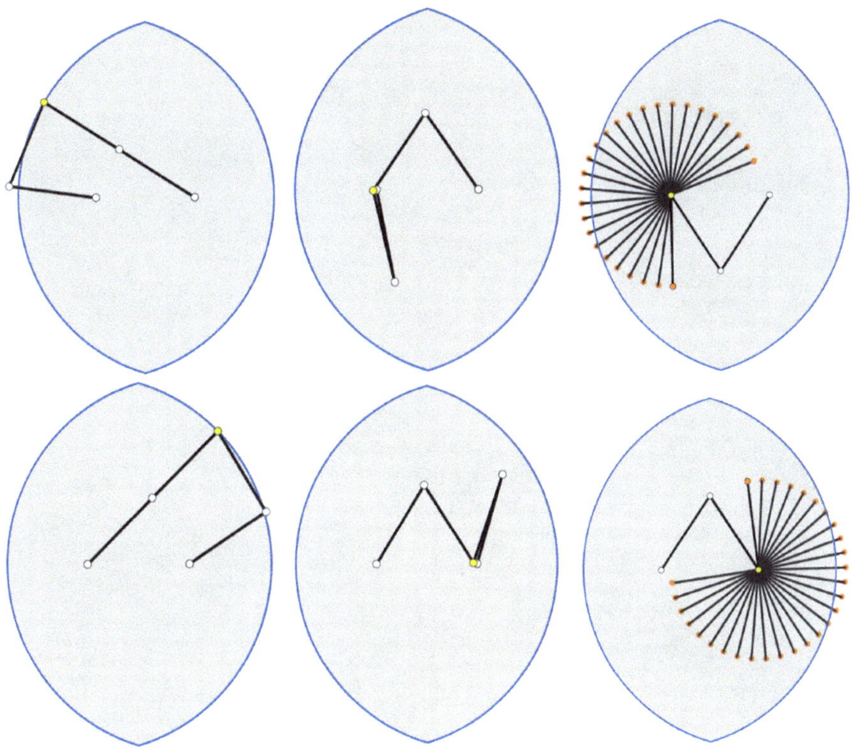

Fig. 10 Left: two configurations with the end-effector at the workspace boundary (right or left arm fully extended). Middle: two configurations with the end-effector at an interior barrier (left or right arm folded back). Right: there is a continuum of configurations for which the end-effector coincides with J_2 or J_5

In fact, the isolated points are curves in configuration space, as there is a one-dimensional continuum of configurations in which J_3 coincides with J_2 or J_5 (Fig. 10, right).

Computing the forward singularity locus: Let us now prepare the `*.cuik` file of the forward singularity locus. We shall extend `cspace.cuik` with new equations forcing the rank deficiency of $\Phi_{\mathbf{y}}(\mathbf{q})$.

First copy `cspace.cuik` to `fwd.cuik` and `cspace.param` to `fwd.param`:

```
cp cspace.cuik fwd.cuik
cp cspace.param fwd.param
```

Then, add the following variables to the `[SYSTEM VARS]` section of `fwd.cuik`:

```
[SYSTEM VARS]
...
xi1:[-1,1]
xi2:[-1,1]
xi3:[-1,1]
xi4:[-1,1]
```

Finally, add the rank deficiency equations to the `[SYSTEM EQS]` section:

```
[SYSTEM EQS]
...
% Forward singularity conditions
-l3*st3*xi3 - l4*st4*xi4 = 0;
 l3*ct3*xi3 + l4*ct4*xi4 = 0;
-xi1 - l3*st3*xi3 = 0;
-xi2 + l3*ct3*xi3 = 0;
 xi1^2 + xi2^2 + xi3^2 + xi4^2 = 1;

% Equation to select just one xi vector for each
singularity
0.0249*xi1 + 0.5923*xi2 + 0.6515*xi3 + 0.4734*xi4 >= 0;
```

Now, save `fwd.cuik` and start the computation of the forward singularity locus:

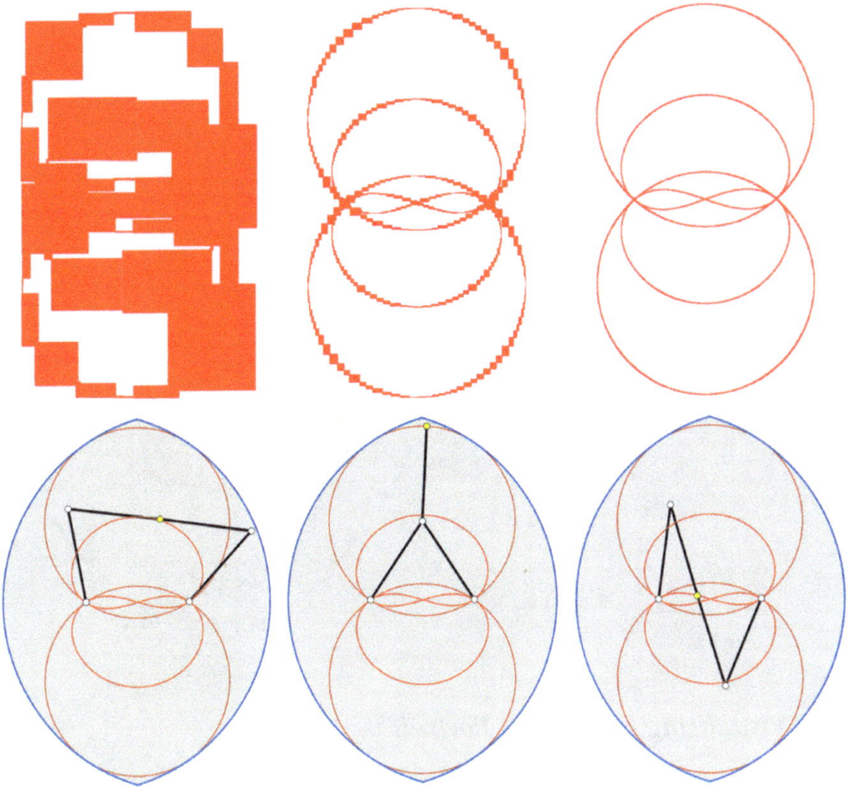

Fig. 11 Top: forward singularity set of the Dextar robot computed at different SIGMA values. From left to right (in parentheses, the computation times needed in seconds): 1 (3), 0.1 (11), 0.01 (69). Bottom: three examples of forward singularities of the Dextar robot

```
cuik fwd
```

When the computation has finished, execute

```
cuikplot3d fwd 1 2 3 0 fwd.gcl
geomview fwd.gcl
```

and use Geomview to obtain the plots in Fig. 11, top.

Each point of the red curve corresponds to a configuration of the mechanism in which J_2, J_3, and J_4 are aligned. The configurations in Fig. 11, bottom, for example, are forward singularities.

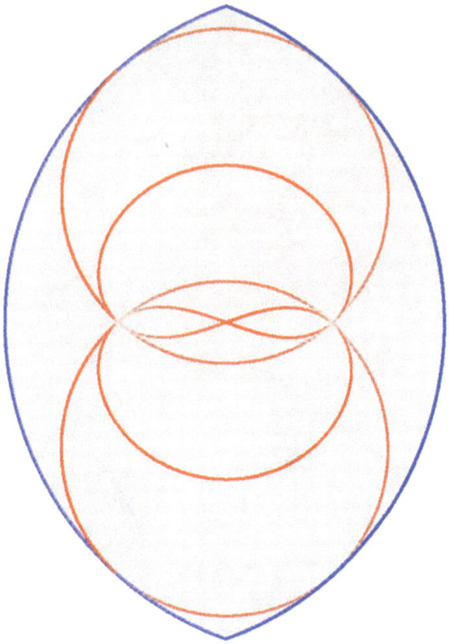

Fig. 12 Output portrait of the Dextar robot

2.7 Visualizing the Output Portrait

The previous examples show Dextar overlaid onto its output portrait (the projection of the C-space and the forward and inverse loci onto the output coordinates). To generate this portrait, run

```
geomview
```

and then load the C-space, inverse and forward singularity loci:

- File → Open, select `cspace.gcl` + click OK;
- File → Open, select `inv.gcl` + click OK;
- File → Open, select `fwd.gcl` + click OK.

By tuning a bit Geomview's output, you will obtain the image in Fig. 12.
To obtain this image, follow these hints after loading the three gcl files:

- Use ortographic projection:
  - Inspect → camera → ortographic
- Center the objects in your window:
  - Select World in "Targets"

  - Use "Zoom" and "Translate"
- Choose a white background:
  - Inspect → camera → background color
- Color the C-space boxes in gray and make them translucent:
  - Select the cspace in "targets"
  - Inspect → appearance and set Shading to "Constant". Also click on [Cf] to select a light gray color for the boxes
  - Inspect → material and click Transparent. Move the Alpha slider to a small value.
- Color the edges of the forward singularity boxes in red:
  - Inspect → appearance → Ce → select red. Also make the edges thicker by setting Line Width to 3 or 4 in the same panel.
- Color the edges of the inverse singularity boxes in blue:
  - Inspect → appearance → Ce → select blue. Also make edges thicker

2.8 *Visualizing the Input Portrait*

A great advantage of computing the singularity sets on the C-space is that they can be projected to any coordinate space. We next show how to project them to the active joint space (θ_2, θ_5).

The boxes in our `*.sol` files only contain the sine and cosine intervals of θ_2 and θ_5, but we can use

```
cuikatan2
```

to replace them by the corresponding angular intervals. Let us do that for `cspace.sol`. We run

```
cuikatan2 cspace 4 3 10 9 cspace_angles
```

since in cspace.cuik st2 and ct2 are declared as the fourth and third variables, and st5 and ct5 are declared as the tenth and ninth variables.

For each box in `cspace.sol`, this instruction substitutes the intervals appearing in the mentioned positions (4, 3, 10, and 9) by the corresponding angle intervals. The new angle intervals are added at the end of the box. The original file `cspace.cuik` is left unaltered, and the results are saved in `cspace_angles.sol`.

Fig. 13 Input portrait of the Dextar robot

Since each box in `cspace.sol` has 10 intervals, we will have 8 intervals in each box of `cspace_angles.sol` (cuikatan2 will have replaced 4 sine–cosine intervals by two angular intervals). The intervals relative to θ_2 and θ_5 will be in the last two positions in `cspace_angles.sol`. Thus, we can run

```
cuikplot3d cspace_angles 8 7 1 0 cspace_angles.gcl
```

to generate the `*.gcl` file corresponding to `cspace_angles.sol`.

To get analogous files for the forward and inverse singularity loci, we run

```
cuikatan inv 4 3 10 9 inv_angles
cuikatan fwd 4 3 10 9 fwd_angles
```

Finally, using Geomview on the created `*.gcl` files, we obtain the input portrait in the active joints space shown in Fig. 13:

```
cuikplot3d inv_angles 12 11 1 0 inv_angles.gcl
cuikplot3d fwd_angles 12 11 1 0 fwd_angles.gcl
```

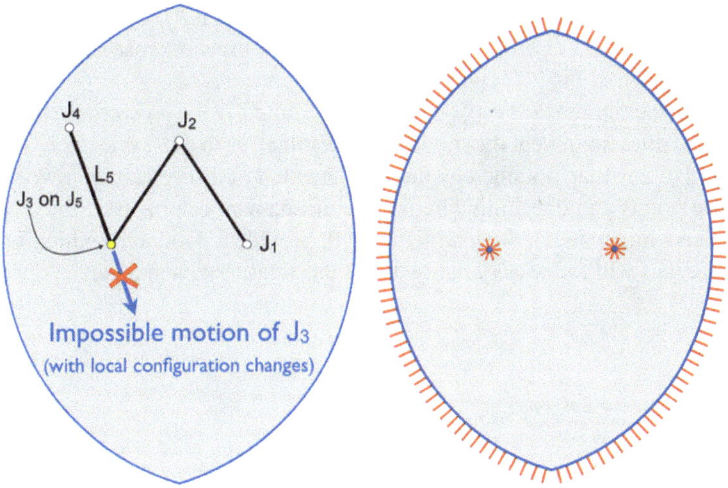

Fig. 14 Left: the shown configuration is an inverse singularity. The end-effector (J_3), coinciding with J_5, cannot move in the direction of link L_5. Right: workspace of Dextar relative to the output coordinates. In the plot, the inverse singularities in blue delimit the workspace boundaries, and small red vectors are shown on the forbidden direction of motion

The horizontal and vertical axes correspond to the θ_5 and θ_2 angles, respectively. The gray area spans the $[-\pi, \pi]$ range in both directions.

The input portrait provides the workspace of the (θ_2, θ_5) angles, which is also useful to a designer. For example, from the previous plot we see that, despite the existence of inner voids, the two angles can take any value in $[-\pi, \pi]$. This implies that Dextar should be equipped with multiturn actuators to exploit its full motion capabilities.

Note also how the roles of the forward and inverse singularity loci are reversed here. While the inverse locus delimited the attainable region in the task space, such region is bounded by the forward locus in the active joint space. However, it is shown in Bohigas et al. (2016) that not all points of the inverse (resp. forward) locus will necessarily project to the boundary of the workspace (resp. active joint) space. The isolated points of the inverse locus of Dextar, explained next, provide one example.

2.9 *Workspace Boundaries and Interior Barriers*

The inverse singularities help to delimit the workspace boundaries relative to the output coordinates, in this case (J_{3x}, J_{3y}). But they can also reveal the existence of interior barriers in the workspace. The isolated blue points that coincide with J_1 and J_5 are an example of such barriers: e.g., if we place J_3 on J_5, J_3 cannot move in the direction of link L_5 (using small variations of the input angles). In other words, there is a motion barrier orthogonal to such direction (Fig. 14, left). Using the techniques of Chap. 4 in Bohigas et al. (2016), we would mark the workspace boundaries and barriers as shown in Fig. 14, right.

Note that our initial workspace map that resulted from projecting the C-space (Fig. 8, left) does not reveal the motion barriers interior to the workspace. The same can be said of any map obtained by the usual method of discretization (sweeping the task space points and checking whether the mechanism can be assembled in them, using inverse kinematics). Such a method will provide a good approximation of the workspace, but will fail to accurately detect the singularities and barriers present in its interior.

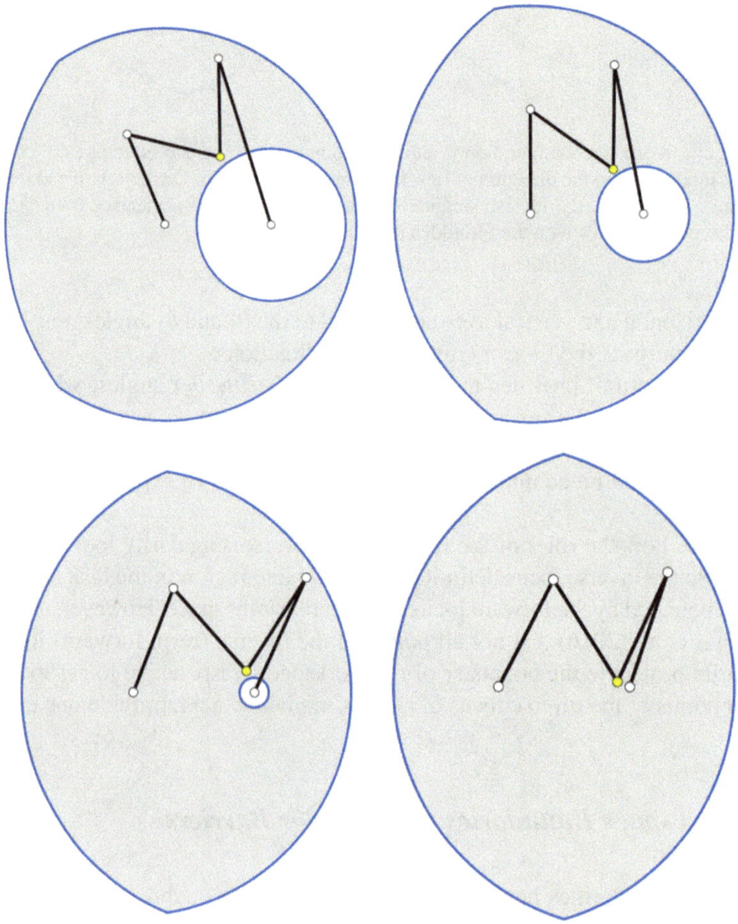

Fig. 15 From left to right and top to bottom, the value of l decreases from 0.7 to zero

The previous punctual barriers can be better understood when seen as the limit case of a family of workspaces. For example, if we consider the family of single-loop five-bar mechanisms with these parameters:

$$l_3 = l_4 = l_5 = 0.9\ ,\ l_2 = l + 0.9$$

and make l evolve from 0.7 to zero, we obtain the workspaces in Fig. 15. It is clear that the circular boundary of the workspace degenerates into an interior barrier with just one point when $l = 0$.

References

O. Bohigas, M. Manubens, L. Ros. *Singularities of Robot Mechanisms: Numerical Computation and Avoidance Path Planning* (Springer, Cham, 2016)

L. Campos, F. Bourbonnais, I. Bonev, P. Bigras, Development of a five-bar parallel robot with large workspace, in *Proceedings of the ASME International Design Engineering Technical Conferences and Computers and Information in Engineering Conference, IDETC/CIE* (Montreal, Quebec, 2010)

Geomview. http://www.geomview.org/

Kinematics and robot design group at the Institut de Robòtica i Informàtica Industrial. http://www.iri.upc.edu/research/kinematics

Mecademic version of Dextar. http://www.mecademic.com

J.M. Porta, L. Ros, F. Thomas, A linear relaxation technique for the position analysis of multi-loop linkages. IEEE Trans. Robot. **25**(2), 225–239 (2009)

J.M. Porta, L. Ros, O. Bohigas, M. Manubens, C. Rosales, L. Jaillet, The CUIK suite: motion analysis of closed-chain multibody systems. IEEE Robot. Autom. Mag. **21**(3), 105–114 (2014)

The CUIK Project Home Page. http://www.iri.upc.edu/cuik

Tutorial Solutions. www.iri.upc.edu/people/ros/srm/Dextar-tutorial.zip

D. Zlatanov, Generalized Singularity Analysis of Mechanisms. Ph.D. thesis, University of Toronto, 1998

Cuspidal Robots

Philippe Wenger

Abstract This chapter is dedicated to the so-called cuspidal robots; i.e., those robots that can move from one inverse geometric solution to another without meeting a singular configuration. This feature was discovered quite recently and has then been fascinating a lot of researchers. After a brief history of cuspidal robots, the chapter provides the main features of cuspidal robots: explanation of the non-singular change of posture, uniqueness domains, regions of feasible paths, identification and classification of cuspidal robots. The chapter focuses on 3-R orthogonal serial robots. The case of 6-dof robots and parallel robots is discussed in the end of this chapter.

1 Introduction

1.1 Preliminaries

This chapter deals with the so-called cuspidal robots. It is focused on serial, open-chain robots. For these robots, the geometric model can be defined by an input map f, such that $\mathbf{X} = f(\mathbf{q})$, where q is the vector of joint variables (the inputs), which are all actuated. Vector $\mathbf{X}$ contains the coordinates of the end-effector pose (the outputs). The robots are assumed non-redundant, namely, the number of joint variables (n) is equal to the number of coordinates (m) that describe the end-effector pose $(n = m)$. The configuration space of these robots can be restricted to their joint space, which is fully defined by the actuated joints. As a consequence, the only singularities are defined by the input singularities. Unless specified, the word singularity will stand for input singularity in this chapter. Figure 1 shows a 3-R robot (made of three revolute joints). The input variables are the three joint coordinates θ_1, θ_2 and θ_3. The output variables are defined by the three Cartesian coordinates x, y and z of a reference point P fixed on the last link. For this robot, $n = m = 3$ and its joint space is a 3-Torus if the joints are unlimited (a 3-D box if they are limited). Generically, the singularities

P. Wenger (✉)
IRCCyN, CNRS, Nantes, France
e-mail: Philippe.Wenger@irccyn.ec-nantes.fr

A. Müller and D. Zlatanov (eds.), *Singular Configurations of Mechanisms and Manipulators*, CISM International Centre for Mechanical Sciences 589,
https://doi.org/10.1007/978-3-030-05219-5_3

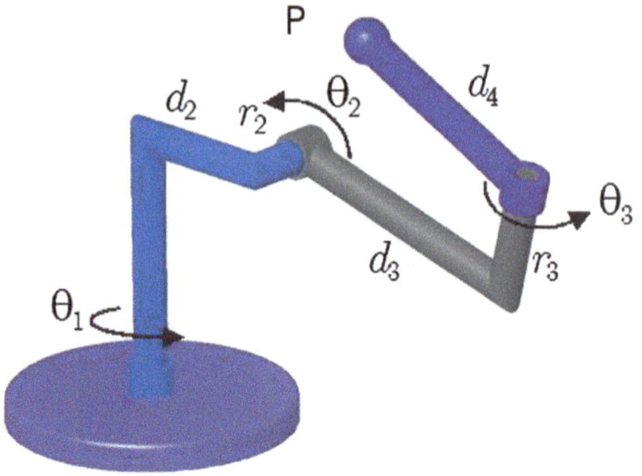

Fig. 1 A serial robot with three revolute joints

form a set of surfaces in the joint space. These surfaces divide the joint space into a set of singularity-free connected regions called aspects, as defined by Borrel and Liegeois (1986).

As it will be shown further, a cuspidal robot is a robot that can move from one of its inverse geometric solution to another without encountering a singularity. This feature was discovered quite recently and has then been fascinating a lot of researchers. Before this discovery, it was thought that any robot should necessarily cross a singularity during a change of solution, like, e.g., when an anthropomorphic robot (a Puma type robot) switches from the elbow up to the elbow down configuration by passing through the fully outstretched (or folded) singular configuration (see Sect. 2.1). A brief history of cuspidal robots is presented next.

1.2 *Brief History of Cuspidal Robots*

In 1986, a mathematical proof was provided in order to confirm the well-established belief that any non-redundant serial robot should cross a singularity when moving from one solution to another (Borrel and Liegeois 1986). This proof was based on the implicit function theorem and some topological arguments. The proof was in fact incorrect but at that time nobody noticed it as it was just confirming a fact that everybody had admitted before. The very first mention of a singularity-free change of solution dates back to 1988, in a talk by V. Parenti-Castelli and C. Innocenti (University of Bologna, Italy) presented during the first conference on Advances in Robot Kinematics that took place in Ljubljana (Parenti and Innocenti 1988). For two different types of 6-R (revolute-jointed) robots, the authors were able to build a singularity-free path joining two inverse solutions. It was such a big challenging of a well-established fact that nobody trusted them at that time. In a separated work conducted in Stanford University, J. W. Burdick showed in its Ph.D. thesis several

3-R robots also able to move from one solution to another without encountering a singularity (Burdick 1988). These results were not published outside the Ph.D. thesis. Nothing appeared during the next four years before this behavior was then formalized in 1992 by Wenger (1992). It was clearly shown in this work why even a simple 3-R robot can move from one of its solutions to another without meeting a singularity. It was not easy to understand why the proof given in (Borrel and Liegeois 1986) was incorrect as it seemed mathematically sound. The main problem was that the proof resorted to a hypothesis that turned out to be not true in all robots. The detailed explanation is a bit more tricky and is not reported there. The belief that a change of solution should necessarily be singular was so strongly rooted in people's mind that it took several years before the research community started to accept the existence of cuspidal robots. The word cuspidal was defined after 1995 when it was shown that the existence of a cusp point in the singularity locus of the robot indicates that this robot can change its posture without encountering a singularity by encircling this cusp point (El Omri and Wenger 1995), see Sects. 2.4 and 4.2.

1.3 Questions of Interest

When the research community started to admit the truth in the second part of the 90s, it was faced with a lot of questions, such as:

- How can a non-singular change of solution be accomplished for a given robot?
- What are the consequences on path-planning?
- Is this ability possible only for specific robots or is it more general?
- Given a new robot, how to know if it will have this ability?
- Is it possible to set some conditions for a robot to have this ability?
- Is it possible to enumerate all the cuspidal robots?

The first two issues are very important for control, while the others are very important for the designer. As it will be shown further, a change in geometric parameters of a non-cuspidal robot may render it cuspidal. Interestingly, this fact can be illustrated with the story of the IRB 6300C robot launched by ABB in 1996. This new manipulator was specially designed for the car industry to minimize the swept volume. The only difference from the Puma was the permutation of the first two link axes, resulting in a manipulator with all its joint axes orthogonal. Commercialization of the IRB 6400C was stopped one year later. Informal interviews with robot customers at that time revealed difficulties in planning off-line trajectories using Robotic CAD systems for this robot. In fact it was shown later that the IRB 6400C robot turns out to be cuspidal.

1.4 Chapter Outline

In the rest of this chapter, the singular and non-singular changes of posture are analyzed in details. Next the feasibility of trajectories in cuspidal robots are investigated, and the maximal regions of feasible paths in the workspace are defined. The identification, enumeration and classification of 3-dof (degree-of-freedom) cuspidal robots and non-cuspidal robots are then addressed. Eventually, the case of 6-dof robots and parallel robots are discussed.

2 Postures

2.1 Definition

A posture is associated with one solution to the inverse geometric problem of a robot. This word refers to the way a given robot places its links in space to reach a given frame. For usual industrial robots, a posture can be easily identified geometrically. For example, it is well known that an anthropomorphic robot (Fig. 2) has at most eight distinct inverse geometric solutions, which are associated with eight postures. These postures can be identified geometrically by the configuration of the elbow (up or down), the shoulder (right or left) and the wrist (flip or no flip). The total number of combinations amount to $2^3 = 8$ distinct postures. Figure 3 shows four of the eight postures obtained in the flip configuration of the wrist. It is well known that the only way for this robot to switch from any of its posture to another is to cross a singular configuration.

2.2 Postures and Aspects Case 1: Non-cuspidal Robots

The aspects are the largest singularity-free connected regions of the joint space (Borrel and Liegeois 1986). They are bounded by the singularity surfaces (or hypersurfaces for a robot with more than 3 joints) and by the joint limits when they exist. Additional boundaries may occur in the presence of obstacles (El Omri 1996). Before cuspidal robots were discovered, one posture was thought to be uniquely associated with one aspect. Indeed, since the robot was thought to cross necessarily a singularity when moving from one posture to another, there could not be more than one posture in each of its aspects. This is true for most usual industrial robots like the anthropomorphic robot and the SCARA robot shown in Figs. 3 and 4, respectively. The SCARA robot is a well-known industrial robot (Fig. 4). This 4-dof robot can produce three translations in space and one rotation about a vertical axis. This motion is known as the Schönflies motion, first studied by the German mathematician Arthur-Moritz Schönflies (1853–1928). This motion is suitable for pick-and-place tasks. The

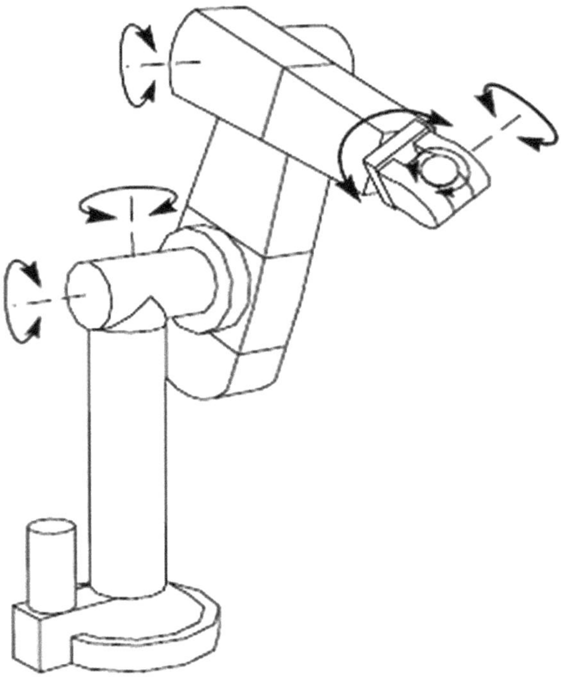

Fig. 2 An anthropomorphic robot

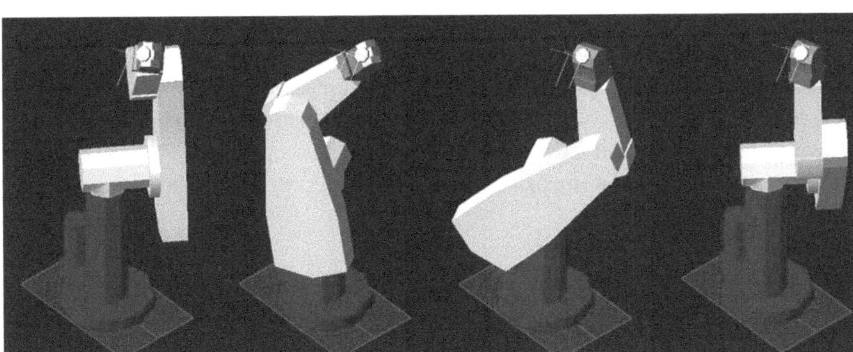

Fig. 3 Four postures for the anthropomorphic robot, from left to right: (elbow up, shoulder left), (elbow up, shoulder right), (elbow down, shoulder right), (elbow down, shoulder left)

determinant of the Jacobian matrix $det(J)$ vanishes whenever $\sin(\theta_2) = 0$, where θ_2 is the second revolute joint variable. Accordingly, a singularity occurs whenever $\theta_2 = 0$ or $\theta_2 = \pi$, namely, when the arm is fully extended or fully folded. This robot has two aspects defined by $\theta_2 > 0$ and $\theta_2 < 0$, respectively. Moreover, the inverse geometric model of this robot admits two solutions, associated with the two

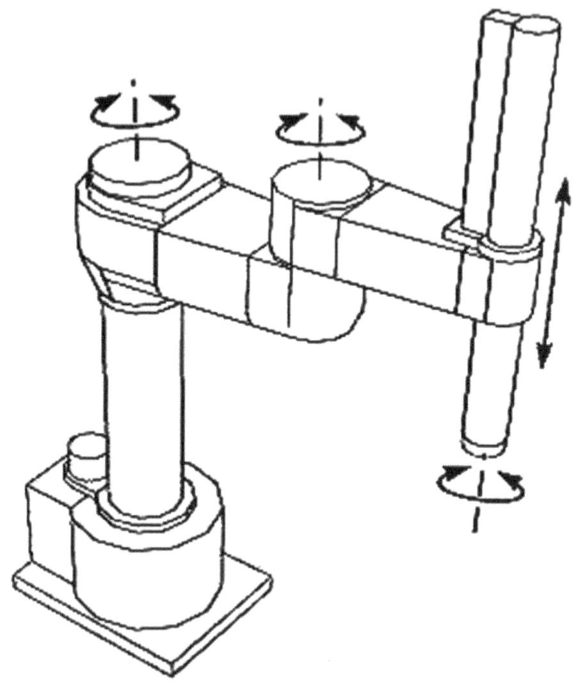

Fig. 4 A SCARA robot

postures elbow up and elbow down, respectively. Fig. 5, left, shows the two aspects in (θ_1, θ_2). The other joint variables can be ignored since they do not play any role in the singularities and aspects. The robot is assumed to have joint limits and its joint space is thus a square in (θ_1, θ_2). Figure 5, right, shows the aspects projected in the workspace and the robot depicted in an elbow down posture. There is only one inverse geometric solution in each aspect; namely, a posture is associated with one aspect.

2.3 *Postures and Aspects Case 2: Cuspidal Robots*

A cuspidal robot has more than one inverse geometric solution in at least one of its aspects. In other words, it can move from one posture to another without meeting a singularity. The 3-R robot with mutually orthogonal joint axes shown in Fig. 1 turns out to be cuspidal because this robot has two aspects and four inverse geometric solutions, two in each aspect. The determinant of the Jacobian matrix **J** of this robot can be written as:

$$det(\mathbf{J}) = (d_3 + c_3 d_4)(c_2(s_3 d_3 - c_3 r_2) + s_3 d_2) \tag{1}$$

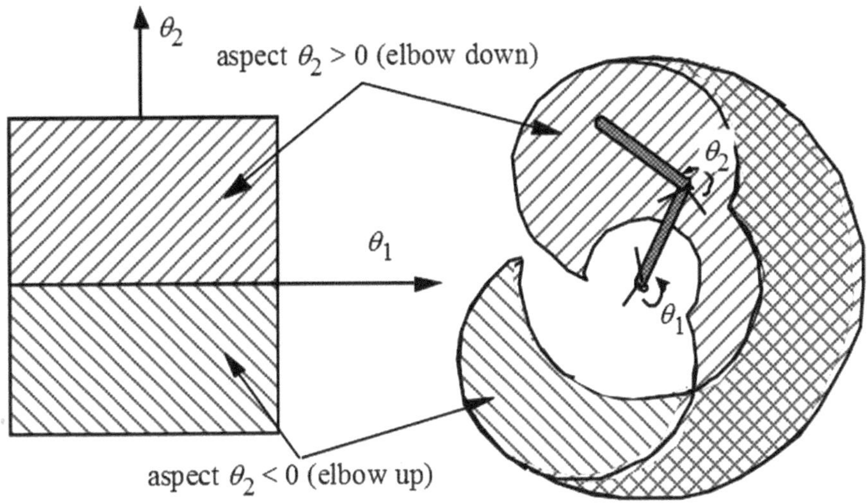

Fig. 5 Aspects of the SCARA robot in the joint space (left) and workspace (right)

where $c_i = \cos(\theta_i)$ and $s_i = \sin(\theta_i)$, $i = 2, 3$. The geometric parameters d_2, d_3, d_4 and r_2, shown in Fig. 1, are the Modified Denavit-Hartenberg as defined in (Khalil and Kleinfinger 1986). It can be noticed that the first joint variable θ_1 does not appear in this determinant. This is always the case in any serial robot whose first joint is revolute since in this case the singularities do not depend on θ_1. The singularity surfaces of our robot, which are the zero sets of $det(\mathbf{J}) = 0$, can then be plotted in (θ_1, θ_2) where they appear as curves. The first factor may vanish only when $d_3 < d_4$, namely when the last link is longer than the second link. If it is not the case, the singularity curves are defined by the zeros of the second factor only. Figure 6 (left) shows the singularity curves obtained when $d_2 = 1$, $d_3 = 2$, $d_4 = 1.5$ and $r_2 = 1$ and for unlimited joints. Note that since the robot is assumed to have no joint limits, its joint space has the structure of a Torus and the singularity curves are defined on the Torus $T(\theta_2, \theta_3)$. For more simplicity, a plane picture of the joint space is shown in Fig. 6, and the reader may retrieve the topology of $T(\theta_2, \theta_3)$ by identifying the opposite sides of the square. Having this important notice in mind, it is easy to see that the singularities form two closed curves S_1 and S_2 and define only two aspects. The singularities can be also plotted in the workspace. For our 3-R robot, the workspace is of dimension 3, but it is sufficient to plot a planar cross section passing through the first joint axis because we know that the singularities do no depend on θ_1. The workspace can then be plotted in the plane (ρ, z), where $\rho = \sqrt{x^2 + y^2}$ (Fig. 6, right). The outside boundary of the workspace WS_2 is the image of S_2 while the internal boundary WS_1 is the image of S_1. The internal boundary is made of four lines that merge at four cusp points. This internal boundary separates two regions: the inner region and the outer region. It can be verified that the robot has four inverse geometric solutions in the inner region but it has only two solutions in the outer region. We solve the inverse geometric model

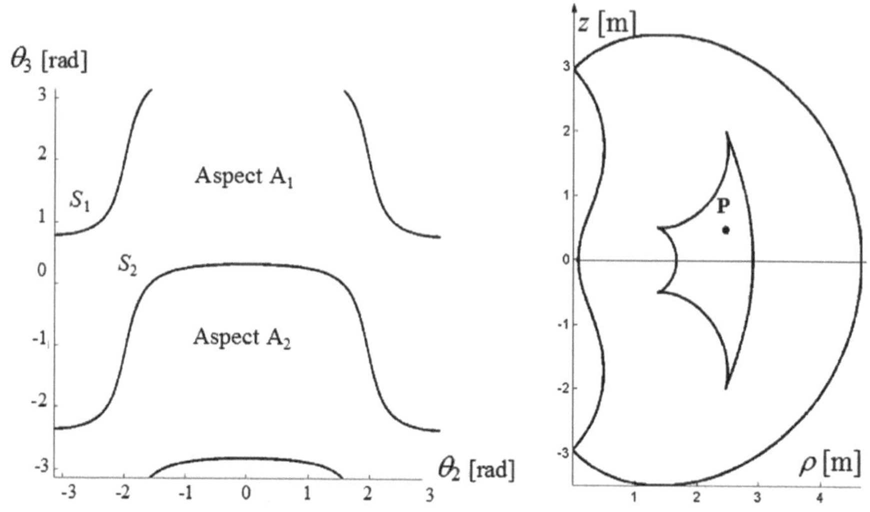

Fig. 6 Singularity curves for the 3-R cuspidal robot shown in Fig. 1(left) and its projection in the workspace (right)

at a point $P(\rho = 2.5, z = 0.5)$ in the inner region. The four solutions are (in radians): $q^{(1)} = [-1.8, -2.8, 1.9]^t$, $q^{(2)} = [-0.9, -0.7, 2.5]^t$, $q^{(3)} = [-2.9, -3, -0.2]^t$ and $q^{(4)} = [0.2, 0.3, 1.9]^t$. It is apparent from Fig. 7 that $q^{(2)}$ and $q^{(3)}$ lie in the same aspect A_1. It is then possible to link $q^{(2)}$ and $q^{(3)}$ by a non-singular path as shown in Fig. 7. Keeping in mind that the opposite sides of the square should be identified, one can see that $q^{(1)}$ and $q^{(4)}$ also lie in one single aspect (A_2). Contrary to their non-cuspidal counterparts, the anthropomorphic robot and the SCARA robot, the postures of this 3-R cuspidal robot are difficult to identify geometrically, as shown in Fig. 8.

2.4 Non-singular Change of Posture Versus Singular Change of Posture

The question of how a cuspidal robot can move from one of its postures to another without encountering a singularity is of primary interest and has puzzled the research community for several years. People thought that the robot should meet a special transitory state between the two postures during a change of posture. If it is not a singularity, what kind of transitory state should it be? In fact no special transitory state is met during a non-singular change of posture and this change is accomplished quite smoothly. To better understand this behavior, it might be helpful to refer to the mappings from a surface onto a plane as formalized by Whitney in 1955. Whitney showed that only two types of stable singularities may occur when a surface is

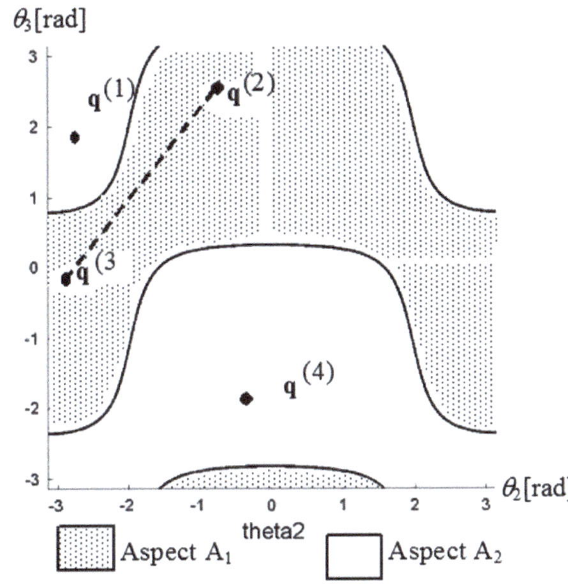

Fig. 7 Two solutions in each aspect and a non-singular path joining two solution in A_1

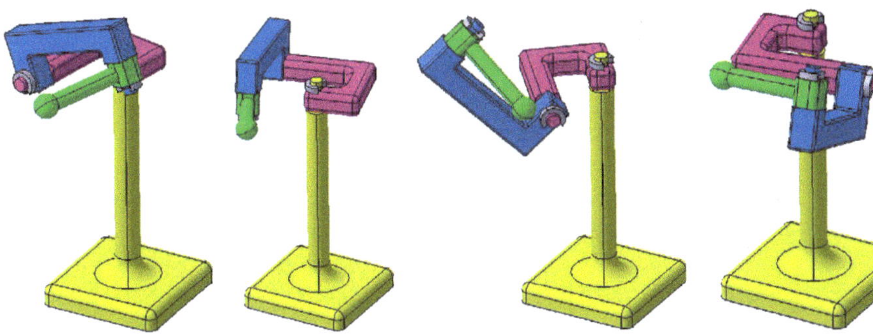

Fig. 8 The four postures of the 3-R cuspidal robot at point P

projected onto a plane: the fold and the cusp (Whitney 1955). A fold arises when the surface is simply folded like in Fig. 9, left. Its projection onto a plane defines a simple line. A cusp arises when the surface is folded twice as in Fig. 9, right. The projection onto a plane gives rise to two fold lines that merge at a cusp point. Now, the non-singular change of posture can be interpreted by considering that the doubly folded surface shown on the right of Fig. 9 is the configuration space $F(\rho, z, t) = 0$ of a cuspidal robot, where ρ and z are the output variables and t is associated to one of the joint variables. The projection onto the plane corresponds to the workspace of the robot. The two red lines merge at a cusp point and define a region where the robot admits three solutions t_1, t_2 and t_3, each one being associated with on one of the

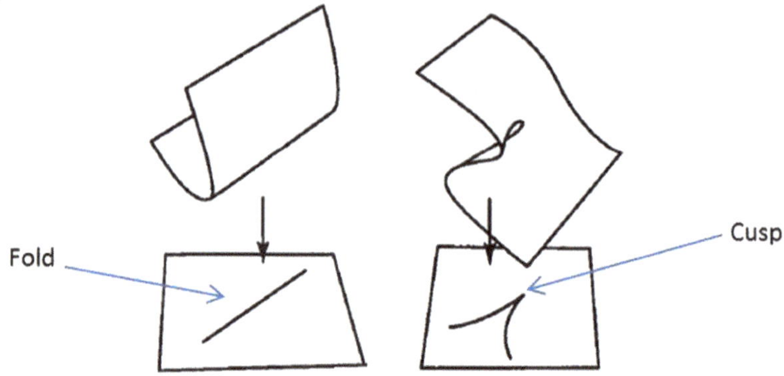

Fig. 9 A fold (left) and a cusp (right). Adapted from (Gibson 2000)

three layers of the doubly folded configuration space surface. Figure 10 shows a path connecting two solutions t_1 and t_3 (from layer 1 to layer 3) at a given point (ρ, z) of this 3-solution region. This path goes smoothly between the solutions and no special transitory configuration is met. Moreover, it turns out that the path goes around the cusp, a fact that will prove of primary importance for the characterization of cuspidal robots. One should be aware that if the projection of the path onto the plane does cross the two red line segments, the path itself never meets the corresponding folds of the configuration space and it is really singularity-free. Coming back to our cuspidal 3-R robot, the non-singular posture changing path shown in Fig. 7 is now projected in the workspace cross section (Fig. 11). Like in Fig. 10, the path encircles a cusp point. The same remark as above also holds here: no singularity is met even if two segments of the inner boundary are crossed. The shape of the configuration space of the robot is a bit more complicated than the one illustrated in Fig. 10. It is easier to cut it along its folds and to show separately its restrictions to the aspects. Figure 12 shows the configuration space restricted to the aspect A_1 (above) and aspect A_2 (below). To get the complete configuration space, these two parts should be glued together along their common edges. In fact the complete configuration space folds along these edges. The fact that there are four inverse geometric solutions in the inner region and only two in the outer region is in accordance with the number of layers of the configuration space in these regions. The non-singular posture changing path of Fig. 11 is shown in red dashed lines. This path goes smoothly from one layer to another like in Fig. 10 by going around a cusp point. Note that in P there are two more solutions in aspect A_2 and a non-singular posture changing path between these two solutions will have to encompass one of the two opposite cusp points. A singular posture changing path can also be defined in P. In the configuration space, such a path should go from one layer of an aspect to a layer of another aspect by moving on a fold. The splitted model of the configuration space shown in Fig. 12 does not make it possible to show this path on the same image. Starting from P in aspect A_1, the path will have to go toward one of the edges. Once on the edge, the path is in fact on the fold that enables it to enter the second aspect A_2 and a singularity is encountered,

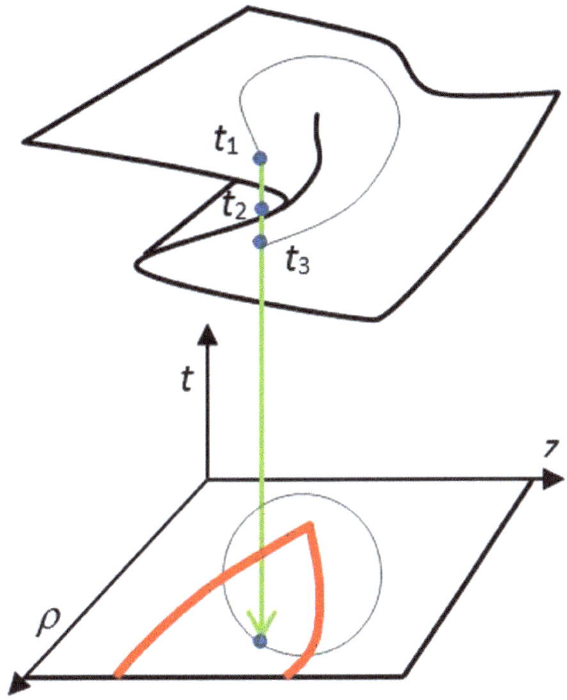

Fig. 10 A non-singular path joining two distinct solutions

where two inverse geometric solutions merge. Physically, the robot behaves like if it was bouncing back against a workspace boundary (think of the SCARA robot that has to bounce back in the fully outstretched configuration when it moves from the elbow up to the elbow down posture).

3 Path Feasibility

3.1 Showing an Example

Another important feature of cuspidal robots is their inability to track certain prescribed paths in their workspace, even in the absence of joint limits or obstacles. Let us consider the straight line path shown in red in Fig. 13. The arrow indicates the direction of motion. The robot is unable to track this path. The reason is that the robot will have to stop on the second crossed boundary line. Indeed, Fig. 12 shows that an edge will be always met, should the robot start from aspect A_1 or A_2. Since when arriving on an edge the robot has no other choice than bouncing back, it cannot track the path further. Note that the path is also infeasible in the reverse direction. In usual non-cuspidal robots met in industrial robotic sites, path infeasibility is usually

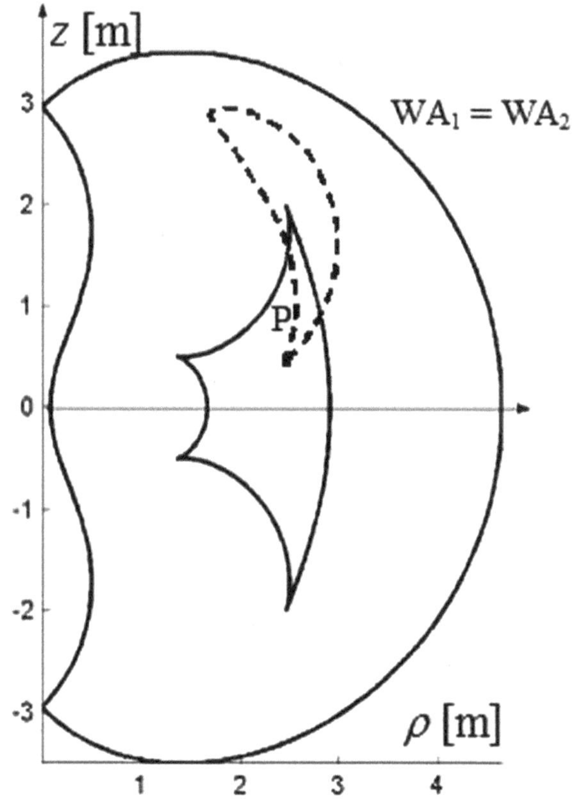

Fig. 11 Non-singular posture changing path in the workspace

due to a physical obstruction such as a joint limit or a collision. The above example shows that a path can be infeasible even if the robot has not joint limits and if there are no collisions. The following section proposes a way of defining the regions of feasible paths in the workspace for cuspidal robots.

3.2 Regions of Feasible Paths for Non-cuspidal Robots

The question of whether a prescribed path in the workspace can be fully tracked or not is of primary interest for many robotic applications such as process tasks. Think of an arc-welding task in which the soldering torch attached to the robot end-effector must realize a continuous welding seam. The welding seam must be fully realized in one step at a constant velocity. This is the reason why P. Borrel introduced the notion of aspects (see Sect. 2.2) when he started to work of the design of a Robotic CAD interface with CATIA for robots off-line programming (Borrel and Liegeois 1986) in 1986. For a non-cuspidal robot, there is only one posture in each aspect. This

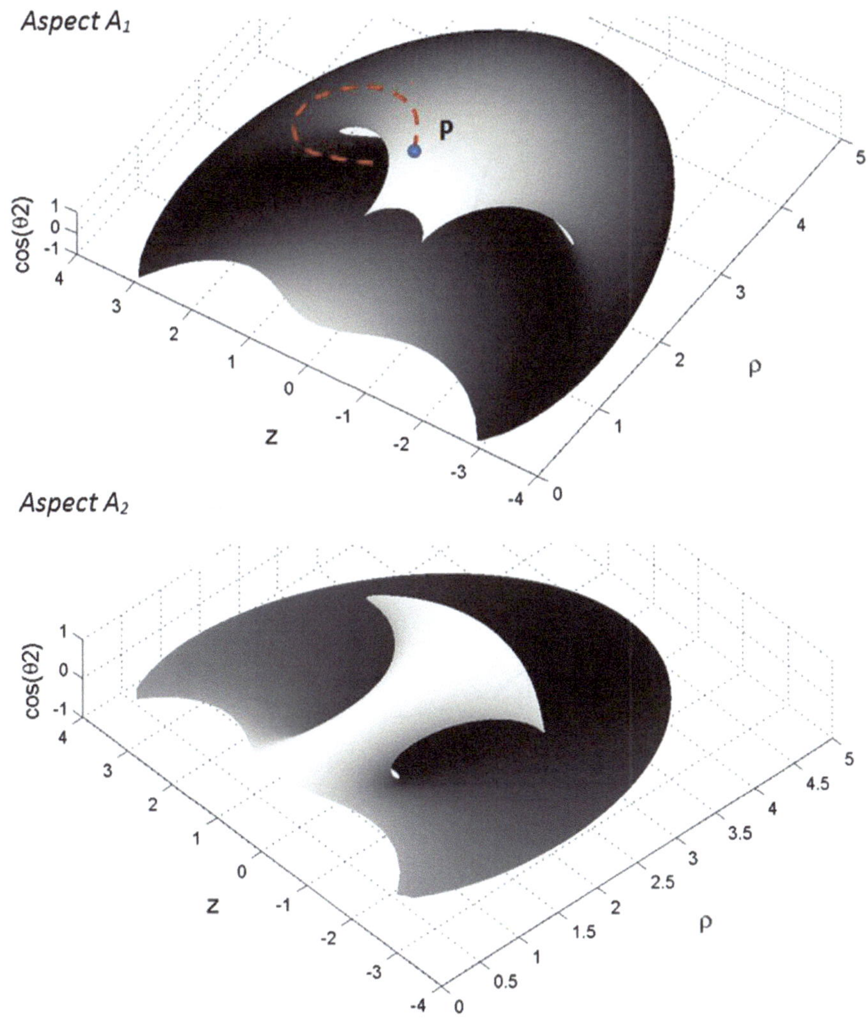

Fig. 12 The splitted configuration space cut along its folds and shown in the two aspects. A non-singular posture changing path is shown in aspect A1 (above)

means that the kinematic map is one-to-one from each aspect onto the workspace. In other words, the aspects are the largest uniqueness domains of the joint space for a non-cuspidal robot. This property is useful to analyze the feasibility of paths. Indeed, since the kinematic map is one-to-one on each aspect, the preimage of any connected set lying in the image of an aspect in the workspace defines a connected set in this aspect. This means that any continuous path lying in the image of an aspect is the image of a continuous path in the joint space. In other words, any path in the image of an aspect is feasible by the robot. This means that the images of the aspects are

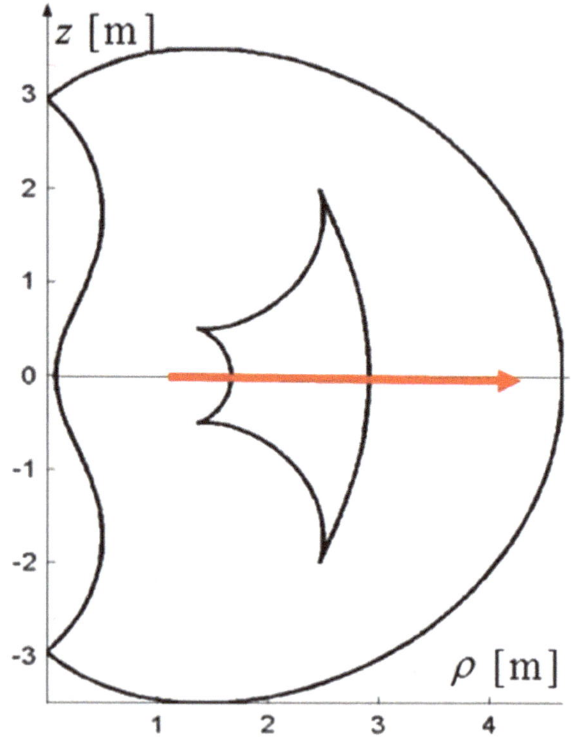

Fig. 13 An infeasible path for the cuspidal robot

the (maximal) regions of feasible paths in the workspace for a non-cuspidal robot. In a robotic CAD system, it is this very convenient to determine the regions of feasible paths in the workspace since the user will then know where to define the prescribed paths during the off-line programming of a robot. Referring to Fig. 5, this SCARA robot has two regions of feasible paths in the workspace that overlap. Suppose that this SCARA robot is used to cut L-shape plates. The robot tool must be able to follow the contour of the shape continuously. Figure 14 shows the two regions of feasible paths along with a suitable and a non-suitable placement of the plates in the workspace. The one that is fully included in the image of an aspect (above) is suitable while the other (below) is not.

For a cuspidal robot, the aspects do not define uniqueness domains anymore. Accordingly, the regions of feasible paths are not the images of the aspects in the workspace. New uniqueness domains must be defined.

3.3 *Characteristic Surfaces*

Since the singular surfaces in the joint space do not separate all the inverse geometric solutions, new separating surfaces should exist. The set obtained by calculating

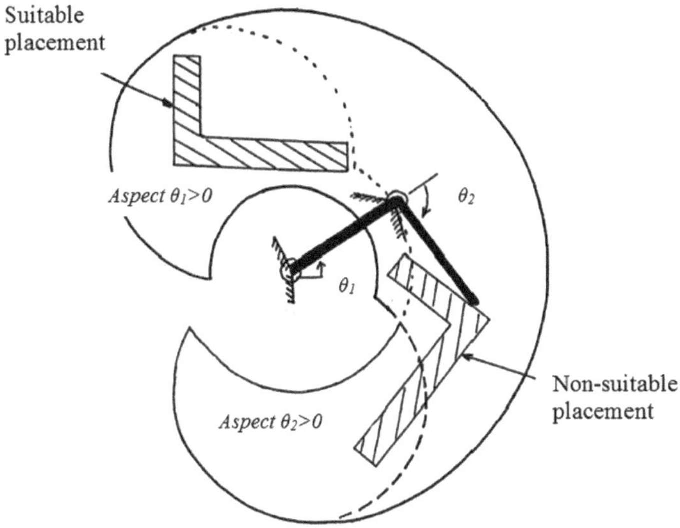

Fig. 14 Use of the aspects for cutting trajectories for a non-cuspidal robot

the non-singular inverse geometric solutions for all points on an internal boundary forms a set of non-singular surfaces in each aspect. These surfaces are called the characteristic surfaces (Wenger 1992). A set of characteristic surfaces is associated with one aspect and they separate the inverse geometric solutions in each aspect, as shown further. A general definition of the characteristic surfaces can be set as follows, which stands for any non-redundant manipulator. Let A_i^* be the boundary of aspect A_i. The characteristic surfaces $\{CS_i\}$ associated with A_i are:

$$\{CS_i\} = f^{-1}(f(A_i^*)) \cap A_i \quad (2)$$

where $f(A_i^*)$ is the image of A_i^* under the forward kinematic map and $f^{-1}(f(A_i^*)) = \{\mathbf{q}/f(\mathbf{q}) \in f(A_i^*)\}$. Note that since an aspect is defined as an open set, A_i does not contain its boundary i.e. $A_i^* \cap A_i = \varnothing$ thus $\{CS_i\}$ might be empty (note that if this is the case for all its aspects, the robot is not cuspidal). Since the general definition of the characteristic surfaces is not algebraic in nature, it is difficult to derive an algebraic expression of $\{CS_i\}$ that would be easy to handle. It should be thus calculated numerically. For the cuspidal robot of Fig. 1 studied in Figs. 6, 7 and 8, the set of characteristic surfaces $\{CS_i\}$ associated with A_i^* can be defined as $\{CS_i\} = f^{-1}(WS_i) \cap A_i$, where $f^{-1}(WS_i) = \{\mathbf{q}/f(\mathbf{q}) \in WS_i\}$. Since $WS_1 = BS_1 \cap BS_2 \cap BS_3 \cap BS_4$ (Fig. 15), we can write $f^{-1}(WS_1) = f^{-1}(BS_1) \cup f^{-1}(BS_2) \cup f^{-1}(BS_3) \cup f^{-1}(BS_4)$. Each set $f^{-1}(BS_j)$ has two components $CS_{1,j} = f^{-1}(BS_j) \cap A_1$ and $CS_{2,j} = f^{-1}(BS_j) \cap A_2$ in A_1 and A_2, respectively. Thus, the two sets of characteristic surfaces can be written as $\{CS_1\} = CS_{1,1} \cup CS_{1,2} \cup CS_{1,3} \cup CS_{1,4}$ and $\{CS_2\} = CS_{2,1} \cup CS_{2,2} \cup CS_{2,3} \cup CS_{2,4}$, respectively

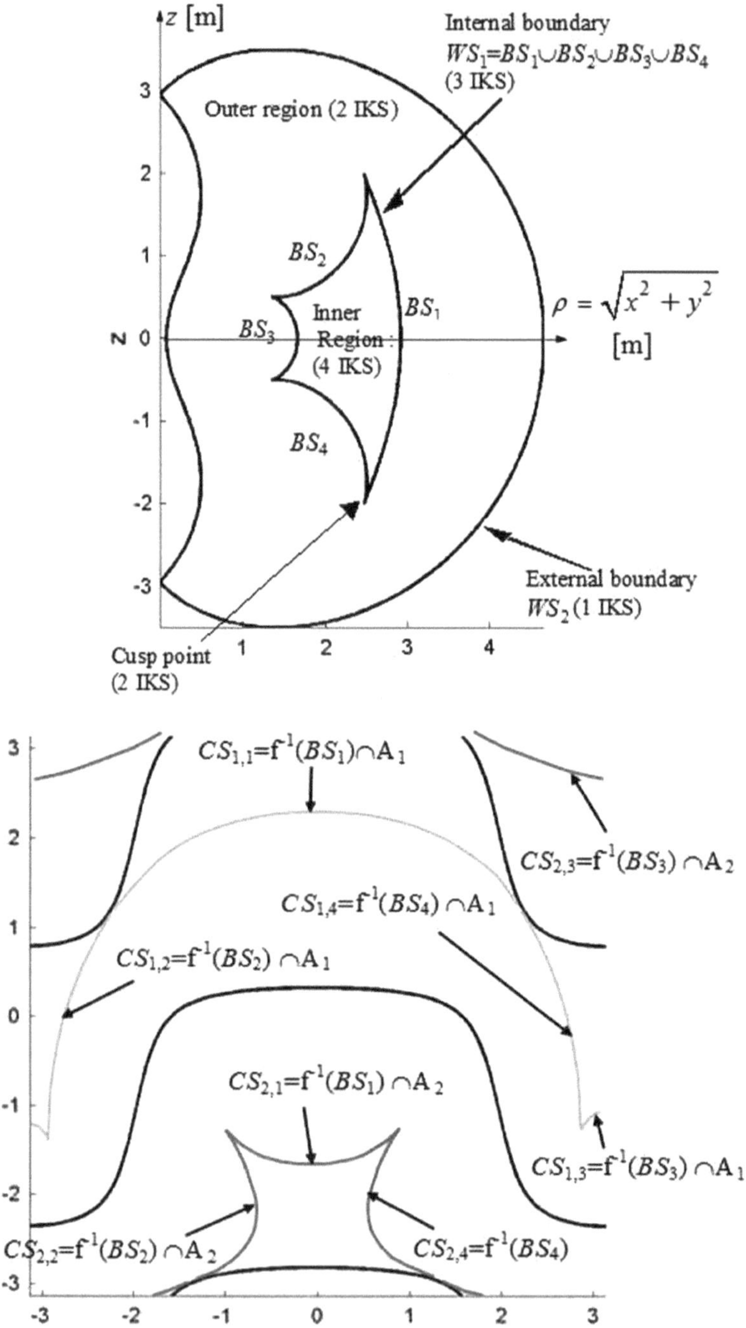

Fig. 15 The characteristic surfaces (below) of the cuspidal robot of Fig. 1 and correspondence with the segments of the internal boundary in the workspace (above)

(Fig. 15). Any two adjacent segments of a set of characteristic surfaces meet at a preimage of a cusp point. This preimage is either a singular configuration or a non-singular one. When the preimage is singular, the segments meet tangentially to a singular surface, whereas a non-singular preimage of a cusp point forms a cusp in an aspect. For example, the segments $CS_{1,1}$ and $CS_{1,2}$ meet tangentially to the singularity surface S_1 at the singular preimage of the cusp point that links BS_1 and BS_2 in the workspace. The remaining non-singular preimage of this cusp point connects the segments $CS_{2,1}$ and $CS_{2,2}$ and forms a cusp in aspect A_2 (Fig. 15).

3.4 *Uniqueness Domains and Regions of Feasible Paths for Cuspidal Robots*

The characteristic surfaces induce a partition of each aspect into smaller sets Ra_{ij}. Also, the internal boundaries induce a partition of the workspace into regions and each such region is associated with several sets Ra_{ij}. For the cuspidal robot at hand, the inner region is associated with the four sets Ra_{11}, Ra_{12}, Ra_{21} and Ra_{22} (in gray in Fig. 16). The first two, Ra_{11} and Ra_{12}, are in A_1, while Ra_{21} and Ra_{22} are in A_2. The outer region is associated with the two sets Ra_{13} and Ra_{23} that belong to A_1 and A_2, respectively. The sets Ra_{ij} are used to determine the maximal uniqueness domains. First, it can be proved that the sets Ra_{ij} are uniqueness domains (Wenger 2004). Second, there exist larger uniqueness domains in the joint space. In effect, Fig. 16 shows that each aspect is made of three sets Ra_{ij}, two of them being associated with the same region in the workspace. If one removes one of these two sets and its boundary from the aspect, the remaining domain is a uniqueness domain (Wenger 2004). Thus, there is still a unique inverse geometric solution in the domain defined by $Qu_1 = A_1\dot{-}C(Ra_{12})$ as well as in $Qu_2 = A_1\dot{-}C(Ra_{11})$ ($\dot{-}$ means the difference between sets, $C(Ra_{ij})$ means the closure of Ra_{ij}). In the same way, $Qu_3 = A_2\dot{-}C(Ra_{22})$ and $Qu_4 = A_2\dot{-}C(Ra_{21})$ are still uniqueness domains. In addition, these uniqueness domains are maximal (Wenger 2004). Figure 17 depicts the four uniqueness domains Qu_1, Qu_2, Qu_3 and Qu_4. The uniqueness domains can be determined numerically. A simple numerical method can be found in (Wenger 2004) and will not be reported here.

Now that we have determined the (maximal) uniqueness domains, it is then possible to obtain the (maximal) regions of feasible paths in the workspace for a cuspidal robot. These regions are the images in the workspace of the uniqueness domains. Figure 18 shows the four regions $Wf_i = f(Qu_i), i = 1, , 4$ of feasible paths for the cuspidal robot at hand. The internal segments that appear in each region indicate sets of points that do not belong to this region and, in turn, segments that the manipulator cannot cross. In fact, each region of feasible paths is the image of an aspect minus one or several segments of the internal boundary surface. This is because the maximal uniqueness domains are defined from the aspects by removing one set Ra_{ij} together with its boundary. By doing so, one removes a part of the boundary surface in the

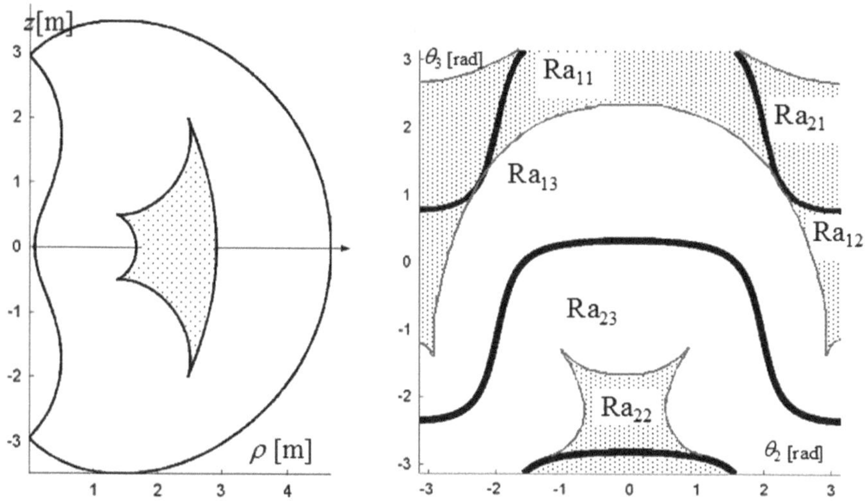

Fig. 16 The characteristic surfaces divide the aspects into smaller sets

Cartesian space. Note that the four regions Wf_i define regions where any arbitrary path is feasible but they do not feature the full model of feasible paths. In effect, it is always possible to define a feasible path that undergoes a non-singular change of posture in aspect A_1 (resp. in aspect A_2). In this case, the path would start in Wf_1 (resp. in Wf_3) and stop in Wf_2 (resp. in Wf_4). In fact, the full model of feasible paths is obtained when Wf_1 and Wf_2 (resp. Wf_3 and Wf_4) are properly glued together. To better visualize this fact, the reader is invited to come back to Fig. 12, which precisely shows the full model of feasible paths.

Note: The above definitions and formula stand for any non-redundant robot, with or without joint limits. In (Wenger 2004), a cuspidal 3-R robot with joint limits has been analyzed with this theory. It is even possible to include the influence of collisions (self-collisions or collisions with the environment) in the calculus, as reported in (Wenger and El Omri 1993). Practically, however, it is difficult to handle robots with more than 3 joints.

4 Identification, Enumeration and Classification of Cuspidal and Non-cuspidal Robots

4.1 Simplifying Geometric Conditions

Because of the more complex behavior of a cuspidal robot and because of the difficulty in modeling its kinematic properties, industrial robots should be preferably non-cuspidal. Thus, before designing an innovative kinematic architecture,

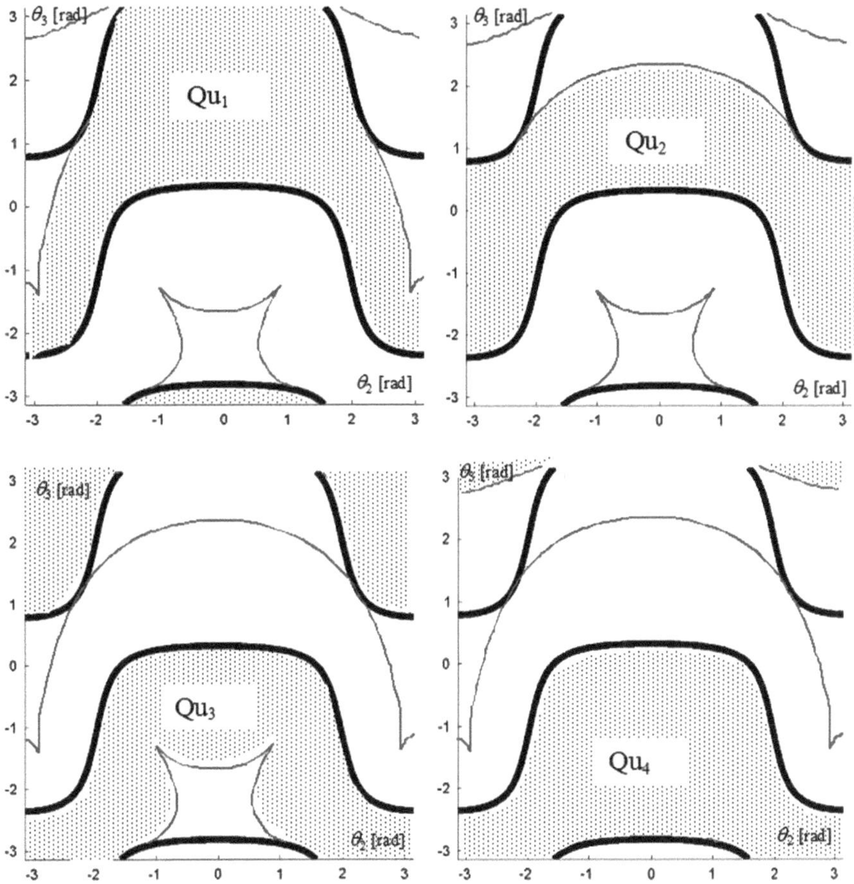

Fig. 17 The four uniqueness domains for the robot of Fig. 1

robot manufacturers should have guidelines and design rules to help them. Why a manipulator with a given geometry is cuspidal on non-cuspidal has long been a very intriguing question. It is worth noting that this question remains not completely solved. One of the pioneer contributors to this problem was J. Burdick, who observed that under simplifying geometric conditions such as intersecting or parallel joint axes, a 3-R manipulator was non-cuspidal (Burdick 1995). Other simplifying conditions were exhibited later (Wenger 1997) as a direct consequence of the condition recalled in the next section. Finally, the following seven geometric conditions were found to define non-cuspidal 3R manipulators:

1. first two joint axes are parallel;
2. last two joint axes are parallel;
3. first two joint axes intersect;
4. last two joint axes intersect;

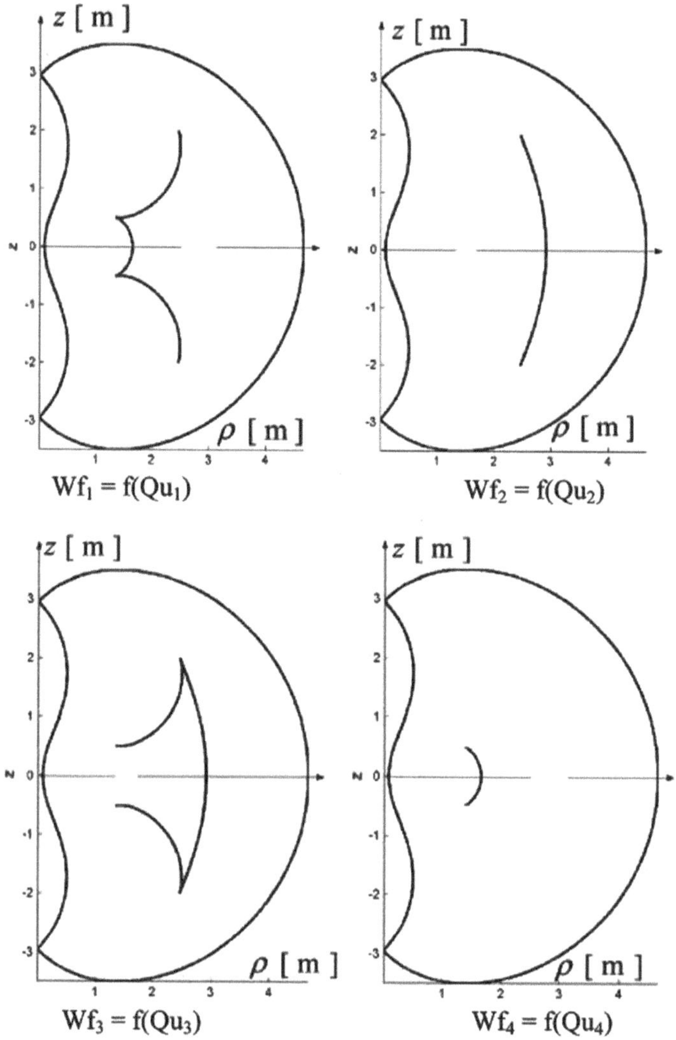

Fig. 18 The four maximal regions of feasible paths for the robot of Fig. 1

5. first two joint axes are orthogonal and all joint offsets vanish;
6. the joint axes are mutually orthogonal and the first joint offset vanishes;
7. the joint axes are mutually orthogonal and $d_4^2 > d_3^2(1 + (\frac{r_2}{d_2 - d_3})^2)$.

These conditions also hold for 6-R manipulators with a spherical wrist because the singularity conditions for the wrist and the regional structure can be decoupled. It is worth noting that conditions 2. and 3. are encountered in most industrial 6-R manipulators. However, the last three conditions are unusual.

A design methodology taking into account the aforementioned geometric conditions was proposed by Wenger (1999).

4.2 Identification of Cuspidal Robots and Classification

The identification of cuspidal robots is one of the most important issues to solve. When a new robot is designed or even when an existing robot is modified, it is necessary to know if the resulting robot will be cuspidal or not before using it. The first important result in the identification of cuspidal robots was found in 1995: the existence of a point in the workspace where three inverse geometric solutions coincide indicates that the robot is cuspidal (El Omri and Wenger 1995). The formal mathematical proof is not detailed here, it relates on the facts reported in Sect. 2.4. A point with three coincident inverse geometric solutions can be identified in a planar cross section of the workspace as a cusp point (hence the word cuspidal robot). Moreover, the existence of a point with three coincident solutions can be verified using the univariate polynomial derived to solve the inverse geometric model. For a 3-R manipulator, the existence of cusps can be determined from its fourth-degree inverse kinematic polynomial $P(t) = at^4 + bt^3 + ct^2 + dt + e$ in $t = tan(\frac{\theta_3}{2})$ whose coefficients are function of the geometric parameters and of the variables $R = x^2 + y^2$ and $Z = z^2$ (see (Kholi and Spanos 1985) for more details on the derivation and properties of this polynomial). The condition for $P(t)$ to have three equal roots can be set as follows:

$$\begin{cases} P(t, d_2, d_3, d_4, \alpha_2, \alpha_3, r_2, R, Z) = 0 \\ \frac{\partial P}{\partial t}(t, d_2, d_3, d_4, \alpha_2, \alpha_3, r_2, R, Z) = 0 \\ \frac{\partial^2 P}{\partial t^2}(t, d_2, d_3, d_4, \alpha_2, \alpha_3, r_2, R, Z) = 0 \end{cases} \tag{3}$$

where $d_2, d_3, d_4, r_2, \alpha_2$ and α_3 are the geometric parameters of the robot. If at least one solution to this system exists, this means that the robot is cuspidal. More interestingly, one can try to derive a condition on the geometric parameters for the above system to have (real) solutions. This is very challenging but by doing so, a condition for a 3-R robot to be cuspidal could be found. One needs to eliminate the three variables t, R and Z. This task is not tractable in the general case but fortunately it becomes feasible (although still complex) once the angle values α_2 and α_3 are assigned to 90°, i.e., in the particular case of orthogonal manipulators. We can go even further by looking for the conditions under which the number of real solutions changes. By doing so, one will obtain a set of bifurcation surfaces in the parameter space of orthogonal 3-R manipulators where the number of cusp points changes. Accordingly, such bifurcating surfaces can be regarded as sets of transition robots that divide the parameter space into domains where all robots have the same number of cusp points. The algebra involved in system (3) is too complex to be handled by commercial computer algebra tools. Corvez and Rouillier (2004) resorted to sophisticated computer algebra tools to solve system (3) by normalizing with $d_2 = 1$ (without loss of generality) and first considering the more particular case $r_3 = 0$ (no offset along the last joint axis like the robot shown in Fig. 1). They used Groebner Bases and Cylindrical Algebraic Decomposition (Lazard and Rouillier 2004), (Collins 1975) to find the equations of the bifurcating surfaces and the number of domains generated by these surfaces. A kinematic interpretation of this theoretical work was conducted by Baili et al. (2003): the authors analyzed global kinematic properties of one representative manipulator in each domain. Finally, only five different cases were found to exist and the true bifurcating surfaces were shown to take on the following explicit form (Baili et al. 2003):

$$C_1: \ d_4 = \sqrt{\frac{1}{2}(d_3{}^2 + r_2{}^2 - \frac{(d_3{}^2 + r_2{}^2)^2 - (d_3{}^2 - r_2{}^2)}{AB})} \tag{4}$$

$$C_2: \ d_4 = \frac{d_3}{1 - d_3} B \ \text{ and } \ d_3 < 1 \tag{5}$$

$$C_3: \ d_4 = \frac{d_3}{d_3 - 1} B \ \text{ and } \ d_3 > 1 \tag{6}$$

$$C_4: \ d_4 = \frac{d_3}{1 - d_3} B \ \text{ and } \ d_3 < 1 \tag{7}$$

where:

$$A = \sqrt{(d_3 + 1)^2 + r_2{}^2} \ \text{ and } \ B = \sqrt{(d_3 - 1)^2 + r_2{}^2} \tag{8}$$

These four surfaces divide the parameter space into five domains with 0, 2 or 4 cusps. Figure 19 shows the plots of the surfaces in a section (d_3, d_4) of the parameter

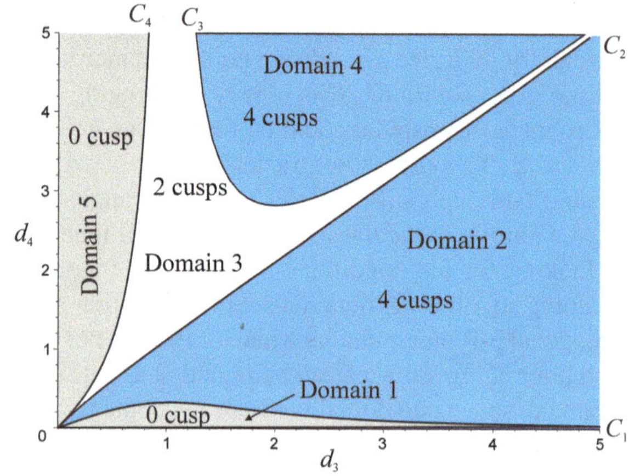

Fig. 19 The four bifurcating surfaces and the five domains in the parameter space (section $r_2 = 1$)

space for $r_2 = 1$. The separating surfaces, which appear as curves there, are labeled with C_i in accordance with the labels in Eqs. (4) to (8). Plotting sections for different values of r_2 changes the size of each region but the general pattern does not change and the number of cells remains the same. There are two domains of non-cuspidal manipulators (domains 1 and 5), two domains of cuspidal manipulators with four cusps (domains 2 and 4) and one domain of cuspidal manipulators with two cusps. Figure 20 shows the cross sections of the workspace and the singular curves in the joint space, for one representative robot in each domain of the partition. The number of inverse geometric solutions in each region of the workspace is indicated. Figure 20 shows that robots in domain 1 have only two inverse geometric solutions. Also, they have a void in their workspace and they are non-cuspidal. In fact, it can be shown that all other robots have 4 inverse geometric solutions. The other non-cuspidal robots are in domain 5. They have a region with 4 inverse geometric solutions and no void. It is interesting to note that robots in domain 5 were in fact already identified in (Wenger 1997), they correspond to the case 7 enumerated in Sect. 4.1.

The partition of the parameter space and the equations of the bifurcating surfaces allow us to define an explicit necessary and sufficient condition for an orthogonal manipulator with no offset along its last joint axis to be cuspidal. Indeed, Fig. 19 shows that a manipulator is cuspidal if and only if it belongs to domains 2, 3 or 4. Thus, an orthogonal manipulator with no offset along its last joint axis is non-cuspidal if and only if:

$$\begin{cases} d_4 > \sqrt{\frac{1}{2}\left(d_3^2 + r_2^2 - \frac{\left(d_3^2+r_2^2\right)^2 - d_2^2\left(d_3^2-r_2^2\right)}{\sqrt{(d_3+d_2)^2+r_2^2}\sqrt{(d_3-d_2)^2+r_2^2}}\right)} \\ \text{and} \\ d_3 > d_2 \text{ or } \left(d_3 < d_2 \text{ and } d_4 < \frac{d_3}{d_2-d_3}\sqrt{(d_3-d_2)^2 + r_2^2}\right) \end{cases} \tag{9}$$

It is worth noting that this necessary and sufficient condition on the geometric parameters is not straighforward and not intuitive at all.

Case $r3 \neq 0$: the case of nonzero offset along the last joint axis has also been studied. The number of parameters to handle is 4 instead of 3 and the bifurcating surface equations get much more complicated: the equation of one of the bifurcating surfaces is a 12th–degree polynomial in the square of the geometric parameters and contains 536 monomials! Because of this complexity, it is difficult to derive an algebraic condition like (9) for a robot to be cuspidal. Figure 21 shows a section of the parameter space at $r_2 = 0.3$ and $r_3 = 0.8$. Numerals indicate the numbers of cusps in the domains. It turns out that a robot with nonzero joint offset along its last joint axis may have up to 8 cusps. Such a robot is shown in Fig. 22.

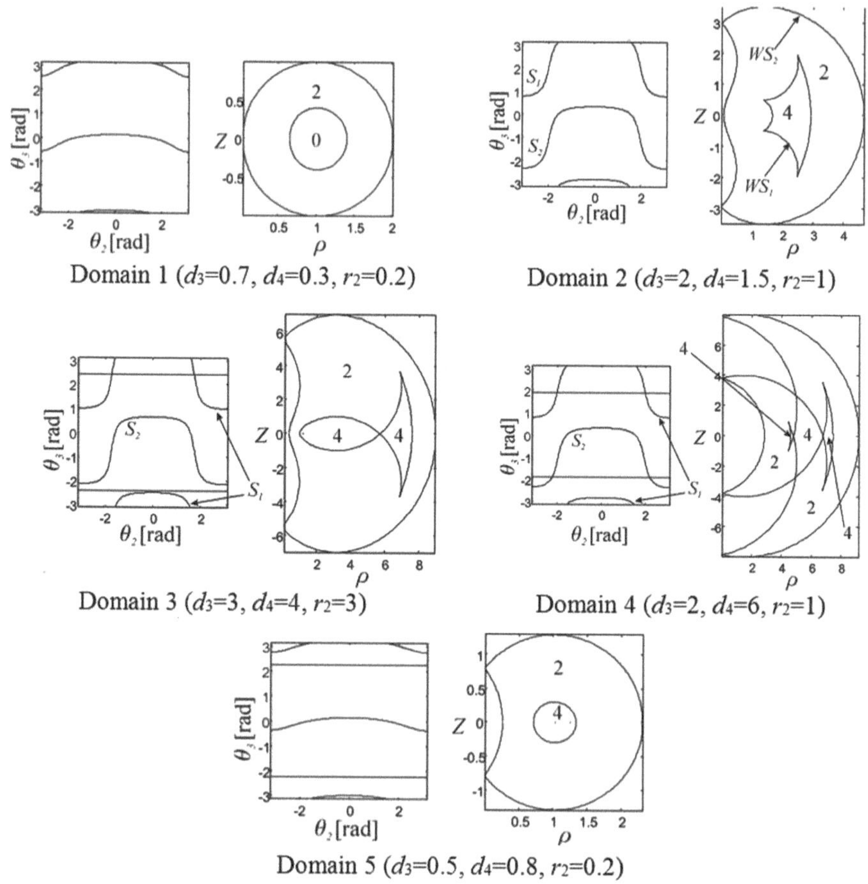

Fig. 20 Workspace topologies in each domain

Note1: The existence of a cusp point is a sufficient condition for a robot to be cuspidal, but it is not necessary in general. The reason is that the presence of a cusp point provides a way of defining a non-singular change of posture in a local way (the encircling path can be made as close as possible to the cusp point). In theory, there could exist a more global way of defining a non-singular change of posture. Figure 23 shows a workspace pattern that could enable a non-singular change of posture in the absence of any cusp point. This global non-singular change of posture can be interpreted using a comparison with a Fig. 8 race track where the crossroads is realized with a bridge; i.e., there are two levels at the crossroads. Each level is associated with a posture. Starting from the crossroads on the bridge (level 1), the track is followed until the crossroads at the same horizontal position but under the bridge is reached (level 0).

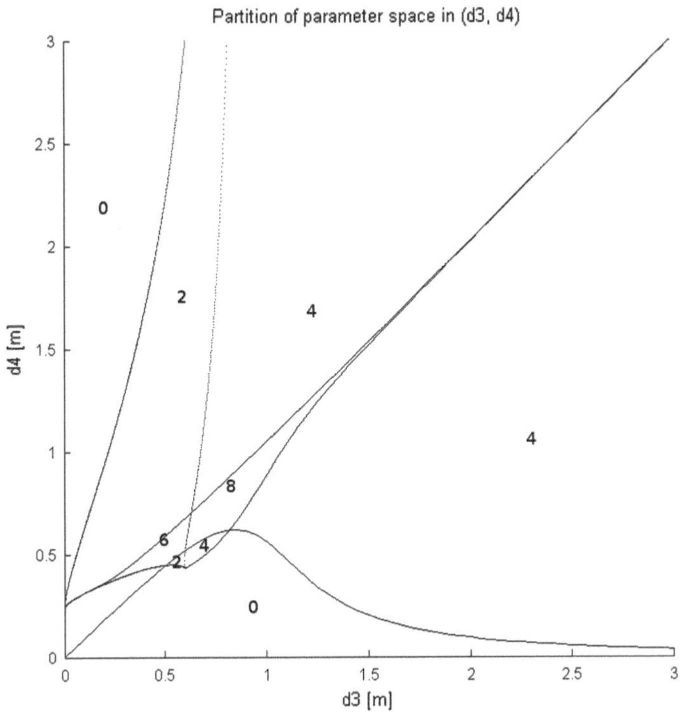

Fig. 21 Partition of the parameter space for 3-R robots with joint offsets, section at $r_2 = 0.3$, $r_3 = 0.8$

In fact, it turns out that the existence of a cusp point is a necessary and sufficient condition for a 3-R orthogonal robot to be cuspial and consequently the pattern shown in Fig. 23 cannot exist for these robots. This can be shown by verifying that for one arbitrary robot in each domain of the parameter space associated with no cusp, there is only one inverse geometric solution in each aspect. It is likable that this result still stands for general 3-R robots (without orthogonal joint axes). On the other hand, a planar parallel cuspidal robot was shown in (Coste et al. 2014) to feature a joint space pattern as in Fig. 23 (see Sect. 5.2).

Note2: 3-dof robots with one prismatic joint can be classified in a similar way. In fact, the equations will be simpler and such robots are more likely to be non-cuspidal because their inverse geometric model can be often solved in a cascade of two quadratics (Pieper 1968). Note that any 3-dof robot with more than one prismatic joint is always non-cuspidal because its inverse geometric model admits two solutions at most (only one for a 3-P robot) (Pieper 1968).

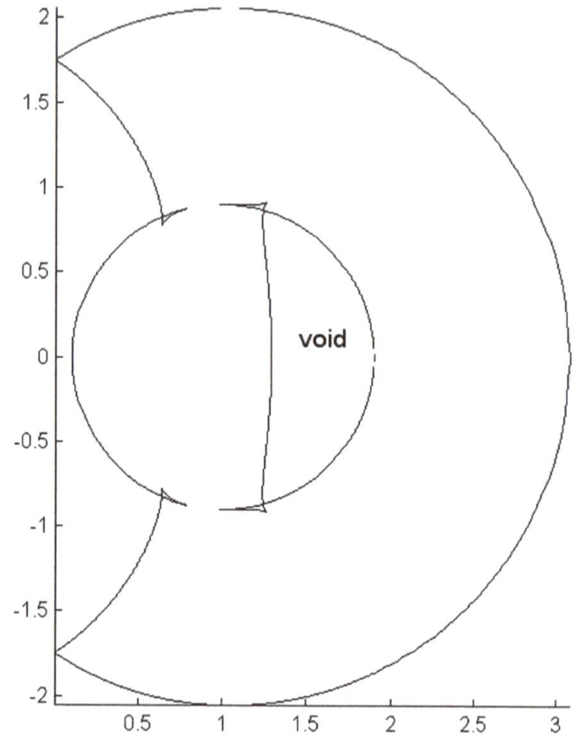

Fig. 22 A 3-R orthogonal robot with 8 cusps ($d_2 = 1$, $d_3 = 0.91$, $d_4 = 0.94$, $r_2 = 0.3$, $r_3 = 0.9$)

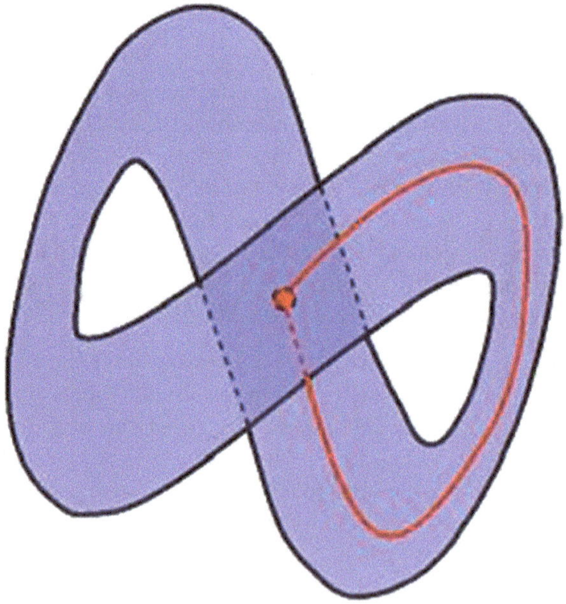

Fig. 23 A workspace pattern that would allow a non-singular posture changing path in the absence of cusps

5 Higher-Degree-of-Freedom Robots and Parallel Robots

5.1 6-dof Serial Robots

The results pertaining to 3-R robots also hold for 6-R robots with a spherical wrist (i.e, with their last three joint axes intersecting at a common point) because the singularity analysis of the wrist can then be decoupled from that of the regional structure. In Sect. 1.3, we reported the story of the IRB 6400C robot. This robot, shown in Fig. 24, has a spherical wrist and its regional structure is an orthogonal 3R robot that can be shown to be cuspidal. Note that the main objective of this new robot design was to save space along the assembly lines and this is why its first joint axis is horizontal instead of vertical (Hemmingson et al. 1996). This was a good idea but at the time the engineers of ABB designed their new robot, the classification results were not published. It would be interesting to attempt a new design, keeping the orthogonal architecture with its first axis horizontal but tuning the length parameters in order that the robot falls in one of the interesting classes of non-cuspidal orthogonal robots described by Zein et al. (2006). On the other hand, there is no general result about the enumeration of cuspidal 6-dof manipulators with non-spherical wrist. One of the reasons is the difficulty in analyzing the singularities of general 6-R robots, which depend on four joint variables instead of two in 3-R robots. We think that 6-R manipulators are very likely to be cuspidal, even if the simplifying geometric conditions listed in Sect. 4.1 are satisfied. This is because the inverse kinematics of most 6-dof manipulators with non-spherical wrist is a polynomial of degree higher than 4, which is more likely to admit triple roots. Further research work is required before stating more definitive results but several examples of simple 6-R cuspidal robots with non-spherical wrist exist. One of these robots is the GMF P150 shown in Fig. 25 used in the automotive industry for car painting (a similar version exists by COMAU). This robot is close to a PUMA robot, the only difference being the presence of a wrist offset. El Omri showed that without taking account the joint limits, this robot has 16 inverse kinematic solutions and only two aspects (El Omri 1996). Thus, it is cuspidal. Another example is the ROBOX painting robot studied by Zoppi (2002) (see Fig. 26). The kinematic architecture is very close to the GFM P150 but the wrist offset is not along the same wrist axis. This cuspidal robot has also 16 inverse kinematic solutions and only two aspects.

If the classification of 6-dof cuspidal and non-cuspidal robots is far from being simple, it is possible to enumerate a set of non-cuspidal robots, namely, those whose inverse geometric polynomial can be solved with quadratics or linear equations in cascade. Such robots were enumerated by Mavroidis and Roth (1994).

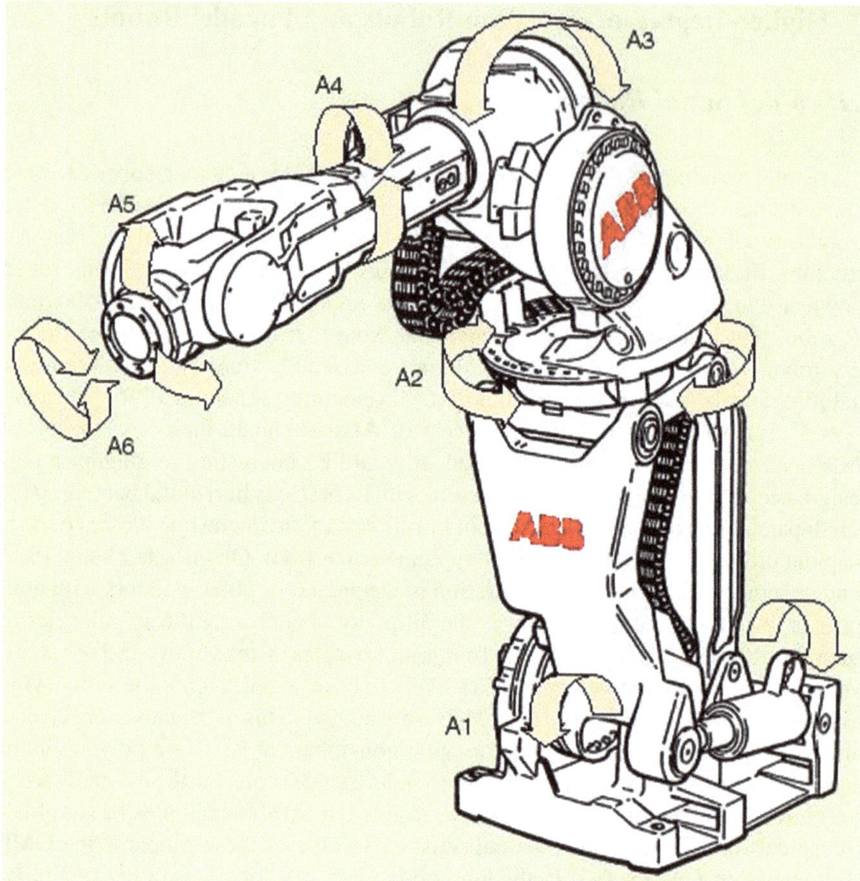

Fig. 24 The IRB 6400C robot

5.2 *Parallel Robots*

A parallel robot may change its assembly-mode without encountering an output singularity. As first observed in 1998, a parallel robot may be cuspidal in the sense that it may change its assembly-mode (an assembly mode is associated with a solution to the direct geometric problem) without crossing an output singularity (Innocenti and Parenti-Castelli 1998), (Wenger and Chablat 1998). As shown by McAree and Daniel (1999) and explained in details by Zein et al. (2008), if a parallel robot has 3 coincident assembly modes, thus defining a cusp point in a section of its joint space (not in the workspace this time), then this robot is cuspidal. A non-singular change of assembly-mode can then be accomplished by encircling a cusp point in a section of the joint space (McAree and Daniel 1999), (Zein et al. 2008). Figure 27 shows the singularity curves and a non-singular assembly-mode changing path in a section

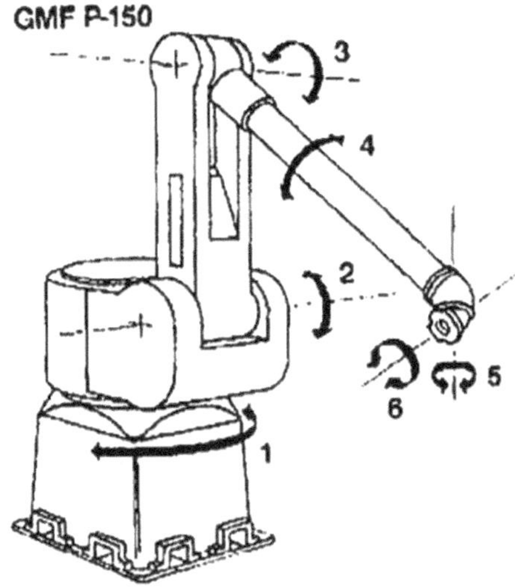

Fig. 25 The GMF P150 robot

Fig. 26 The ROBOX robot

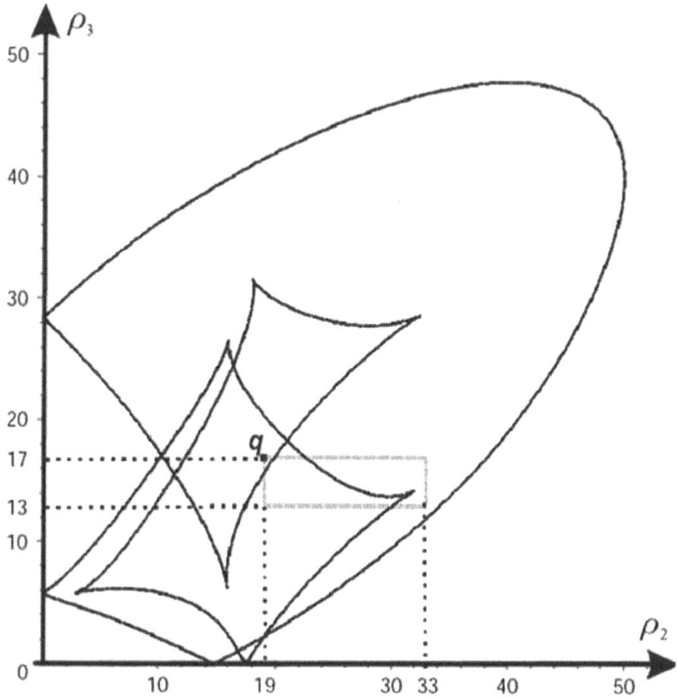

Fig. 27 A non-singular assembly-mode changing path encircling a cusp point in a section of the joint space for a 3-RPR planar parallel robot

$\rho_1 = 17$ of the joint space for the 3-RPR robot (the underlined letter refers to the actuated joint) analyzed in (Innocenti and Parenti-Castelli 1998) and (McAree and Daniel 1999).

Because the kinematic equations of a parallel manipulator are very complex, it seems very difficult to derive general geometric conditions for a parallel robot to be cuspidal. However, some results are available for planar parallel robots. One of these results claims that to be cuspidal, a 3-RPR robot parallel planar robot should not have similar platform and base triangles (McAree and Daniel 1999), (Kong and Gosselin 2000) but this is false if the legs are RRR instead of RPR (Wenger 2004). Another interesting result claims that a generic 3-RPR robot has always two aspects (Husty 2009), (Coste 2012). Since such robots have up to six assembly-modes, this shows that they have more than one assembly-mode in one of their aspects, which thus allows them to accomplish non-singular assembly-mode motions. The last interesting result is a proof that a planar parallel robot can be cuspidal while there are no cusp points at all (Coste et al. 2014), see **Note1** in Sect. 4.2 and Fig. 23. More details on cuspidal planar parallel robots can be found in (Zein et al. 2007) and (Zein et al. 2008).

Quite few results exist for 6-dof parallel robots. Out of the octahedral Gough-Stewart platform studied by McAree and Daniel (1999) and J.P. Merlet who showed the existence of non-singular assembly-mode changing motions for this robot, the only other available result, to the author's knowledge, was published in (Caro et al. 2012). In this paper, a non-singular assembly-mode changing path was generated for a 6-dof 3-PPPS parallel robot.

If the interest of designing a cuspidal serial robot is questionable, a parallel robot that is cuspidal is interesting since the robot can thus move in a larger part of its workspace without crossing any output singularity (Wenger 2004), (Zein et al. 2008).

6 Conclusions

The results presented in this chapter can be summarized as follows. Cuspidal robots, which were first discovered in 1988, have multiple inverse geometric solutions that are not separated by a singular surface. In the joint space, additional surfaces called the characteristic surfaces, divide the aspects and separate the solutions. These surfaces are used to define new uniqueness domains and regions of feasible paths in the workspace. The definitions are general and stand for any serial, non-redundant robot with or without joint limits. For 3-dof robots, it is possible to calculate and plot these characteristic surfaces, uniqueness domains and regions of feasible paths. If the first joint is revolute and unlimited, 2-dimensional plots are sufficient. Because there is no simple algebraic definition, these sets must be calculated numerically.

A 3-R robot is non-cuspidal as soon as its first two or last two joint axes are parallel or intersect, or if the three joint axes are mutually orthogonal and the first joint offset is equal to zero. But an orthogonal manipulator with its last joint offset equal to zero may be cuspidal.

A robot is cuspidal if there is at least one point where three inverse geometric solutions coincide. For 3-dof robots, such points appear as cusp points in a cross section of the workspace. In general, the existence of a cusp point is only sufficient for a robot to be cuspidal but for the particular case of 3-R orthogonal robots, this condition is also necessary. The full partition of the parameter space of 3-R orthogonal robots is possible. Bifurcating surfaces divide the parameter space into domains where all robots have the same number of cusp points.

For 3-R orthogonal robots with no joint offset along their last joint axis, it is possible to derive an explicit DH-parameter based necessary and sufficient condition for a robot to be cuspidal. For general 3-R orthogonal manipulators, the classification is much more complex and does not lend itself to explicit conditions.

Little research work has been conducted on 6-R cuspidal robots. It appears that 6-R robots with non-spherical wrist are very likely to be cuspidal, even if two joint axes intersect or are parallel. However, there is still much work to do before having definitive geometric conditions for general 6-R robots. Resorting to some transversality theorems used by singularity theorists would help going further, providing that we remain in the generic case (Pai and Leu 1991), (Donelan and Gibson 1999), (Donelan and Müller 2010). So the first step would be to enumerate all 6-R generic robots.

Finally, the issue of classifying cuspidal and non-cuspidal parallel robots is still much more difficult and remains open. As shown in (Coste et al. 2014), a parallel robot may be cuspidal without having any cusp, so that we lack a necessary and sufficient condition for the classification of parallel robots.

References

M. Baili, P. Wenger, D. Chablat, Classification of one family of 3R positioning manipulators, in *Proceedings of 11th International Conference on Advanced Robotics* (2003)

P. Borrel, A. Liegeois, A study of manipulator inverse geometric solutions with application to trajectory planning and workspace determination, in *Proceedings of IEEE International Conference Robotics and AUT*, pp. 1180–1185 (1986)

J.W. Burdick, *Kinematic analysis and design of redundant manipulators*. Ph.D. thesis, Standford University (1988)

J.W. Burdick, A classification of 3R regional manipulator singularities and geometries. Mech. Mach. Theor. **30**(1), 71–89 (1995)

S. Caro, P. Wenger, D. Chablat, Non-singular assembly mode changing trajectories of a 6-dof parallel robot, in *In Proceedings of ASME Design Engineering Technical Conferences and Computers and Information in Engineering Conference* (2012)

G.E. Collins, Quantifier elimination for real closed fields by cylindrical algebraic decomposition. in *Spring lecture, Notes in Computer Science* vol. 3, pp. 515–532 (1975)

S. Corvez, F. Rouillier, Using computer algebra tools to classify serial manipulators, in *Automated Deduction in Geometry, Lectures Notes in Computer Science*, pp. 31–43 (2004)

M. Coste, A simple proof that generic 3-RPR manipulators have two aspects. *ASME J. Mech. Robot.* **4**(1) (2012)

M. Coste, D. Chablat, P. Wenger, Non-singular change of assembly mode without any cusp, in *Advances in Robot Kinematics*, ed. J. Lenarčič, O. Khatib (Springer, 2014)

P.S. Donelan, C.G. Gibson, Singular phenomena in kinematics, in *Singularity Theory*, ed. by B. Bruce, D. Mond (Cambridge University Press, 1999)

P.S. Donelan, A. Müller, Singularities of regional manipulators revisited, in *Advances in Robot Kinematics, Motion in Man and Machine*, ed. by J. Lenarčič, M. Staničić (Springer, 2010)

J. El Omri, *Kinematic analysis of robot manipulators (in French)*. Ph.D. thesis, Ecole Centrale de Nantes (1996)

J. El Omri, P. Wenger, How to recognize simply a nonsingular posture changing 3-dof manipulator, in *Proceedings of 7th International Conference on Advanced Robotics*, pp. 215–222 (1995)

C.G. Gibson, Kinematics from the singular viewpoint, in *Geometrical Foundation of Robotics*, ed. J.M. Selig (World Scientific Press, 2000)

E. Hemmingson, S. Ellqvist, J. Pauxels, *New Robot Improves Cost-Efficiency of Spot Welding* (ABB Review, Spot Welding Robots, 1996)

M. Husty, Non-singular assembly mode change in 3-RPR parallel manipulators, in *Singularity Theory*, ed. A. Müller, A. Kecskemthy (Springer, 2009)

C. Innocenti, V. Parenti-Castelli, Singularity-free evolution from one configuration to another in serial and fully-parallel manipulators. ASME J. Mech. Des. **120**, 73–99 (1998)

W. Khalil, J. Kleinfinger, A new geometric notation for open and closed loop robots, in *Proceedings of IEEE International Conference Robotic and AUT*, pp. 1174–1179 (1986)

D. Kholi, J. Spanos, Workspace analysis of mechanical manipulators using polynomial discriminant. ASME J. Mech. Transm. Autom. Des. **107**, 209–215 (1985)

X. Kong, C.M. Gosselin, Determination of the uniqueness domains of 3-RPR planar parallel manipulators with similar platforms, in *Proceedings of ASME Design Engineering Technical Conferences and Computers and Information in Engineering Conference* (2000)

D. Lazard, F. Rouillier. *Solving parametric polynomial systems*. INRIA Technical Report (2004)
C. Mavroidis, B. Roth, Structural parameters which reduce the number of manipulator configurations. ASME J. Mech. Des. **116**, 3–10 (1994)
P.R. McAree, R.W. Daniel, An explanation of never-special assembly changing motions for 3–3 parallel manipulators. Int. J. Robot. Res. **18**(6), 556–574 (1999)
D.K. Pai, M.C. Leu, Genericity and singularities of robot manipulators. IEEE Trans. Robot. Autom. **20**(4), 545–559 (1991)
C.V. Parenti, C. Innocenti, Position analysis of robot manipulators: Regions and sub-regions, in *Proceedings of International Conference on Advances in Robot Kinematics*, pp. 150–158 (1988)
B. Pieper. *The Kinematics of Manipulators Under Computer Control*. Ph.D. thesis, Stanford University (1968)
P. Wenger, A new general formalism for the kinematic analysis of all non-redundant manipulators, in *Proceedings of IEEE International Conference Robotic and AUT*, pp. 442–447 (1992)
P. Wenger, Design of cuspidal and noncuspidal manipulators, in *Proceedings of IEEE International Conference Robotic and AUT*, pp. 2172–2177 (1997)
P. Wenger, Some guidelines for the kinematic design of new manipulators. Mechanisms Mach. Theor. **35**(3), 437–449 (1999)
P. Wenger, Uniqueness domains and regions of feasible paths for cuspidal manipulators. IEEE Trans. Robot. **20**(4), 750–754 (2004)
P. Wenger, D. Chablat, Workspace and assembly-modes in fully parallel manipulators: a descriptive study, in *Advances in Robot Kinematics*, ed. by J. Lenarčič, M. Husty (Kluwer Academic Publisher, 1998)
P. Wenger, J. El Omri, Modeling kinematic properties of type-2 regional manipulators using octrees, in *Proceedings of IEEE International Conference Man and Cybernetics*, pp. 183–188 (1993)
H. Whitney, On singularities of mappings of euclidean spaces 1. mappings of the plane into the plane. Ann. Math. **62**(3), 374–410 (1955)
M. Zein, P. Wenger, D. Chablat, An exhaustive study of the workspace topologies of all 3R orthogonal manipulators with geometric simplifications. Mechanisms Mach. Theor. **41**(8), 971–986 (2006)
M. Zein, P. Wenger, D. Chablat, Singular curves in the joint space and cusp points of 3-RPR parallel manipulators. Robotica Spec. Issue Geometry Robot. Sens. **25**(6), 714–724 (2007)
M. Zein, P. Wenger, D. Chablat, Non-singular assembly-mode changing motions for 3-RPR parallel manipulators. Mechanisms Mach. Theor. **43**(4), 480–490 (2008)
M. Zoppi, Effective backward kinematics for an industrial 6Rrobot, in *Proceedings of ASME Design Engineering Technical Conferences and Computers and Information in Engineering Conference* (2002)

Mechanism Constraints and Singularities—The Algebraic Formulation

Manfred L. Husty and Dominic R. Walter

Abstract Various mathematical formulations are used to describe mechanism and robot kinematics. This mathematical formulation is the basis for kinematic analysis and synthesis, i.e., determining displacements, velocities and accelerations, on the one hand, and obtaining design parameters on the other hand. Vector/matrix formulation containing trigonometric functions is arguably the most favored approach used in the engineering research community. A less well-known but nevertheless very successful approach relies on an algebraic formulation. This involves describing mechanism constraints with algebraic (polynomial) equations and solving these equation sets that pertain to some given mechanism or robot, with the powerful tools of algebraic and *numerical* algebraic geometries. In the first section of this chapter, the algebraic formulation of Euclidean displacements using Study coordinates (or dual quaternions) will be recalled. Then it will be shown how constraint equations of different kinematic chains can be derived using either geometric insight, elimination methods or the linear implicitization algorithm (LIA). LIA will be described in detail because it can be used even without deep kinematic and geometric insight into the properties of the kinematic chain. In case the kinematic chain consists of only elementary joints, the constraint equations consist of a set of polynomial equations. In the language of algebraic geometry, the corresponding polynomials form an ideal which can be treated with almost classical methods that might not be well known in the engineering community. Therefore in the third section, these methods will be recalled and it will be shown how they can be used to derive the properties of the kinematic entities. In the last section, we apply the introduced algebraic methods to the complete analysis of the 3-UPU-TSAI parallel manipulator.

M. L. Husty (✉) · D. R. Walter
University of Innsbruck, Innsbruck, Austria
e-mail: manfred.husty@uibk.ac.at

A. Müller and D. Zlatanov (eds.), *Singular Configurations of Mechanisms and Manipulators*, CISM International Centre for Mechanical Sciences 589,
https://doi.org/10.1007/978-3-030-05219-5_4

1 Introduction

The mathematical description of mechanisms or robots needs essentially two ingredients: the description of Euclidean displacements and the mathematical description of the mechanical device itself. The literature (e.g., Bottema and Roth 1979) knows many different representations of the Euclidean displacement group $SE(3)$ such as Euler angles, Rodrigues parameters, Euler parameters, Study parameters, quaternions and dual quaternions. Within this chapter, the algebraic description of mechanisms is sought and therefore Study parameterization will be used throughout. Concerning the description of the mechanisms, we adopt the following point of view: No matter how complicated the mechanism might be, it consists of one or more kinematic chains. The kinematic chain is a serial assembly of joints and links, and the mechanism or the robot is a serial, parallel or hybrid assembly of kinematic chains. Vector/matrix formulation containing trigonometric functions is arguably the most favored approach used in the engineering research community. Most of the time, vector loop equations are used to describe the kinematic chains, and very often, only a single numerical solution to the basic tasks like inverse or direct kinematics is obtained. Complete analysis and synthesis of mechanisms needs all solutions. Therefore, the use of algebraic constraint equations is proposed as to be able to use strong methods and algorithms from algebraic geometry or numerical algebraic geometry (see Sommese and Wampler 2005). An important task is to find the simplest algebraic constraint equations, that describe the chains, because after establishing these equations they have to be processed with different solution algorithms. Geometric and algebraic preprocessing is needed before elimination, Groebner base computation or numerical solution processes start. An important feature of algebraic constraint equations is that they provide a complete description of the overall motion behavior of the mechanism. This chapter is therefore a contribution to **Global Kinematics**.

2 Parameterization of $SE(3)$

Euclidean three space is the three-dimensional real vector space $\mathbb{R}^3$ together with the usual scalar product $\mathbf{x}^T\mathbf{y} = \sum_{i=1}^{3} x_i y_i$. A Euclidean displacement is a mapping

$$\gamma : \mathbb{R}^3 \to \mathbb{R}^3, \quad \mathbf{x} \mapsto \mathbf{M}_R\mathbf{x} + \mathbf{a} \tag{1}$$

where $\mathbf{M}_R \in \mathrm{SO}(3)$ is a proper orthogonal three-by-three matrix and $\mathbf{a} \in \mathbb{R}^3$ is a vector. The entries of $\mathbf{M}_R$ fulfill the well-known orthogonality condition $\mathbf{M}_R^T \cdot \mathbf{M}_R = \mathbf{I}_3$, where $\mathbf{I}_3$ is the three-by-three identity matrix.

The group of all Euclidean displacements is denoted by SE(3). It is a convenient convention to write Eq. (1) as product of a four-by-four matrix and a four-dimensional vector according to[1]

[1]Note that homogeneous coordinates in this chapter are written in the European notation, with homogenizing coordinate on first place.

$$\begin{bmatrix}1\\ \mathbf{x}\end{bmatrix} \mapsto \mathbf{M}\cdot\begin{bmatrix}1\\ \mathbf{x}\end{bmatrix} = \begin{bmatrix}1 & \mathbf{o}^T\\ \mathbf{a} & \mathbf{M}_R\end{bmatrix}\cdot\begin{bmatrix}1\\ \mathbf{x}\end{bmatrix}. \tag{2}$$

The displacements Eq. (2) can be parameterized in different ways. In many applications, Euler angles or Rodrigues parameters are used to parameterize the matrix $\mathbf{M}_R$. In the aim of using algebraic tools, an algebraic version of the complete 4×4 displacement operator is adopted. This parameterization of the spatial Euclidean transformation matrix $\mathbf{M} \in SE(3)$ is named after *E. Study* (Study 1903) and can be written as follows (for detailed information on this approach, see Husty and Schröcker 2012):

$$\mathbf{M} = \begin{pmatrix} x_0^2+x_1^2+x_2^2+x_3^2 & \mathbf{0}^\top \\ \mathbf{a} & \mathbf{M}_R \end{pmatrix}, \quad \mathbf{a} = \begin{pmatrix} 2(-x_0y_1+x_1y_0-x_2y_3+x_3y_2) \\ 2(-x_0y_2+x_1y_3+x_2y_0-x_3y_1) \\ 2(-x_0y_3-x_1y_2+x_2y_1+x_3y_0) \end{pmatrix}$$

$$\mathbf{M}_R = \begin{pmatrix} x_0^2+x_1^2-x_2^2-x_3^2 & 2(x_1x_2-x_0x_3) & 2(x_1x_3+x_0x_2) \\ 2(x_1x_2+x_0x_3) & x_0^2-x_1^2+x_2^2-x_3^2 & 2(x_2x_3-x_0x_1) \\ 2(x_1x_3-x_0x_2) & 2(x_2x_3+x_0x_1) & x_0^2-x_1^2-x_2^2+x_3^2 \end{pmatrix} \tag{3}$$

The vector $\mathbf{a}$ represents the translational part, and $\mathbf{M}_R$ represents the rotational part of the transformation operator $\mathbf{M}$. The parameters $x_0, x_1, x_2, x_3, y_0, y_1, y_2, y_3$ which appear in the matrix $\mathbf{M}$ are called *Study* parameters of the transformation $\mathbf{M}$, and we can consider them as a projective point in $\mathbb{P}^7$. The mapping

$$\kappa\colon SE(3) \to P \in \mathbb{P}^7$$
$$\mathbf{M}(x_i, y_i) \mapsto (x_0 : x_1 : x_2 : x_3 : y_0 : y_1 : y_2 : y_3)^T \neq (0:0:0:0:0:0:0:0)^T \tag{4}$$

is called *kinematic mapping* and maps each Euclidean displacement of $SE(3)$ to a point P on a quadric $S_6^2 \subset \mathbb{P}^7$. In this way, every projective point $(x_0 : x_1 : x_2 : x_3 : y_0 : y_1 : y_2 : y_3) \in \mathbb{P}^7$ represents a spatial Euclidean transformation if it fulfills the following equation of S_6^2:

$$x_0y_0 + x_1y_1 + x_2y_2 + x_3y_3 = 0 \tag{5}$$

and the inequality:

$$x_0^2 + x_1^2 + x_2^2 + x_3^2 \neq 0. \tag{6}$$

Equation (6) is a normalization term, and points with $x_0 = x_1 = x_2 = x_3 = 0$ do not represent Euclidean transformations; they form the 3-dimensional *exceptional generator* contained in S_6^2. The points on S_6^2 are called *kinematic image points* of the corresponding displacement, and the seven-dimensional projective space is called *kinematic image space*. More information on this parameterization can be found in Husty and Schröcker (2012), where especially the relation of the Study parameters to dual quaternions is highlighted.

Very often the practical question arises how to compute the Study parameters when an other parameterization is given. Let therefore $\mathbf{A}$ be an orthogonal 3×3 matrix and $\mathbf{a} = (a_1, a_2, a_3)^T$ a three-dimensional vector that determines a displacement Eq. (2). Then the entries of Matrix $\mathbf{M}_R$ can be found by one of the four relations:

$$\begin{aligned} x_0 : x_1 : x_2 : x_3 &= 1 + a_{11} + a_{22} + a_{33} : a_{32} - a_{23} : a_{13} - a_{31} : a_{21} - a_{12} \\ &= a_{32} - a_{23} : 1 + a_{11} - a_{22} - a_{33} : a_{12} + a_{21} : a_{31} + a_{13} \\ &= a_{13} - a_{31} : a_{12} + a_{21} : 1 - a_{11} + a_{22} - a_{33} : a_{23} + a_{32} \\ &= a_{21} - a_{12} : a_{31} + a_{13} : a_{23} - a_{32} : 1 - a_{11} - a_{22} + a_{33} \end{aligned} \tag{7}$$

In general, all four proportions of Eq. (7) yield the same result. If, however, $1 + a_{11} + a_{22} + a_{33} = 0$, the first proportion yields $0 : 0 : 0 : 0$ and is invalid. We can use the second proportion instead as long as $a_{22} + a_{33}$ is different from zero. If this happens, we can use the third proportion unless $a_{11} + a_{33} = 0$. In this last case, we resort to the last proportion which yields $0 : 0 : 0 : 1$. Having computed the first four Study parameters, the remaining four parameters $y_0 : y_1 : y_2 : y_3$ can be computed from

$$\begin{aligned} 2y_0 &= \quad a_1x_1 + a_2x_2 + a_3x_3, \\ 2y_1 &= -a_1x_0 + a_3x_2 - a_2x_3, \\ 2y_2 &= -a_2x_0 - a_3x_1 + a_1x_3, \\ 2y_3 &= -a_3x_0 + a_2x_1 - a_1x_2. \end{aligned} \tag{8}$$

Example 1 A rotation about the x-axis with angle φ is described by the matrix

$$\mathbf{Q} = \begin{bmatrix} 1 & 0 & 0 & 0 \\ 0 & 1 & 0 & 0 \\ 0 & 0 & \cos\varphi & -\sin\varphi \\ 0 & 0 & \sin\varphi & \cos\varphi \end{bmatrix}. \tag{9}$$

Its kinematic image, computed via (7) and (8), is

$$\mathbf{r} = [1 + \cos\varphi : \sin\varphi : 0 : 0 : 0 : 0 : 0 : 0]. \tag{10}$$

As φ varies in $[0, 2\pi)$, $\mathbf{r}$ describes a straight line on the Study quadric which reads after algebraization with half-tangent substitution

$$\mathbf{r}_x = [1 : u : 0 : 0 : 0 : 0 : 0 : 0]. \tag{11}$$

It is easy to see that the other two elementary rotations about $y-$ and $z-$axis can be written in Study parameters as:

$$\mathbf{r}_y = [1 : 0 : v : 0 : 0 : 0 : 0 : 0], \mathbf{r}_z = [1 : 0 : 0 : w : 0 : 0 : 0 : 0]. \tag{12}$$

A second example, which is already a little bit more sophisticated, shows the Study representation (or the kinematic image) of a simple $2 - R$ chain.

Example 2 Let **M** and **K** be the elementary rotation about the $x-$ and the $y-$axis. Let furthermore **N** be a translation along the $z-$ axis. The concatenation of these three transformations yields a $2 - R$ chain, and its representation is given by $\mathbf{L} = \mathbf{M} \cdot \mathbf{N} \cdot \mathbf{K}$;

$$\mathbf{M} = \begin{bmatrix} 1 & 0 & 0 & 0 \\ 0 & 1 & 0 & 0 \\ 0 & 0 & \cos(t) & -\sin(t) \\ 0 & 0 & \sin(t) & \cos(t) \end{bmatrix}, \quad \mathbf{N} = \begin{bmatrix} 1 & 0 & 0 & 0 \\ 0 & 1 & 0 & 0 \\ 0 & 0 & 1 & 0 \\ 2a & 0 & 0 & 1 \end{bmatrix},$$
$$\mathbf{K} = \begin{bmatrix} 1 & 0 & 0 & 0 \\ 0 & \cos(s) & 0 & -\sin(s) \\ 0 & 0 & 1 & 0 \\ 0 & \cos(s) & 0 & \sin(s) \end{bmatrix},$$

$$\mathbf{L} = \mathbf{M} \cdot \mathbf{N} \cdot \mathbf{K} = \begin{bmatrix} 1 & 0 & 0 & 0 \\ 0 & \cos(s) & 0 & -\sin(s) \\ -2\sin(t)\,a & -\sin(t)\cos(s) & \cos(t) & -\sin(t)\sin(s) \\ 2\cos(t)\,a & \cos(t)\cos(s) & \sin(t) & \cos(t)\sin(s) \end{bmatrix}.$$

Its kinematic image, computed via (7) and (8), is

$$\mathbf{l} = \begin{bmatrix} 1 + \cos(s) + \cos(t) + \cos(t)\sin(s) \\ \sin(t) + \sin(t)\sin(s) \\ -\sin(s) - \cos(t)\cos(s) \\ -\sin(t)\cos(s) \\ -\sin(t)\,a\,(-\sin(s) - \cos(t)\cos(s)) - \cos(t)\,a\sin(t)\cos(s) \\ \cos(t)\,a\,(-\sin(s) - \cos(t)\cos(s)) - (\sin(t))^2\,a\cos(s) \\ \sin(t)\,a\,(1 + \cos(s) + \cos(t) + \cos(t)\sin(s)) - \cos(t)\,a\,(\sin(t) + \sin(t)\sin(s)) \\ -\cos(t)\,a\,(1 + \cos(s) + \cos(t) + \cos(t)\sin(s)) - \sin(t)\,a\,(\sin(t) + \sin(t)\sin(s)) \end{bmatrix}. \tag{13}$$

It seems that this is much more complicated than the matrix **L**, but after algebraization with half-tangent substitution this yields:

$$\mathbf{l} = [1 : u : v : uv : -uav : av : ua : -a]. \tag{14}$$

Equation (14) is the parametric representation of the $2 - R$ chain in Study coordinates.

Remark 1 In the Study representation of $SE(3)$, planar and spherical displacements are contained via setting $x_1 = x_2 = y_0 = y_3 = 0$ for planar displacements and $y_0 = y_1 = y_2 = y_3 = 0$ for spherical displacements. Pure translations are obtained in setting $x_0 = 1,\ x_1 = 0,\ x_2 = 0,\ x_3 = 0$.

3 Constraint Equations

In the kinematic analysis or synthesis of a mechanism, it is crucial to find a mathematical description of the overall motion capability. In this section, we will show different possibilities how a complete set of equations can be found. In this chapter, we are interested to find algebraic constraint equations. This restricts first of all the design possibilities of the mechanism, as helical joints cannot be described by algebraic (polynomial) equations. As we have seen already in Example 2 the parametric description of a serial chain is straightforward. The question is now, how can we find the corresponding algebraic equations? Generally, there are three methods to derive these algebraic constraint equations:

- Using geometric properties of the mechanism. These properties can be, for example, one point of the moving system (end effector system) is bound to move on a line, a circle, a sphere or a plane.
- Elimination method: Given the parametric description of the motion of the moving system of the mechanism, one can use resultant methods or dialytic elimination methods to derive the algebraic equations. The disadvantage of these methods is that they introduce "spurious" solutions that come from multiple projections onto coordinate spaces. In simple cases, this method can be very efficient.
- Linear implicitization algorithm. This algorithm was introduced in Walter and Husty (2010) and guarantees a complete solution of the elimination. The algorithm essentially solves an overconstrained linear system which can be very large in case of high-degree polynomial constraint equations.

It is of utmost importance to start any solution process with the simplest set of algebraic equations. There exists always a best-adapted coordinate system for a mechanism or at least for one kinematic chain in a more complicated mechanism. When a kinematic chain is represented in its "best"-adapted coordinate system, then it is called *canonical chain*. Furthermore, geometric preprocessing, whenever possible, is highly recommended. Both issues will be explained in detail in the next subsections where we will discuss all three methods with representative examples.

3.1 Geometric Constraint Equations

The use of geometric properties to derive constraint equations will be demonstrated with a planar 3-RRR parallel manipulator. This manipulator has been discussed by many authors. Many details can be found in Bonev (2002).

The manipulator consists of three anchor points P_1, P_2, P_3 in the base and three points p_1, p_2, p_3 in a moving system (Fig. 1). The three pairs of points are connected via three planar dyads. The links in the dyads are denoted by $l_i, k_i, i = 1, \ldots 3$. The revolute joints at the base points P_1, P_2, P_3 are active. All other revolute joints are passive. For the mathematical description, we use the planar displacement subgroup parameterized by Study parameters:

$$\mathbf{M} = \begin{bmatrix} {x_0}^2 + {x_3}^2 & 0 & 0 & 0 \\ -2\,x_0\,y_1 + 2\,x_3\,y_2 & {x_0}^2 - {x_3}^2 & -2\,x_0\,x_3 & 0 \\ -2\,x_0\,y_2 - 2\,x_3\,y_1 & 2\,x_0\,x_3 & {x_0}^2 - {x_3}^2 & 0 \\ 0 & 0 & 0 & {x_0}^2 + {x_3}^2 \end{bmatrix}. \tag{15}$$

Without loss of generality, we can cancel the last row and the last column and obtain the 3×3 planar transformation matrix.

$$\mathbf{A} = \begin{bmatrix} {x_0}^2 + {x_3}^2 & 0 & 0 \\ -2\,x_0\,y_1 + 2\,x_3\,y_2 & {x_0}^2 - {x_3}^2 & -2\,x_0\,x_3 \\ -2\,x_0\,y_2 - 2\,x_3\,y_1 & 2\,x_0\,x_3 & {x_0}^2 - {x_3}^2 \end{bmatrix}. \tag{16}$$

From Fig. 1, right it is obvious that, in case of fixed input parameters α, β, γ, the platform points p_1, p_2, p_3 are bound to be on three circles having the "knee joints" (e.g., J in Fig. 1) as centers and radii k_i. Using homogeneous coordinates $(X_0, X_1, X_2)^T$ in the base coordinate system, a circle with center $(1, m, n)^T$ can be written

$$X_1^2 + X_2^2 - 2mX_0X_1 - 2nX_0X_2 + (m^2 + n^2 - r^2)X_0^2 = 0 \tag{17}$$

Let $\mathbf{p} = (1, x, y)^T$ be the coordinates of a point in the moving system, then its coordinates in the fixed base system are given by

$$\mathbf{P} = \mathbf{A}\mathbf{p} = \begin{bmatrix} x_0^2 + x_3^2 \\ -2x_0y_1 + 2x_3y_2 + (x_0^2 - x_3^2)x - 2x_0x_3y \\ -2x_0y_2 - 2x_3y_1 + 2x_0x_3x + (x_0^2 - x_3^2)y \end{bmatrix}. \tag{18}$$

Substituting Eq. (18) into the circle equation, Eq. (17) yields after factoring and canceling $x_0^2 + x_3^2$ the general circle constraint equation

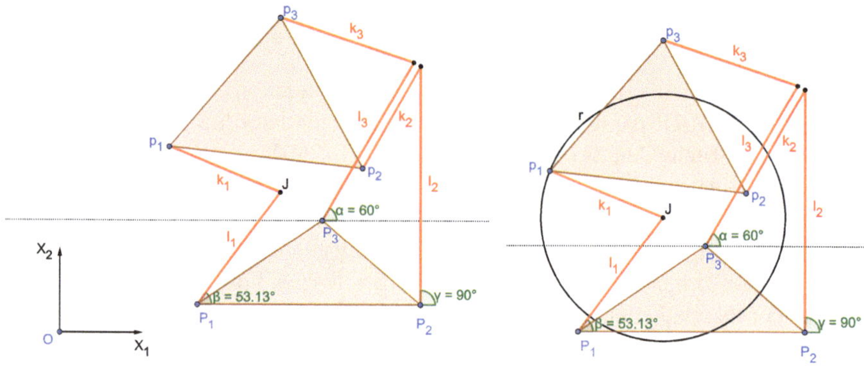

Fig. 1 3-RRR parallel manipulator

Fig. 2 Constraint hyperboloids in kinematic image space

$$(x^2 + y^2 + m^2 - 2mx + n^2 - 2ny - r^2)x_0^2 + 4(my - nx)x_0x_3 + 4(m - x)x_0y_1 + 4(n - y)x_0y_2 + (x^2 + y^2 + m^2 + 2mx + n^2 + 2ny - r^2)x_3^2 + 4(y + n)x_3y_1 - 4(x + m)x_3y_2 + 4y_1^2 + 4y_2^2 = 0. \quad (19)$$

Note that in Eq. (19), all expressions in round brackets are design constants. Therefore, the circle constraint equation allows a simple geometric interpretation in the three-dimensional kinematic image space of planar kinematics (Fig. 2): It represents a hyperboloid which passes, independently of the choice of the design constants, through the points $J_1(0, 1, I)^T$ and $J_2(0, 1, -I)^T$ (see Bottema and Roth 1979, pp. 409). A further observation is that the expression $4y_1^2 + 4y_2^2$ in Eq. (19) does not contain any design variable.

An important next step is to use the best-adapted coordinate system. Without loss of generality, P_1, p_1 can be the origins of fixed and moving coordinate systems and P_2, p_2 can be on the X_1 resp. x-axis of both systems. Therefore, we have the following assignments

$$P_1 = [1, 0, 0]^T, \quad P_2 = [1, A_2, 0]^T, \quad P_3 = [1, A_3, B_3]^T,$$
$$p_1 = [1, 0, 0]^T, \quad p_2 = [1, a_2, 0]^T, \quad p_3 = [1, a_3, b_3]^T.$$

In the next step, one can observe that the knee joints also move on circles. These circles can be parameterized as follows:

$$\begin{array}{lll} m_1 = l_1 \frac{1-u^2}{1+u^2}, & m_2 = l_2 \frac{1-v^2}{1+v^2} + A_2, & m_3 = l_3 \frac{1-w^2}{1+w^2} + A_3, \\ n_1 = l_1 \frac{2u}{1+u^2}, & n_2 = l_2 \frac{2v}{1+v^2}, & n_3 = l_3 \frac{2w}{1+w^2} + B_3. \end{array}$$

Substituting all assignments into the general circle constraint equations using $r = k_i, i = 1 \ldots 3$ for the respective legs, we obtain three constraint equations which completely describe the 3-RRR parallel mechanism:

$$\begin{aligned}
h_1 :& (l_1^2 - k_1^2)(x_0^2 + x_3^2) + 4l_1 \left(\frac{1-u^2}{1+u^2}(x_0 y_1 - x_3 y_2) + \frac{2u}{1+u^2}(x_0 y_2 + x_3 y_1) \right) + \\
& 4(y_1^2 + y_2^2) = 0, \\
h_2 :& \left(\frac{r_1 r_2 v^2 + r_3 r_4)}{v^2+1} \right) x_0^2 + \left(\frac{r_5 r_6 v^2 + r_7 r_8)}{v^2+1} \right) x_3^2 - 4a_2(x_0 y_1 + x_3 y_2) + \\
& 4(l_2 \frac{1-v^2}{1+v^2} + A_2)(x_0 y_1 - x_3 y_2) + 4l_2 \frac{2v}{1+v^2}(a_2 x_0 x_3 + x_0 y_2 + x_3 y_2) + 4(y_1^2 + y_2^2) = 0, \\
& r_1 := A_2 - k_2 - a_2 - l_2,\ r_2 := A_2 + k_2 - a_2 - l_2,\ r_3 := A_2 - k_2 - a_2 + l_2, \\
& r_4 := A_2 + k_2 - a_2 + l_2,\ r_5 := A_2 + k_2 + a_2 - l_2,\ r_6 := A_2 - k_2 + a_2 - l_2, \\
& r_7 := A_2 + k_2 + a_2 + l_2,\ r_8 := A_2 - k_2 + a_2 + l_2, \qquad (20) \\
h_3 :& \frac{(q_1^2 + q_2)w^2 + 4l_3(B_3 - b_3)w + q_4^2 + q_2 q_3}{1+w^2} x_0^2 + \\
& \left(4 \left(\frac{l_3(1-w^2)}{w^2+1} + A_3 \right) b_3 - (4(\frac{2wl_3}{1+w^2} + B_3))a_3) \right) x_0 x_3 \\
& \left(-4a_3 + 4l_3 \frac{1-w^2}{w^2+1} + 4A_3 \right) x_0 y_1 + \left(-4b_3 + \frac{8wl_3}{(w^2+1} + 4B_3 \right) x_0 y_2 + \\
& \left(4b_3 + \frac{8wl_3}{w^2+1} + 4B_3 \right) x_3 y_1 + \left(-4a_3 - 4l_3 \frac{1-w^2}{w^2+1} - 4A_3 \right) x_3 y_2 \\
& \frac{(q_5^2 + q_6 q_7)w^2 + 4l_3(B_3 b_3)w + q_8^2 + q_6 q_7}{1+w^2} x_3^2 + 4(y_1^2 + y_2^2) = 0, \\
& q_1 := A_3 - a_3 - l_3,\ q_2 := -b_3 + B_3 - k_3,\ q_3 := -b_3 + B_3 + k_3,\ q_4 := A_3 - a_3 + b_3, \\
& q_5 := A_3 - l_3 + a_3,\ q_6 := b_3 + B_3 + k_3,\ q_7 := b_3 + B_3 - k_3,\ q_8 := A_3 + a_3 + l_3.
\end{aligned}$$

Using the three equations h_1, h_2, h_3 and a normalization condition, one can solve the direct kinematics (DK), the inverse kinematics (IK), and the forward and the inverse singularities completely. These are just the most important kinematic tasks which will be shown in the sequel, but the equations can be used also for other tasks.

Direct Kinematics In the direct kinematics, all design variables and the three input parameters are given and the pose of the moving platform has to be computed. Although the univariate polynomial of the direct kinematics of this manipulator can be computed without assigning values to the design and input variables, we will show the algorithm for the example that was already used to produce Fig. 1. Before values assigned to the design parameters, one should have a closer look at the constraint equations. It is important to know that several manipulations are allowed, such as addition, subtraction, multiplication with arbitrary polynomials and constants without changing the solution set of the constraint equations. A stringent mathematical formulation of this statement will be given in Sect. 4. In case of the quadratic constraint equations h_1, h_2, h_3, one can observe that a subtraction $h_{1j} := h_j - h_1, j = 2, 3$ creates two new equations that are linear in the variables y_1, y_2. It would be tempting to create one more difference equation, but it is easy to show that a third difference equation would be redundant and would not contain any new information. At least one of the three original equations has to be kept. The detailed solution process will be shown in the following example.

Example 3 The following design variables are assigned to a 3-RRR planar parallel manipulator:

$$A_2 = 16, A_3 = 9, B_3 = 6, a_2 = 14, a_3 = 7, b_3 = 10, l_1 = 10, l_2 = 17, l_3 = 13,$$
$$k_1 = \sqrt{75}, k_2 = \sqrt{70}, k_3 = 10.$$

Furthermore, the three input variables are given by

$$u = \frac{1}{2}, v = 1, w = \frac{\sqrt{3}}{3}.$$

The three constraint equations simplify considerably

$$\begin{aligned} h_1 &: 25x_3^2 + 32x_3y_1 - 24x_3y_2 + 4y_1^2 + 4y_2^2 + 24y_1 + 32y_2 + 25 = 0, \\ h_2 &: 1119x_3^2 + 68x_3y_1 - 120x_3y_2 + 4y_1^2 + 4y_2^2 - 952x_3 + 8y_1 + 68y_2 + 223 = 0, \\ h_3 &: 620\,x_3 + \frac{2025\,x_3{}^2}{4} - 130\,\sqrt{3} - \frac{191}{4} + 40\,y_1\,x_3 + 34\,y_1 - 90\,x_3\,y_2 - 40\,y_2 + 4\,y_1{}^2 + \\ &4\,y_2{}^2 + \left(20x_3{}^2 + 4\,y_1x_3 - 28x_3 + 4y_2\right)\left(\frac{13\sqrt{3}}{2} + 6\right) + \left(x_3{}^2 + 1\right)\left(\frac{13\sqrt{3}}{2} + 6\right)^2 = 0. \end{aligned} \tag{21}$$

In this set of equation, $x_0 = 1$ has been used as normalization. This normalization excludes all 180° turns. For a complete analysis, one has to check if $x_0 = 0$ is a solution. In this example, it is easy to see that this is not the case. First the equations h_{12} and h_{13} are solved for y_1 and y_2, and the result is substituted into h_1. This yields the univariate polynomial:

$$1012018158645001\,x_3^6 + 373126531431576\,\sqrt{3}x_3^5 + 828170897821956\,\sqrt{3}x_3^4$$
$$- 1870238901095276\,x_3^5 - 3830372502668712\,\sqrt{3}x_3^3 - 309592552617273\,x_3^4-$$
$$1367698801300104\,\sqrt{3}x_3^2 + 5703740216839288\,x_3^3 + 2552443644341760\,\sqrt{3}x_3+$$
$$2666944473586507\,x_3^2 - 584052482710476\,\sqrt{3} - 4438269370622172\,x_3+$$
$$1009620776386125 = 0. \tag{22}$$

This polynomial has four real and two complex roots for x_3, which is the rotation angle of the moving system:

$$- 0.01155649481 - .8571792684\,\mathrm{I}, \quad -0.01155649481 + .85717926841\,I,$$
$$- 0.05446878513, \quad .17472650281, \quad .3874512485, \quad .7248336963. \tag{23}$$

These six values have to be substituted into the solution for y_1 and y_2, which then yields the pose variables for the six solutions of the DK of this manipulator. This procedure is shown for the first real solution. Substituting this solution into the solutions for y_1 and y_2 yields

$$y_1 = 1.3224316875, \qquad y_2 = -5.5627826306. \tag{24}$$

Now all pose parameters are determined and can be substituted into the transformation matrix Eq. (18):

$$\mathbf{A} = \begin{bmatrix} 1.002966848554 & 0 & 0 \\ -2.038867351357 & 0.997033151446 & 0.1089375702689 \\ 11.26962775615 & -0.10893757026889 & 0.99703315144598 \end{bmatrix}.$$

Note that this matrix is not normalized (has no 1 at the leftmost entry of the first row). This is due to the fact that in the image space, the normalization $x_0 = 1$ was used. By dividing the matrix with this entry, one obtains the normalized matrix

$$\mathbf{A} = \begin{bmatrix} 1 & 0.0 & 0.0 \\ -2.032836234114 & 0.9940838551976 & 0.1086153250488 \\ 11.23629138131 & -0.1086153250488 & 0.9940838551976 \end{bmatrix}.$$

Multiplying $\mathbf{A}$ with the coordinates of the points p_1, p_2, p_3, one obtains their coordinates in the base system and the pose of the end effector is given as solution of the DK problem.

$$p_1^b = [1, -2.032836234114, 11.2362913813]^T,$$
$$p_2^b = [1, 11.8843377387, 9.715676830627]^T,$$
$$p_3^b = [1, 6.01190400276, 20.41682265796]^T.$$

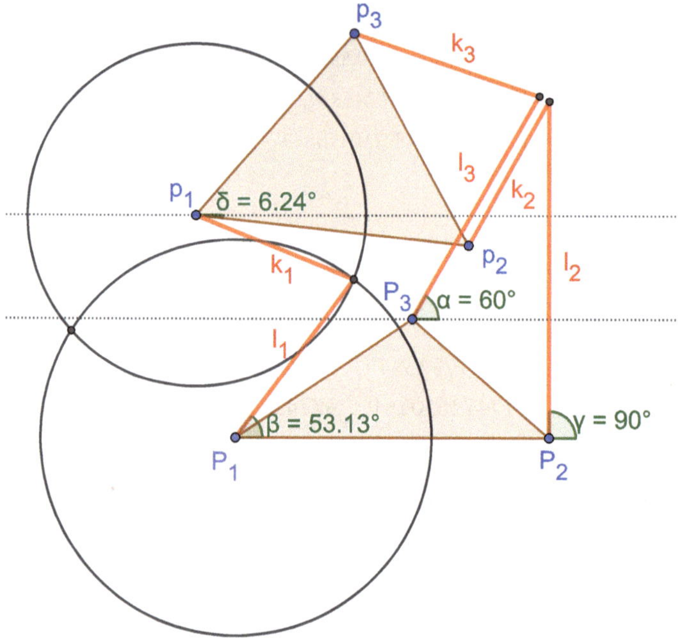

Fig. 3 DK solution 3-RPR

Figure 3 shows this solution.

The rotation angle of the platform system can be calculated via the inverse of the half-tangent substitution

$$\cos\varphi = \frac{1-x_3^2}{1+x_3^2} \wedge \sin\varphi = \frac{2x_3}{1+x_3^2} \wedge x_3 = -0.05446878513$$
$$\Rightarrow \varphi = -6.23550128120984°.$$

Note that the input angles were

$$u = \frac{1}{2} \equiv 53.13010235°,\ v = 1 \equiv 90°,\ w = \frac{\sqrt{3}}{3} \equiv 60°.$$

Inverse Kinematics In the inverse kinematics (IK), the design variables and the pose parameters are given. The same constraint equations can be used. We continue with Example 3.

Example 4 (Continuation of Example 3) Substituting the design variables into Eq. (20) yields

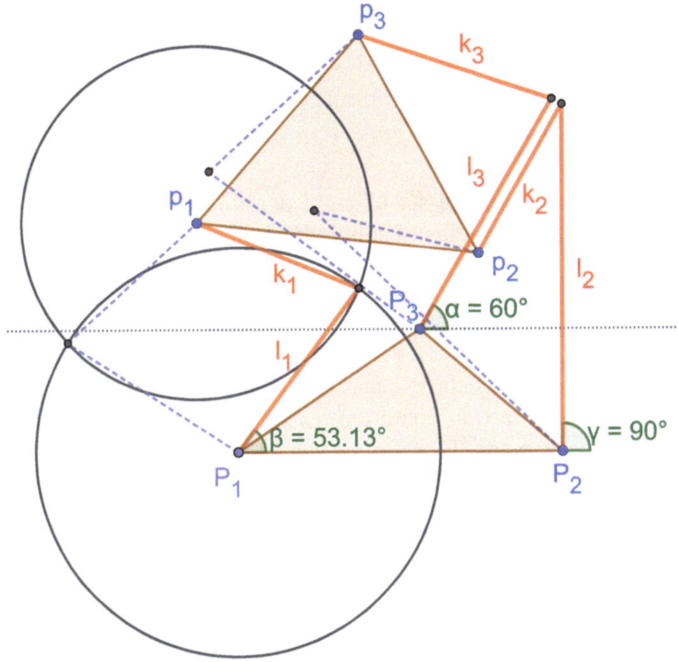

Fig. 4 IK solution of 3-RRR manipulator

$$\begin{aligned}
h_1 :&25u^2x_3^2 + 40u^2x_3y_2 + 4u^2y_1^2 + 4u^2y_2^2 - 40u^2y_1 + 80ux_3y_1 + 25u^2 + 80uy_2 + 25x_3^2 - \\
&40x_3y_2 + 4y_1^2 + 4y_2^2 + 40y_1 + 25 = 0 \\
h_2 :&99v^2x_3^2 - 52v^2x_3y_2 + 4v^2y_1^2 + 4v^2y_2^2 - 60v^2y_1 + 136vx_3y_1 + 155v^2 - 1904vx_3 + \\
&136vy_2 + 2139x_3^2 - 188x_3y_2 + 4y_1^2 + 4y_2^2 + 76y_1 + 291 = 0 \qquad (25)\\
h_3 :&165w^2x_3^2 + 64w^2x_3y_1 - 12w^2x_3y_2 + 4w^2y_1^2 + 4w^2y_2^2 - 328w^2x_3 - 44w^2y_1 - 16w^2y_2 + \\
&832wx_3^2 + 104wx_3y_1 + 37w^2 - 728wx_3 + 104wy_2 + 997x_3^2 + 64x_3y_1 - 116x_3y_2 + 4y_1^2 + \\
&4y_2^2 - 208w + 712x_3 + 60y_1 - 16y_2 + 14 = 0.
\end{aligned}$$

This system of equations in the unknowns u, v, w allows eight different solutions when a set of pose parameters are given. We use the results of Eqs. (22) and (23) to compute the inverse kinematics of the pose with pose parameters

$$x_0 = 1, \quad x_3 = -0.05446878513, \quad y_1 = 1.3224316875, \quad y_2 = -5.5627826306.$$

The solutions for the resulting set of equations are

$u = .500000, v = 1.00000, w = .57735,\ u = .500000, v = 1.00000, w = 3.0254,$
$u = 3.41747, v = 1.00000, w = .57735,\ u = 3.41747, v = 1.00000, w = 3.0254,$
$u = .500000, v = 2.46987, w = .57735,\ u = .500000, v = 2.46987, w = 3.0254,$
$u = 3.41747, v = 2.46987, w = .57735,\ u = 3.41747, v = 2.46987, w = 3.0254.$

Figure 4 shows the complete set of eight possible solutions, consisting of all possible combinations of the pairs of solutions of each equation $h_i, i = 1, 2, 3$. The second solutions for each dyad are drawn in dashed lines. Each solution yields a working mode of the manipulator. The first set of solutions (the first working mode) is the one we have started with in the DK.

Singularities The system of three constraint equations $\mathcal{S} = \{h_1, h_2, h_3, n\}$ augmented with the normalization condition $n : x_0^2 + x_3^2 = 1$ (Eq. 20) comprises the forward and the inverse maps. Therefore, it can be used to compute the singularities of both maps. Referring to (Gosselin and Angeles 1990), the singularities of both maps are given by

$$\mathbf{J}_o \dot{\mathbf{y}} + \mathbf{J}_i \dot{\mathbf{t}} = 0, \tag{26}$$

where

$$\mathbf{J}_o = \begin{bmatrix} \frac{\partial n}{\partial x_0} & \frac{\partial n}{\partial x_3} & 0 & 0 \\ \frac{\partial h_1}{\partial x_0} & \frac{\partial h_1}{\partial x_3} & \frac{\partial h_1}{\partial y_1} & \frac{\partial h_1}{\partial y_2} \\ \frac{\partial h_2}{\partial x_0} & \frac{\partial h_2}{\partial x_3} & \frac{\partial h_2}{\partial y_1} & \frac{\partial h_2}{\partial y_2} \\ \frac{\partial h_3}{\partial x_0} & \frac{\partial h_3}{\partial x_3} & \frac{\partial h_3}{\partial y_1} & \frac{\partial h_3}{\partial y_2} \end{bmatrix}, \quad \mathbf{J}_i = \begin{bmatrix} 0 & 0 & 0 & 0 \\ 0 & \frac{\partial h_1}{\partial u} & 0 & 0 \\ 0 & 0 & \frac{\partial h_2}{\partial v} & 0 \\ 0 & 0 & 0 & \frac{\partial h_3}{\partial w} \end{bmatrix},$$

$$\dot{\mathbf{y}} = [\dot{x}_0, \dot{x}_3, \dot{y}_1, \dot{y}_2]^T, \quad \dot{\mathbf{t}} = [0, \dot{u}, \dot{v}, \dot{w}]^T.$$

$\mathbf{J}_o$ is the Jacobian matrix of the forward map and $\mathbf{J}_i$ is the Jacobian matrix of the inverse map of the manipulator. In the literature, the singularities of the inverse map are sometimes called type 1 singularities and those of the forward map are called type 2 singularities (Bonev 2002).[2] The redundant motion which exists either on the platform or in the joints can be infinitesimal or finite. The infinitesimal case is treated in the following sections. The finite case, which occurs only at the presence of specific design or pose parameters, can be detected in the solution of the direct or inverse kinematics described in the previous sections. In the solution process, it happens that the elimination stops before a univariate polynomial can be found. Then the manipulator allows a self-motion, i.e., the platform or an input chain will move without active input. An example of this phenomenon will be shown later in Sect. 4 on page 153.

[2] A refined classification, taking into account special situations which can occur for special design parameters of the manipulator, was given in Zlatanov et al. (1994).

Forward singularities From kinematic point of view, a forward singularity occurs when a parallel manipulator allows an infinitesimal (local) mobility when all the inputs are locked. Mathematically, this corresponds to the condition $\dot{\mathbf{t}} = [0, 0, 0, 0]^T$ and substitution of this condition into Eq. (26) yields

$$\mathbf{J}_o \dot{\mathbf{y}} = 0. \tag{27}$$

Equation (27) represents a homogeneous system of linear equations. From linear algebra, it is well known that this system allows a nontrivial solution if and only if $\det \mathbf{J}_o = 0$ holds. Computing this determinant without assigning design variables is feasible but without practical use. One has to realize that this determinant still contains all pose parameters x_0, x_3, y_1, y_2 and all input parameters u, v, w. Adding $\det \mathbf{J}_o = 0$ to the system of constraint equations $\mathscr{S}$ yields a system of five equations. In this system, one can, for example, eliminate the input parameters u, v, w to obtain an equation in the pose parameters x_0, x_3, y_1, y_2 and a normalization condition that represent in the kinematic image space all poses which are forward singular. The disadvantage of this equation is that the working modes cannot be distinguished. Elimination of the pose parameters yields the singular set in the joint space. The practicality of the latter is although questionable, because once this equation is obtained one has to perform the direct kinematics again and only one of the obtained six solutions yields a singular pose!

Example 5 (Continuation of Example 3) Taking the system of constraint equations (25) and computing the determinant of $\mathbf{J}_o$ yield a polynomial h_4 of degree 10 in the unknowns $x_0, x_3, y_1, y_2, u, v, w$. Together with the original constraint equations h_1, h_2, h_3, one obtains a system of four algebraic equations that allow to derive a single polynomial in the Study parameters or in the joint parameters which will comprise all forward singular poses of the manipulator. A suitable normalization is still possible. The zero set of this polynomial describes a surface in the kinematic image space or the joint space depending on which of the variables have been eliminated. One can eliminate either the input or the output variables. Due to the above-mentioned fact that computing the singularity surface in the joint space needs an additional computation of the forward kinematics, only the elimination of the input parameters will be shown. To obtain a single polynomial, we compute the following sequence of resultants of equations: $h_{14} = res(h_1, h_4, u)$ is the resultant of h_1 and h_4 that eliminates u. Then we take $h_{142} = res(h_{14}, h_2, v)$, which eliminates v and finally we take $h_{1423} = res(h_{142}, h_3, w)$ which eliminates w to arrive at a polynomial of degree 44 in the remaining pose parameters x_0, x_3, y_1, y_2. As spurious solutions might occur, it is recommended to use a different sequence of elimination and compare the two final polynomials. In case they are different, one has to look for their greatest common divisor. In case of the example, the second sequence $h_{34} = res(h_3, h_4, w) \rightarrow h_{342} = res(h_{34}, h_2, v) \rightarrow h_{3421} = res(h_{342}, h_1, u)$ was used and this elimination yielded the same polynomial of degree 44.[3] The degree of the polynomial is so high because it describes all forward singularities of all working modes.

---

[3]The degree of this polynomial disagrees with the results found in the thesis (Bonev 2002).

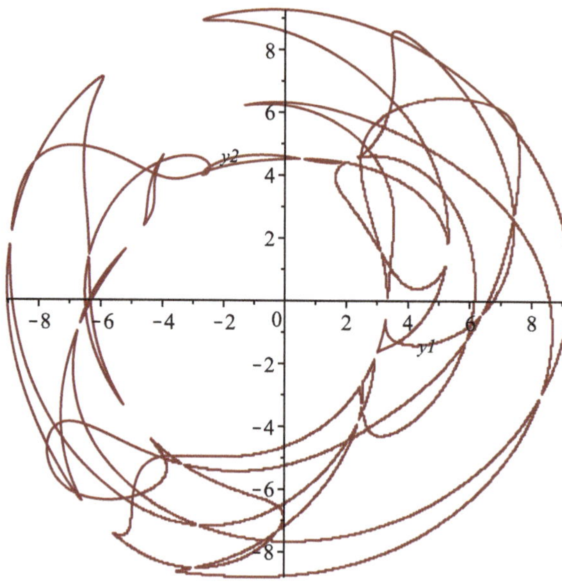

Fig. 5 Section of forward singularity surface at $x_0 = 1, x_3 = 0$ in image space

In the kinematic image space, the polynomial describes a surface of order 44. Figure 5 shows a section of this surface at $x_0 = 1$, $x_3 = 0$. To show a single singular pose of the manipulator, we take the uppermost point of this curve on the y_2 axis of Fig. 5. Substituting $x_0 = 1$, $x_3 = 0$, $y_1 = 0$ into the polynomial of degree 44 yields a univariate polynomial which is to big to be displayed completely

$$\begin{aligned}&72113027247051332648 96 y_2^{44} - 8474147875931485844275 2 y_2^{43} - \dots\\&- 12050792400144740726958497665280369335937500000000 0 y_2 +\\&21775525730952737167082108233173573318481445312500 = 0.\end{aligned}$$

This polynomial has twelve real roots as can be seen also in Fig. 5:

$$\begin{aligned}&-8.773883747, -7.680052573, -7.189867449, -6.457249958,\\&-5.253827948, -4.649535025, 4.507174754, 4.592908938,\\&6.218593708, 6.317560304, 8.566805064, 9.256203129.\end{aligned}$$

Figure 6 shows the singular pose corresponding to the root $y_2 = 9.256203129$. The three lines f_1, g_1, h_1 covering three end links of the dyads intersect in one point E, which is the instantaneous center of the redundant infinitesimal rotation in this singular pose. The pose belongs to a working mode consisting of one red and two blue dashed input dyads. Note that the moving platform is below the fixed platform in the displayed pose!

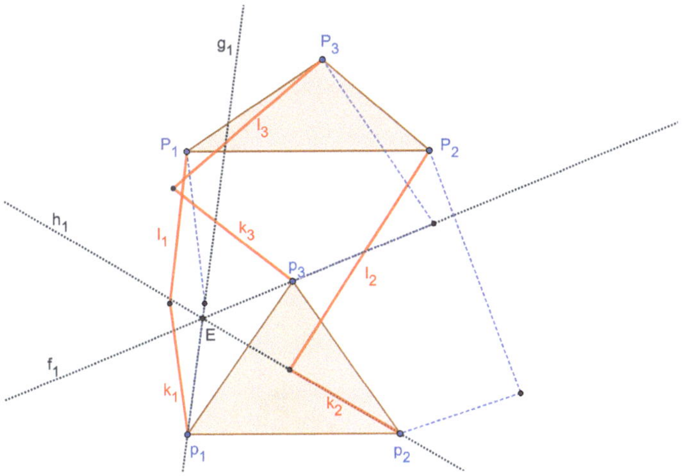

Fig. 6 Forward singularity 3-RRR

Inverse singularities From kinematic point of view, an inverse singularity occurs when the manipulator allows an infinitesimal (local) mobility of an input chain when the platform is not moving. Mathematically, this corresponds to the condition $\dot{\mathbf{y}} = [0, 0, 0, 0]^T$ and substitution of this condition into Eq. (26) yields

$$\mathbf{J}_i \dot{\mathbf{t}} = 0. \tag{28}$$

It is quite obvious from Eq. (26) that this determinant factors into three parts:

$$\begin{aligned} h_5 : &\big[(B_3x_0^2 + B_3x_3^2 - 2a_3x_0x_3 - b_3x_0^2 + b_3x_3^2 + 2x_0y_2 + 2x_3y_1)w^2 + \\ &(2A_3x_0^2 + 2A_3x_3^2 - 2a_3x_0^2 + 2a_3x_3^2 + 4b_3x_0x_3 + 4x_0y_1 - 4x_3y_2)w \\ &-2x_0y_2 - 2x_3y_1 - B_3x_0^2 - B_3x_3^2 + 2a_3x_0x_3 + b_3x_0^2 - b_3x_3^2\big]\, l_3 \cdot \\ &\big[(-a_2x_0x_3 + x_0y_2 + x_3y_1)v^2 + (A_2x_0^2 + A_2x_3^2 - a_2x_0^2 + a_2x_3^2)v + \\ &a_2x_0x_3 + 2vx_0y_1 - 2vx_3y_2 - x_0y_2 - x_3y_1)l_2\big] \cdot \\ &\big[(u^2x_0y_2 + u^2x_3y_1 + 2ux_0y_1 - 2ux_3y_2 - x_0y_2 - x_3y_1)l_1\big] = 0. \end{aligned} \tag{29}$$

Each factor can be treated separately to determine the input singularities for the corresponding leg. For example, when the third factor is used together with equation h_1, the input parameter u can be eliminated to derive all poses of the moving platform in which the first leg is singular. After elimination, the following equation is obtained:

$$\begin{aligned} &(y_1^2 + y_2^2)(x_0^2 + x_3^2)(k_1^2x_0^2 + k_1^2x_3^2 - 2k_1l_1x_0^2 - 2k_1l_1x_3^2 + l_1^2x_0^2 + l_1^2x_3^2 - 4y_1^2 - 4y_2^2)\cdot \\ &(k_1^2x_0^2 + k_1^2x_3^2 + 2k_1l_1x_0^2 + 2k_1l_1x_3^2 + l_1^2x_0^2 + l_1^2x_3^2 - 4y_1^2 - 4y_2^2) = 0. \end{aligned} \tag{30}$$

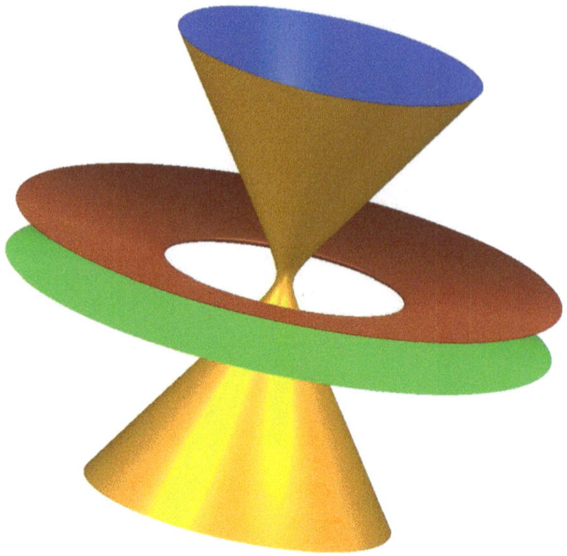

Fig. 7 Inverse singularity surfaces in kinematic image space

The first two factors in this equation have only complex solutions and can be neglected, and the third and the fourth factors yield two surfaces in the kinematic image space which turns out to be two hyperboloids (Fig. 7). One of the hyperboloids represents all poses where the first leg is totally stretched and the other represents the poses where the first leg is folded. Additionally, these hyperboloids represent poses where the manipulator is on the boundary of its workspace.

Example 6 (Continuation of Example 3) Substituting the design parameters of Example 3 into Eq. (30) yields the equations

$$\begin{aligned}&(16y_1^2\sqrt{3}+16y_2^2\sqrt{3}+25x_0^2+25x_3^2-28y_1^2-28y_2^2)=0\\&(16y_1^2\sqrt{3}+16y_2^2\sqrt{3}-25x_0^2-25x_3^2+28y_1^2+28y_2^2)=0.\end{aligned} \tag{31}$$

To derive a single inverse singular pose, a point on one of the hyperboloids is chosen by setting $x_0 = 1$, $y_1 = -1$, $x_3 = -\frac{1}{4}$. This yields the equation

$$16\sqrt{3}+16y_2^2\sqrt{3}-\frac{23}{16}-28y_2^2=0.$$

One of the solutions of this equations is

$$y_2=\frac{1}{8}\,\frac{\sqrt{\left(7-4\sqrt{3}\right)\left(256\sqrt{3}-23\right)}}{4\sqrt{3}-7}.$$

Substituting this value and the chosen parameters $x_0 = 1$, $y_1 = -1$, $x_3 = -\frac{1}{4}$ into the matrix Eq. (16), one obtains the transformation matrix

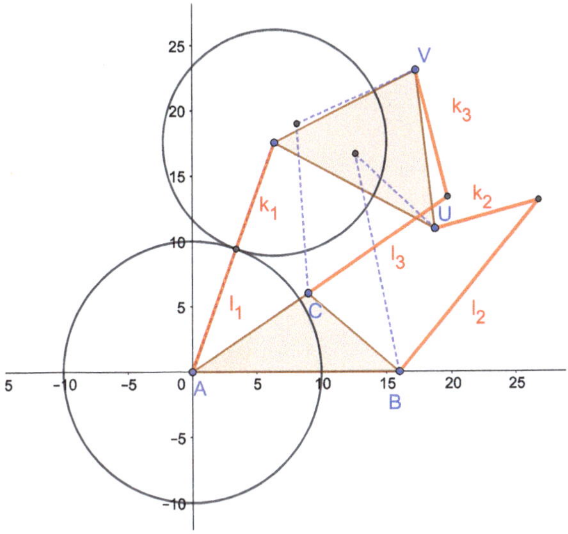

Fig. 8 Inverse singularity of a 3-RRR parallel manipulator

$$\mathbf{A} = \begin{bmatrix} \frac{17}{16} & 0 & 0 \\ 2 - \frac{1}{16}\frac{\sqrt{(7-4\sqrt{3})(256\sqrt{3}-23)}}{4\sqrt{3}-7} & \frac{15}{16} & \frac{1}{2} \\ -\frac{1}{4}\frac{\sqrt{(7-4\sqrt{3})(256\sqrt{3}-23)}}{4\sqrt{3}-7} - \frac{1}{2} & -\frac{1}{2} & \frac{15}{16} \end{bmatrix}.$$

This transformation yields the singular pose of Fig. 8.

In case of the inverse singularities, we will also compute the singularity surface in the joint space, to exemplify the statement made above about this surface. We have to start with the system of equations $\mathscr{S} = \{h_1, h_2, h_3, h_5\}$ consisting of the three constraint equations and the equation of the determinant of the Jacobian $\mathbf{J}_i$. Now the pose parameters x_0, x_3, y_1, y_2 have to be eliminated as to obtain a single equation in the joint parameters u, v, w. To simplify the computation, we use the normalization condition $x_0 = 1$ and compute y_1 and y_2 from the difference equations $h_{12} = h_1 - h_2$ and $h_{13} = h_1 - h_3$ and substitute the solutions for y_1 and y_2 into h_1 and the third factor of h_5. This yields two equations in u, v, w and x_3 from which x_3 can be eliminated using resultant methods. The result is an equation of degree 28 in the input parameters u, v, w representing the singularity surface $\mathscr{F}$ in joint space, which is much to big to be displayed.

Example 7 (Continuation of Example 3) By substituting the design parameters of Example 3 into the equation of $\mathscr{F}$ a picture of the singularity surface can be shown (Fig. 9).

Now a single inverse singular pose is specified by setting $v = 1, w = 1$. Substituting these values into the equation of $\mathscr{F}$ yields a polynomial of degree 12:

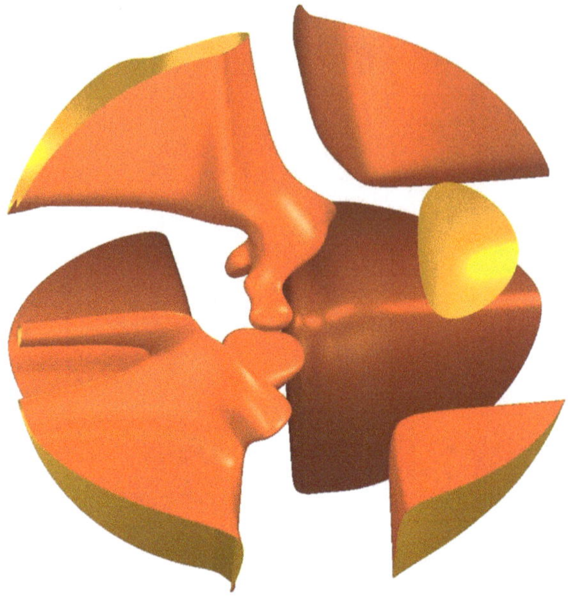

Fig. 9 Inverse singularity surface in joint space

$$
\begin{aligned}
&-818544231103537077\,u^{12} + 3944558869211116800\,u^{11} - 9479437871294979126\,u^{10} + \\
&15757734038472565344\,u^{9} - 19686222594094813627\,u^{8} + 18937447175451975936\,u^{7} - \\
&14224778429264845268\,u^{6} + 8254317529677310144\,u^{5} - 3535122247251754347\,u^{4} + \\
&1029296909802009856\,u^{3} - 175954826762074230\,u^{2} + 12130149459227104\,u + \\
&220928936206427 = 0,
\end{aligned}
$$

which has six real roots for u

$$
\begin{aligned}
&-.01476395296624513,\ .2723109282032673,\ .5270689907577652,\\
&.6265818046074013,\ .6645283032344548,\ 1.402004907351409.
\end{aligned}
$$

We take the root $u = .5270689907577652$ which determines together with $v = 1, w = 1$ a point on $\mathscr{F}$. Now the input for the direct kinematics of this manipulator is determined, and we can apply the procedure developed above to compute the direct kinematics. The computation yields six real poses, one of them is the singular one! The mapping from joint space to the Cartesian space is 1 to 6! Figs. 10, 11 and 12 show all six solutions. In solution Nr. 2, the first leg is singular. It must be noted that the whole computation was done only for the first leg, because the solutions for y_1 and y_2 were substituted into h_1. One also could have used h_2 or h_3 at this step. A complete inverse singularity analysis in joint space yields three surfaces, each of them corresponds to the singularities of one leg.

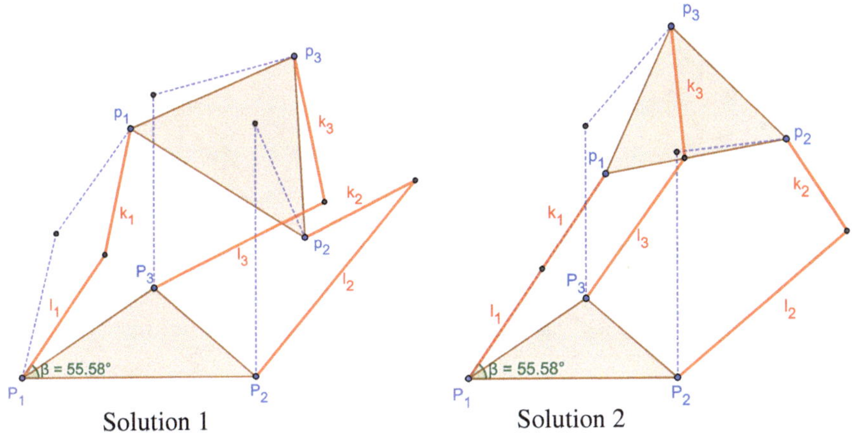

Fig. 10 Solution 1 Solution 2

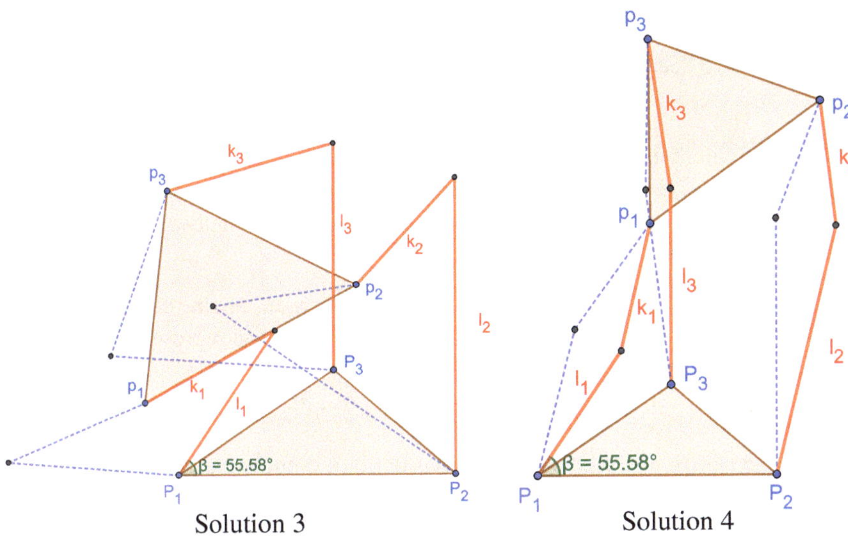

Fig. 11 Solution 3 Solution 4

3.2 Elimination Method

This method follows a simple recipe: Write the forward kinematics of the kinematic chain and then eliminate the motion parameters. When n is the degree of freedom of the kinematic chain, then the number m of constraint equations (in general) to be expected is $m = 6 - n$. Because of projection roots, that are introduced in every single elimination step, the degree of the resulting constraint equations is generally very high. Because of this fact, this method is recommended only for simple chains.

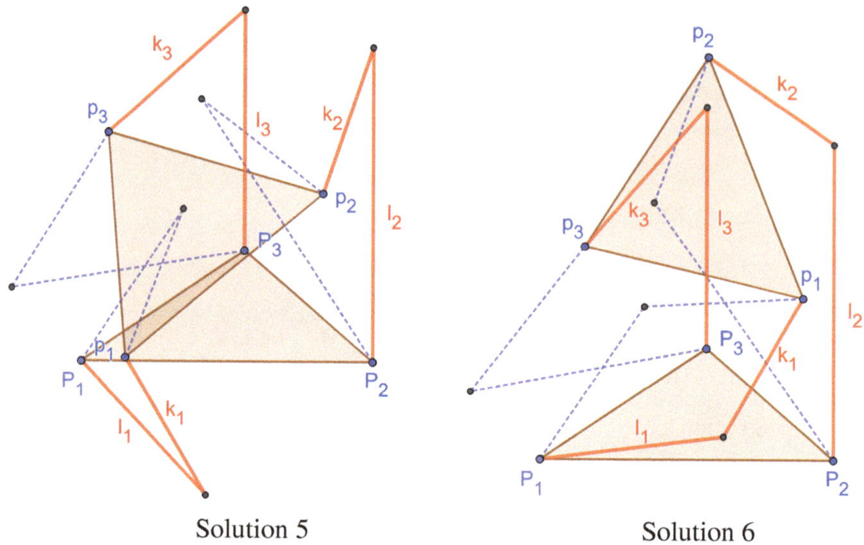

Fig. 12 Solution 5 Solution 6

An example where this method immediately yields a result are the constraint equations of a spatial 2-R chain. It was already shown in Selig (2005) that a 2-R chain can be represented by four linear equations in Study coordinates. We demonstrate this statement using Example 2.

Example 8 In Example 2, it was shown that a 2-R chain, having the x-axis and a parallel line to the y-axis as rotation axes, has the parametric representation

$$\mathbf{l} = [1 : u : v : uv : -uav : av : ua : -a].$$

This homogeneous vector equation consists of eight component equations:

$$\rho x_0 = 1,\ \rho x_1 = u,\ \rho x_2 = v,\ \rho x_3 = uv,\ \rho y_0 = -auv,\ \rho y_1 = av,\ \rho y_2 = au,\ \rho y_3 = -a.$$

The first equation immediately yields $\rho = 1$ and the homogenizing equation $x_0 = 1$. The second and the third equations can be used to eliminate the motion parameters u and v. This yields five equations which seem to be too much because we have to expect four equations:

$$x_3 - x_1x_2 = 0,\quad y_0 + ax_1x_2 = 0,\quad y_1 - ax_2 = 0,\quad y_2 - ax_1 = 0,\quad y_3 + a = 0.$$

But it is allowed to manipulate the equations, and one has to keep in mind that the Study quadric equation has to be fulfilled! Substituting the first equation into the second simplifies the second equation to $y_0 - ax_3 = 0$. Therefore, we have four linear equations and one quadratic equation. But the quadratic equation is essentially the

Study quadric, because substituting the four linear equations into the Study quadric equation yields exactly this first equation. The 2-R chain is therefore represented by the four linear equations

$$y_0 - ax_3 = 0, \quad y_1 - ax_2 = 0, \quad y_2 - ax_1 = 0, \quad y_3 + a = 0.$$

In the kinematic image space, these four equations determine a linear 3-space and all poses the 2-R chain can attain are the points of the intersection of this 3-space with the Study quadric (see also Selig 2005 and Pfurner 2006 where this method was used for deriving a fast algorithm for the inverse kinematics of general 6-R chains).

3.3 Linear Implicitization

To derive the constraint equations of a kinematic chain using geometric properties or elimination has several disadvantages. The first method needs geometric insight into the motion of the chain, and the second method introduces "spurious" solutions that have to be identified and removed from the equations. The question arises if there exists an algorithm that derives "automatically" from a parametric representation of the (allowed) kinematic chain a minimal set of implicit equations that completely describes this kinematic chain. Such an algorithm was developed in Walter and Husty (2010) and called "linear implicitization algorithm" (LIA). It was shown that the parametric representation of any kinematic chain can be transformed into a set of polynomials. The basic idea of this algorithm is that first of all a kinematic chain built from only revolute and prismatic joints can be represented by a set of polynomials. Secondly, one can observe that the parametric expressions have to fulfill the polynomial equations. But unfortunately the degree of the polynomials is unknown from beginning. The algorithm will be explained in detail but at first, a simple example is shown to demonstrate the basic idea.

Example 9 In this example, the implicit equation of a sphere is derived. The sphere is given by its parametric expressions

$$\begin{aligned} X &= \cos(\phi)\cos(\psi) - 3 \\ Y &= \cos(\phi)\sin(\psi) - 2 \\ Z &= \sin(\phi) - 1. \end{aligned} \tag{32}$$

Half-tangent substitution yields

$$\begin{aligned} X &= \frac{(1-u^2)(1-t^2)}{(u^2+1)(t^2+1)} - 3 \\ Y &= \frac{2(1-u^2)t}{(u^2+1)(t^2+1)} - 2 \\ Z &= \frac{2u}{(u^2+1)} - 1. \end{aligned} \tag{33}$$

Actually, one would have to check if a linear polynomial fulfills the parametric equation. We skip this step because we know that the implicit equation needs a quadratic polynomial. In the next step, a quadratic ansatz polynomial is defined

$$Ap: \; ax^2 + by^2 + cz^2 + dxy + exz + fyz + gx + hy + iz + j = 0. \tag{34}$$

The ten unknown coefficients of Ap have to be determined such that a substitution of the parametric expressions Eq. (33) fulfills the equation Eq. (34). Substitution of the parametric expressions Eq. (33) into Eq. (34) yields, after computing a common denominator and taking the numerator, a polynomial of degree 8 in u and t which has to be identical zero for every u and t. This can be only the case when the coefficients of every powerproduct in this equation vanishes. Although this polynomial is quite large, it is displayed

$$\begin{aligned}
&4d + 2f + c + 2e + 4a - 2h - 2g - i + 4b + j + \\
&(4a + 4b + c + 4d + 2e + 2f - 2g - 2h - i + j)t^4u^4 + \\
&(-4c - 4e - 4f + 2i)t^4u^3 + (16a + 8b + 6c + 12d + 6e + 4f - 6g - 4h - 2i + 2j)t^4u^2 + \\
&(-4c - 8e - 4f + 2i)t^4u + (8b + 4d + 2f - 2h)t^3u^4 + \\
&(16a + 12b + 2c + 12d + 6e + 4f - 6g - 4h - 2i + 2j)t^2u^4 + (-8c - 12e - 8f + 4i)t^2u^3 + \\
&(40a + 8b + 12c + 24d + 12e + 8f - 12g - 8h - 4i + 4j)t^2u^2 + (4i - 8c - 12e - 8f)t^2u + \\
&(8b + 8d + 2f - 2h)tu^4 + (-8b - 4d - 2f + 2h)t + \\
&(16a + 4b + c + 8d + 4e + 2f - 4g - 2h - i + j)u^4 + (-4c - 8e - 4f + 2i)u^3 + \\
&(16a + 8b + 6c + 12d + 6e + 4f - 6g - 4h - 2i + 2j)u^2 + (-4c - 4e - 4f + 2i)u + \\
&(16a + 4b + c + 8d + 4e + 2f - 4g - 2h - i + j)t^4 + (-8b - 8d - 2f + 2h)t^3 + \\
&(16a + 12b + 2c + 12d + 6e + 4f - 6g - 4h - 2i + 2j)t^2 - 4ft^3u^3 + 4dt^3u^2 + \\
&4ft^3u - 4ftu^3 - 4dtu^2 + 4ftu = 0.
\end{aligned} \tag{35}$$

One can observe that each coefficient of the powerproducts yields linear equations in the ten unknowns $a, b, c, d, e, f, g, h, i, j$. As there are 25 equations, we have a highly overconstrained system of linear equations. Nevertheless, this system has to have a unique solution which reads

$$a = 1, b = 1, c = 1, d = 0, e = 0, f = 0, g = 6, h = 4, i = 2, j = 13, \tag{36}$$

and from this solution, we obtain the implicit equation of the sphere

$$x^2 + y^2 + z^2 + 6x + 4y + 2z + 13 = 0. \tag{37}$$

An easy computation shows that the parametric expressions Eqs. (32) and (33) fulfill the implicit equation Eq. (37). This simple example already shows the disadvantage of this algorithm. The linear system will always be overconstrained and can become very large even for relatively simple chains.

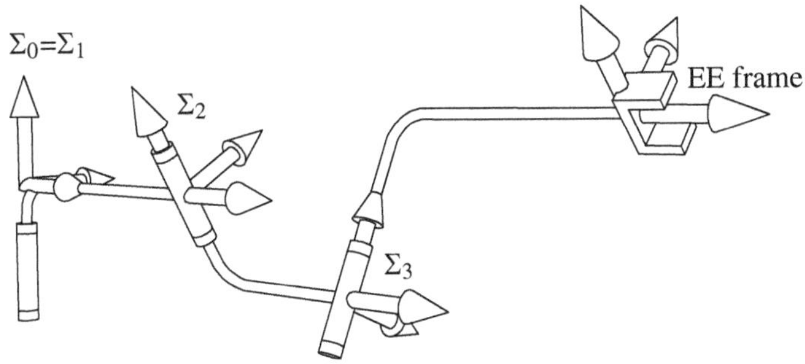

Fig. 13 Canonical 3R-chain

Remark 2 In case the linear system has no solution or only the trivial solution, one has to repeat the algorithm with an ansatz polynomial of higher degree.

We proceed now to explain the LIA for a general kinematic chain. In the first step of the algorithm, one has to compute the forward transformation of the kinematic chain to obtain a parametric expression.

If the relative position of two rotation axes is described by the usual Denavit–Hartenberg parameters (α_i, a_i, d_i), the coordinate transformation between the coordinate systems attached to the rotation axes is given by

$$\mathbf{G}_i = \begin{bmatrix} 1 & 0 & 0 & 0 \\ a_i & 1 & 0 & 0 \\ 0 & 0 & \cos(\alpha_i) & -\sin(\alpha_i) \\ d_i & 0 & \sin(\alpha_i) & \cos(\alpha_i) \end{bmatrix}. \tag{38}$$

Using this transformation, we assume the axes of an nR-chain to be in a canonical start position, where all the axes are parallel to a plane, the first rotation axis is the z-axis of the base coordinate system and the x-axis is the common normal of first and second revolute joints. A simple consideration shows that this is always possible and no restriction of generality (Pfurner 2006; Husty et al. 2007b). As shown in Fig. 13 the motion axes are always the z-axes of the coordinate systems, they are either rotations or translations and can be written

$$\mathbf{M}_i = \begin{bmatrix} 1 & 0 & 0 & 0 \\ 0 & \cos(u_i) & -\sin(u_i) & 0 \\ 0 & \sin(u_i) & \cos(u_i) & 0 \\ 0 & 0 & 0 & 1 \end{bmatrix}, \quad \mathbf{M}_i = \begin{bmatrix} 1 & 0 & 0 & 0 \\ 0 & 1 & 0 & 0 \\ 0 & 0 & 1 & 0 \\ t_i & 0 & 0 & 1 \end{bmatrix}. \tag{39}$$

Following this sequence of transformations, the end effector pose will have the following description:

$$\mathbf{D} = \mathbf{M}_1 \cdot \mathbf{G}_1 \cdot \mathbf{M}_2 \cdot \mathbf{G}_2 \cdot \dots \cdot \mathbf{M}_n, \tag{40}$$

where n is the number of motion axes in the chain. From this parametric representation of the chain, the parametric representation in the kinematic image space has to be computed using Eqs. (7) and (8). Half-tangent substitution transforms the rotation angles u_i into algebraic parameters t_i, and one ends up with eight parametric equations of the form:

$$\begin{aligned} x_0 &= f_0(t_1, \dots t_n), \\ x_1 &= f_1(t_1, \dots t_n), \\ &\vdots \\ y_3 &= f_8(t_1, \dots t_n). \end{aligned} \tag{41}$$

These equations will be rational having a denominator of the form $(1 + t_1^2) \cdot \ldots \cdot (1 + t_n^2)$ which can be canceled because the Study parameters x_i, y_i are homogeneous. The same can be done with a possibly appearing common factor of all parametric expressions. After these allowed manipulations, the simplest possible parametric representation of the kinematic chain in the kinematic image space has been found.

It is well known that there exists a one-to-one correspondence from all spatial transformations to the Study quadric which lives in $\mathbb{P}^7$. Particularly this means that a tuple of Study parameters describing a transformation is a projective point and consequently always only unique up to scalar multiples. If we have a transformation parameterized by n parameters $t_1, \dots, t_n$, we obtain by kinematic mapping a set of corresponding points in $\mathbb{P}^7$ and we ask now for the smallest variety $\mathscr{V} \in \mathbb{P}^7$ (with respect to inclusion) which contains all these points. What do we know about this variety? What can be said definitely is that its describing set of polynomials consists of homogeneous polynomials and contains $x_0y_0 + x_1y_1 + x_2y_2 + x_3y_3$, i.e., the equation for the Study quadric S_6^2. In the following, it is shown how additional equations can be computed which are necessary to describe $\mathscr{V}$. It should be noted that the minimum number of polynomials necessary to describe $\mathscr{V}$ corresponds to the number of parameters, which in turn correspond to the degrees of freedom (DOF) of the kinematic chain. If the number of generic parameters is n, then $m = 6 - n$ polynomials are necessary to describe $\mathscr{V}$. This is of course a rough statement, because different numbers can appear when special situations (e.g., redundant DOFs, special design parameters) are in place.

Now we are searching for homogeneous polynomials which vanish on all points that can be obtained from the parameterization of the kinematic chain, i.e., polynomials in $x_0, x_1, x_2, x_3, y_0, y_1, y_2, y_3$ which vanish when the expressions of the parameterization are substituted. One possibility to find such polynomials is the following: A general ansatz of a homogeneous polynomial in $x_0, x_1, x_2, x_3, y_0, y_1, y_2, y_3$ with given degree n is made, and then the Study parameters of the parametric representation are substituted. The resulting expression f is treated as a polynomial $f(t_1, \dots t_n)$. Due to the fact that f has to vanish for all values of the t_i, it has to be the zero polynomial. As it already was demonstrated in Example 9, all coefficients of f have to vanish. This means that, after extraction of these coefficients, one obtains a system of linear equations where the unknowns are the $\binom{n+7}{n}$ coefficients from the general ansatz. This system can be solved (assuming that the design parameters a_i, d_i and α_i are generic), and the solution can be substituted into the ansatz. The result is an expression r describing all homogeneous polynomials of degree n which vanish on the points of $\mathscr{V}$. An important point is that if the solution of the linear system is positive dimensional, the corresponding parameters also appear in the final expression, i.e., the expression r itself is parameterized.

In the following, this important part of the algorithm is explained in more detail: The simplest homogeneous ansatz polynomial would be a linear one. Therefore at first, a general linear polynomial is generated:

$$ansatz: \quad C_1y_3 + C_2x_3 + C_3y_2 + C_4x_2 + C_5y_1 + C_6x_1 + C_7y_0 + C_8x_0, \qquad (42)$$

with unknown coefficients C_i. The question is now if there exist values for these unknowns such that Eq. (42) vanishes identically for points fulfilling the parametric representation of the given kinematic chain. To test this, the parametric expressions Eq. (41) are substituted into Eq. (42) which yields a polynomial in t_i with coefficients in C_i and the Denavit–Hartenberg parameters a_i, d_i and α_i. We collect with respect to the powerproducts of the t_i and extract their coefficients. This yields a set of linear equations in C_i, a_i, d_i and α_i. The number of equations depends on the particular design of the chain. In general, the system will consist of more equations than unknowns because in general there are more powerproducts than unknowns C_i. This does not mean that there is no solution, because the equations can be dependent. Already in the Example 9, we have seen that they have to be dependent, at least if the degree of the ansatz polynomial is high enough, because the constraint variety will have some algebraic degree.

In the next step, the system of linear equations is solved. If this system has a solution, then there exists a linear constraint polynomial which is an element of the set of polynomials describing $\mathscr{V}$. If this system has no solution, which means only the null vector solves it, one has to proceed to degree two. A general ansatz polynomial of degree two is created, and then one has to follow the same steps as above. Note that a general quadratic polynomial in Study coordinates has 36 coefficients. Depending on the design of the chain, one obtains a system of linear equations in 36 unknowns. For many kinematic chains, the second step already yields solutions. For example, a chain consisting of a universal joint and a spherical joint (as it is the case for one leg of a Stewart–Gough platform, when the actuated P-joint is locked), the second step comes immediately up with the quadratic constraint equation for this leg. This case will be shown in the next example explicitly.

Note that as many polynomials have to be obtained as the chain has constraints c. From this follows that the algorithm has to be continued increasing the degree of the ansatz polynomial until the number of necessary equations is obtained.

In the following, several examples will be presented. We follow the general outline of the algorithm, and we will only show how the constraint equations of canonical chains are found.

Table 1 Denavit–Hartenberg parameters of the UPS chain

	α_i	a_i	d_i
$\mathbf{G}_1$	$\frac{\pi}{2}$	0	0
$\mathbf{G}_2$	0	L	0
$\mathbf{G}_3$	$\frac{\pi}{2}$	0	0
$\mathbf{G}_4$	$\frac{\pi}{2}$	0	0

Example 10 (Canonical leg of a Stewart–Gough platform (UPS chain)) It is well known that the Stewart–Gough parallel manipulator (SGP) consists of a base and a platform connected by six identical legs (Husty 1996). Each of these legs is a serial chain that can be modeled as a UPS serial linkage. To compute the direct kinematics, the actuated prismatic joint is considered to be locked. The resulting canonical linkage therefore consists of five revolute joints, and the forward kinematics of this chain is described by the coordinate transformations:

$$\mathbf{D} = \mathbf{M}_1 \cdot \mathbf{G}_1 \cdot \mathbf{M}_2 \cdot \mathbf{G}_2 \cdot \mathbf{M}_3 \cdot \mathbf{G}_3 \cdot \mathbf{M}_4 \cdot \mathbf{G}_4 \cdot \mathbf{M}_5. \tag{43}$$

The necessary entries of the coordinate transformation matrices are displayed in Table 1. By computing the forward kinematics Eq. (43) and then transforming to Study parameters using Eqs. (7) and (8) and performing half-tangent substitution, one will arrive at the parametric representation of the variety representing this linkage in $\mathbb{P}^7$:

$$\begin{aligned}
x_0 = & -1 + t_5t_1 - t_5t_2 - t_5t_1t_2t_3 + t_2t_5t_1t_4 - t_1t_4 - t_5t_4 - t_5t_3 + t_4t_1t_2t_3 + t_4t_2t_3t_5 - t_1t_2 - \\
& t_4t_3 - t_4t_2 + t_4t_1t_3t_5 - t_1t_3 + t_2t_3 \\
x_1 = & -t_4t_1t_2t_3 - t_5t_1t_2t_3 - t_2t_5t_1t_4 - t_1t_2 - t_4t_1t_3t_5 - t_1t_3 + t_1t_4 + t_5t_1 + t_4t_2t_3t_5 - t_2t_3 \\
& - t_4t_2 + t_5t_2 - t_4t_3 + 1 + t_5t_3 - t_5t_4 \\
x_2 = & t_1 + t_2 - t_1t_2t_3 - t_4t_1t_2 + t_1t_4t_2t_3t_5 - t_4 + t_5t_1t_2 + t_3 + t_2t_5t_4 + t_4t_2t_3 + t_5t_2t_3 - t_4t_1t_3 \\
& - t_5 + t_5t_1t_3 - t_5t_1t_4 + t_4t_3t_5 \\
x_3 = & -t_1 + t_2 + t_1t_2t_3 - t_5t_1t_3 - t_4t_1t_2 + t_1t_4t_2t_3t_5 + t_4 - t_5t_1t_2 + t_3 - t_5t_1t_4 - t_4t_2t_3 - \\
& t_4t_1t_3 - t_5 + t_5t_2t_3 - t_2t_5t_4 - t_4t_3t_5
\end{aligned}$$

$$\begin{aligned}
y_0 = & -\frac{1}{2}L(-1 + t_1t_4 - t_4t_1t_3t_5 + t_4t_1t_2t_3 + t_1t_2 + t_5t_1 + t_4t_3 + t_5t_1t_2t_3 - t_4t_2 + t_2t_5t_1t_4 - \\
& t_1t_3 - t_2t_3 + t_5t_2 + t_4t_2t_3t_5 - t_5t_3 + t_5t_4) \\
y_1 = & -\frac{1}{2}L(-1 + t_1t_4 - t_4t_1t_3t_5 - t_5t_3 + t_4t_1t_2t_3 - t_1t_2 - t_4t_3 - t_5t_1t_2t_3 + t_4t_2 + t_1t_3 - \\
& t_2t_3 - t_4t_2t_3t_5 - t_5t_1 + t_5t_2 - t_5t_4 + t_2t_5t_1t_4) \\
y_2 = & \frac{1}{2}L(t_1 - t_2 + t_1t_2t_3 + t_5t_1t_3 - t_4t_1t_2 + t_1t_4t_2t_3t_5 + t_4 - t_5t_1t_2 + t_3 + t_5t_1t_4 + \\
& t_4t_2t_3 + t_4t_1t_3 - t_5 - t_5t_2t_3 + t_2t_5t_4 - t_4t_3t_5) \\
y_3 = & -\frac{1}{2}L(-t_1 - t_2 - t_1t_2t_3 - t_5t_1t_3 - t_4t_1t_2 + t_1t_4t_2t_3t_5 - t_4 + t_5t_1t_2 + t_3 + t_5t_1t_4 - \\
& t_4t_2t_3 + t_4t_1t_3 - t_5 - t_5t_2t_3 - t_2t_5t_4 + t_4t_3t_5).
\end{aligned} \tag{44}$$

As this chain has five parameters, we can expect one equation that will describe the constraint variety $\mathscr{V}$ together with the equation for S_6^2. One may see immediately that an elimination process will be very tedious, maybe hopeless. Definitely, the elimination will blow up the degree of the resulting equations. We will show that the proposed algorithm yields a result after two steps: In the first step, the parametric equations Eq. (44) are substituted into the linear ansatz equation Eq. (42). The resulting equation is collected with respect to the monomials in t_i. We obtain an expression with 32 terms. Only the beginning and the end of this expression are displayed:

$$(C_3L + C_1L + 2C_4 - 2C_2)t_1 + (-C_7L + 2C_6 + C_5L + 2C_8)t_4t_2t_3t_5 \\ + (C_7L + C_5L + 2C_8 - 2C_6)t_4t_1t_3t_5 + \ldots + (C_3L + C_1L + 2C_4 - 2C_2)t_4t_2t_3 = 0. \tag{45}$$

It turns out that this system of 32 linear equations in 8 unknowns has no solution. Therefore, no linear equation describing the constraint variety exists. In the next step, we build a general second-order equation and proceed as before. Now we obtain a linear system with 234 equations in 32 unknowns. This system is solved, and it yields a two-dimensional solution vector, which is back substituted into the quadratic ansatz equation:

$$(y_0^2 + y_1^2 + y_2^2 + y_3^2 - \frac{1}{4}L^2(x_0^2 + x_1^2 + x_2^2 + x_3^2))\lambda + (x_0y_0 + x_1y_1 + x_2y_2 + x_3y_3)\mu = 0. \tag{46}$$

This surprisingly simple result shows that the constraint variety of the canonical leg of the Stewart–Gough platform is described by the Study quadric equation and the second quadratic equation in Eq. (46) and linear combinations of these two equations.

One question remains to be solved: How do we obtain the constraint equations of a chain in general position from the constraint equations of a kinematic chain in canonical position? The answer to this question was given in Pfurner (2006) and will be recalled briefly here. Suppose that $\alpha : \mathbf{x} \mapsto \mathbf{y} = \mathbf{A}\mathbf{x} + \mathbf{a}$ is a Euclidean displacement. The vectors $\mathbf{x}$ and $\mathbf{y}$ are elements of $\mathbb{R}^3$, but in kinematics, it is advantageous to consider them as elements of two distinct copies of $\mathbb{R}^3$, called the *moving space* and the *fixed space*. The description of α in Study parameters depends on the choice of coordinate frames—*moving frame* and *fixed* or *base frame*—in both spaces. In kinematics, the moving frame is the space attached to a mechanism's output link, and the fixed space is the space where the mechanism itself is positioned. Both types of transformations induce transformations of the Study quadric and thus impose a geometric structure on $\mathbb{P}^7$. Kinematic mapping is constructed such that these transformations act linearly on the Study parameters (i.e., they are projective transformations in $\mathbb{P}^7$). We are going to compute their coordinate representations.

Consider a Euclidean displacement described by a four-by-four transformation matrix $\mathbf{X}$, as in Eq. (3). It maps a point $(1, \mathbf{a})^T$ to $(1, \mathbf{a}')^T = \mathbf{X} \cdot (1, \mathbf{a})^T$. Now we change coordinate frames in fixed and moving spaces and compute the matrix $\mathbf{Y}$ such that $(1, \mathbf{b}')^T = \mathbf{Y} \cdot (1, \mathbf{b})^T$ is the representation of the displacement in the new fixed coordinate frame and the *old* moving coordinate frame. This is slightly different from the typical change of coordinates known from linear algebra where one describes the new transformation in terms of new coordinates in *both* spaces but more suitable for application in kinematics, in particular for describing the position of the end effector tool or for concatenation of different joint displacements in kinematic chains. When the changes of coordinates in fixed and moving frames are described by

$$(1, \mathbf{a})^T = \mathbf{M} \cdot (1, \mathbf{b})^T, \quad (1, \mathbf{b}')^T = \mathbf{F} \cdot (1, \mathbf{a}')^T, \tag{47}$$

we have $\mathbf{Y} = \mathbf{F} \cdot \mathbf{X} \cdot \mathbf{M}$. Denote now by $\mathbf{y}$, $\mathbf{x}$, $\mathbf{f} = [f_0, \ldots, f_7]^T$ and $\mathbf{m} = [m_0, \ldots, m_7]^T$ the corresponding Study vectors. Straightforward computation (see Pfurner 2006) yields

$$\mathbf{y} = \mathbf{T}_f \mathbf{T}_m \mathbf{x}, \quad \mathbf{T}_m = \begin{bmatrix} \mathbf{A} & \mathbf{O} \\ \mathbf{B} & \mathbf{A} \end{bmatrix}, \quad \mathbf{T}_f = \begin{bmatrix} \mathbf{C} & \mathbf{O} \\ \mathbf{D} & \mathbf{C} \end{bmatrix}, \tag{48}$$

where

$$\mathbf{A} = \begin{bmatrix} m_0 & -m_1 & -m_2 & -m_3 \\ m_1 & m_0 & m_3 & -m_2 \\ m_2 & -m_3 & m_0 & m_1 \\ m_3 & m_2 & -m_1 & m_0 \end{bmatrix}, \quad \mathbf{B} = \begin{bmatrix} m_4 & -m_5 & -m_6 & -m_7 \\ m_5 & m_4 & m_7 & -m_6 \\ m_6 & -m_7 & m_4 & m_5 \\ m_7 & m_6 & -m_5 & m_4 \end{bmatrix}, \tag{49}$$

$$\mathbf{C} = \begin{bmatrix} f_0 & -f_1 & -f_2 & -f_3 \\ f_1 & f_0 & -f_3 & f_2 \\ f_2 & f_3 & f_0 & -f_1 \\ f_3 & -f_2 & f_1 & f_0 \end{bmatrix}, \quad \mathbf{D} = \begin{bmatrix} f_4 & -f_5 & -f_6 & -f_7 \\ f_5 & f_4 & -f_7 & f_6 \\ f_6 & f_7 & f_4 & -f_5 \\ f_7 & -f_6 & f_5 & f_4 \end{bmatrix}, \tag{50}$$

and $\mathbf{O}$ is the four-by-four zero matrix.

Example 11 (Continuation of Example 10) The computation of the constraint equations of a kinematic chain in general position from the canonical equations will be demonstrated for the SGP manipulator. To move the canonical leg from the best-adapted position, which has the origin of the fixed frame in the intersection point of axes of the U-joint and the origin of the moving frame in the center of the spherical joint, into a general position, two translations, one in the fixed and one in the moving frame, are needed. Both transformation matrices of Eq. (47) become

$$\mathbf{F} = \begin{bmatrix} 1 & 0 & 0 & 0 \\ A_1 & 1 & 0 & 0 \\ B_1 & 0 & 1 & 0 \\ C_1 & 0 & 0 & 1 \end{bmatrix}, \quad \mathbf{M} = \begin{bmatrix} 1 & 0 & 0 & 0 \\ -a_1 & 1 & 0 & 0 \\ -b_1 & 0 & 1 & 0 \\ -c_1 & 0 & 0 & 1 \end{bmatrix}. \tag{51}$$

Transforming into Study coordinates and constructing the matrices $\mathbf{T}_f$ and $\mathbf{T}_m$ yield

$$
\mathbf{T}_m\mathbf{T}_f = \begin{bmatrix}
4 & 0 & 0 & 0 & 0\,0\,0\,0 \\
0 & 4 & 0 & 0 & 0\,0\,0\,0 \\
0 & 0 & 4 & 0 & 0\,0\,0\,0 \\
0 & 0 & 0 & 4 & 0\,0\,0\,0 \\
0 & -2\,a1+2\,A_1 & -2\,b_1+2\,B_1 & -2\,c_1+2\,C_1 & 4\,0\,0\,0 \\
2\,a_1-2\,A_1 & 0 & 2\,c_1+2\,C_1 & -2\,b_1-2\,B_1 & 0\,4\,0\,0 \\
2\,b_1-2\,B_1 & -2\,c_1-2\,C_1 & 0 & 2\,a_1+2\,A_1 & 0\,0\,4\,0 \\
2\,c_1-2\,C_1 & 2\,b_1+2\,B_1 & -2\,a_1-2\,A_1 & 0 & 0\,0\,0\,4
\end{bmatrix}. \tag{52}
$$

We denote both Study vectors in Eq. (48) by $\mathbf{x} = (x_0, x_1, x_2, x_3, y_0, y_1, y_2, y_3)^T$ and $\mathbf{y} = (x_0', x_1', x_2', x_3', y_0', y_1', y_2', y_3')^T$ and realize that we have to use the inverse of the transformation Eq. (52) to obtain expressions in the coordinates of $\mathbf{x}$

$$
\mathbf{x} = (\mathbf{T}_m\mathbf{T}_f)^{-1}\mathbf{y}
$$

so that these expressions can be substituted into the canonical constraint equation of the leg in optimal position with respect to both coordinate systems. One arrives at the equation

$$
\begin{aligned}
&\left((A_1 - A_1)\,x_1' + (b_1 - B_1)\,x_2' + (c_1 - C_1)\,x_3' + 2\,y_0'\right)^2 + \\
&\left((-a_1 + A_1)\,x_0' + (-c_1 - C_1)\,x_2' + (b_1 + B_1)\,x_3' + 2\,y_1'\right)^2 + \\
&\left((-b_1 + B_1)\,x_0' + (c_1 + C_1)\,x_1' + (-a_1 - A_1)\,x_3' + 2\,y_2'\right)^2 + \\
&\left((-c_1 + C_1)\,x_0' + (-b_1 - B_1)\,x_1' + (a_1 + A_1)\,x_2' + 2\,y_3'\right)^2 - \\
&\frac{1}{4}\,L^2\left(4\,x_0'^2 + 4\,x_1'^2 + 4\,x_2'^2 + 4\,x_3'^2\right) = 0.
\end{aligned} \tag{53}
$$

Equation (53) is the constraint equation of kinematic chain corresponding to leg of the SGP in general position. It is obvious that the equation is still quadratic, and additionally, it can be seen that all $y_i'^2$ are free of design parameters. This is an analogous fact to the planar case which was discussed in Example 3.

Example 12 (3-R chain) In this example, the constraint variety of a 3-R chain is discussed. This variety is interesting because it has been used in a semi-parametric form in Husty et al. (2007a), Husty et al. (2007b), Pfurner (2006) to efficiently compute the inverse kinematics of general 6-R serial manipulators. In this case, we have to expect three constraint equations because the dimension of the constraint variety must be three.

As in the two previous cases, the linear ansatz does not give a constraint equation. On the other hand, the quadratic ansatz yields nine quadratic equations. All nine equations fulfill the parametric equations of this chain. Any three of these nine equations plus the equation for S_6^2 (which is contained in the set) can be taken for further computations. It is guaranteed that the variety generated by them will contain all points that correspond to poses of the end effector of the 3-R chain. But it might be that the variety is bigger than necessary. This means that the other six quadratic equations can (or better should) be taken for solution verification.

More details on LIA can be found in Walter and Husty (2010), where especially some remarks are made to special situations that can occur during the algorithm.

3.4 Singularities in Serial Chains

Algebraic methods are of course not restricted to the analysis of parallel manipulators. As the basic element of these methods are kinematic chains, serial manipulators can be analyzed with the same tool. But one has to be aware that serial and parallel manipulators in some sense behave dual to the basic kinematic tasks. As we have seen in the last subsection, the forward kinematics of a kinematic chain is straightforward (Eq. 40). The inverse kinematics of serial manipulators is more complicated, but using the algebraic methods this task even for the most general 6-R robot can be solved very efficiently (Pfurner 2006). The algorithm is so efficient that a C++ implementation solves almost all cases in real time (Angerer 2016). The application of the algebraic methods for serial manipulators is shown in the computation of all singularities of an industrial robot.

Example 13 (Singularities of a 6-R robot) The geometry of the robot under investigation is given by the following list of DH parameters.

$$
\begin{aligned}
&a_1 = 0, a_2 = 85/2, a_3 = 39, a_4 = 0, a_5 = 0, a_6 = 0, \alpha_1 = 1, \alpha_2 = 0, \alpha_3 = 0, \\
&\alpha_4 = -1, \alpha_5 = 1, \alpha_6 = 0, d_1 = 9, d_2 = 0, d_3 = 0, d_4 = 10, d_5 = 9, d_6 = 8. \quad (54)
\end{aligned}
$$

Note that the angles are given by their algebraic values. The most efficient way to compute the Jacobian of a serial manipulator is given by the fact that its columns contain the Plücker coordinates of the instantaneous position of its axes. To perform these operations, one needs the transformation matrix that transforms lines in a Euclidean transformation. Let $\mathbf{p} = (p_1, p_2, p_3, p_4, p_5, p_6)^T$ and $\mathbf{q} = (q_1, q_2, q_3, q_4, q_5, q_6)^T$ be the Plücker coordinates of two lines, and then the transformation of $\mathbf{p}$ with a transformation as in Eq. (2) is given by

$$
\mathbf{q} = \begin{bmatrix} \mathbf{M}_R & \mathbf{0} \\ \mathbf{a}^\times \mathbf{M}_R & \mathbf{M}_R \end{bmatrix} \mathbf{p}, \tag{55}
$$

where $\mathbf{a}^\times$ is the skew symmetric operator belonging to the vector $\mathbf{a} = (a_1, a_2, a_3)^T$:

$$
\mathbf{a}^\times = \begin{pmatrix} 0 & -a_3 & a_2 \\ a_3 & 0 & -a_1 \\ -a_2 & a_1 & 0 \end{pmatrix}.
$$

Performing this operation for every axis of the manipulator yields the columns Y_i of the Jacobian matrix $\mathbf{J}$. The singularities of the manipulator are the given by $\det \mathbf{J} = 0$. Computing this determinant for the manipulator given by the DH parameters Eq. (54) yields

$$(t_1^2+1)^5(t_2^2+1)^3(t_4^2+1)(t_3^2+1)t_3t_5(7t_2^2t_3^2t_4^2+36t_2^2t_3^2t_4+36t_2^2t_3t_4^2+ \\ 36t_2t_3^2t_4^2+7t_2^2t_3^2+163t_2^2t_4^2+312t_2t_3t_4^2-7t_3^2t_4^2-36t_2^2t_3-36t_2^2t_4- \\ 36t_2t_3^2-144t_2t_3t_4-36t_2t_4^2-36t_3^2t_4-36t_3t_4^2+163t_2^2+312t_2t_3-7t_3^2- \\ 163t_4^2+36t_2+36t_3+36t_4-163)=0. \quad (56)$$

The t_i in this equation are the algebraic values of the joint parameters in the six axes. The first four factors in the equation can be canceled because they can never become real. Then one can observe that first and sixth rotation angles do not contribute to the singularities. This well-known fact applies to all 6-R robots. $t_3 = 0$ means that the second, third and fourth axes are parallel, $t_5 = 0$ means that the fourth and fifth axes are parallel. In both cases, the manipulator is in a singular pose. The remaining equation is of degree 6 in t_2, t_3 and t_4 which describes all remaining singular poses of the manipulator. As it contains only three input variables, it even can be visualized (Fig. 14).

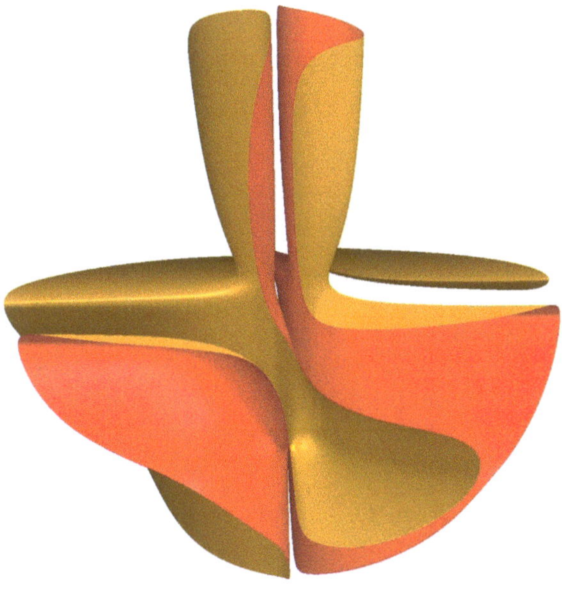

Fig. 14 Singularity surface of a 6-R robot with DH parameters Eq. (54)

4 Algebraic Basics

In this introduction, only the most important definitions and theorems from algebraic geometry are recalled. It should serve the interested reader as a first introduction to this topic. The main emphasis is put on those issues of algebraic geometry that have turned out to be of a certain relevance to the kinematic analysis of mechanisms and robots. This section is based on the book "*Ideals, Varieties, and Algorithms*" by Cox et al. (2007), which is an excellent introduction to algebraic geometry. More detailed descriptions and also the proofs for the theorems can be found there.

Examples were computed in *Maple 16* mostly using the packages `Groebner` and `PolynomialIdeals`. The first package contains the low-level commands, and the second package is newer and contains the more sophisticated ones. There are of course also other software packages, e.g., *Mathematica*, *Singular* or *Macaulay 2*, that could be used for such computations. In most of the examples, the Maple output is not shown, and the reader is advised to test the Maple commands by typing the displayed commands into the system.

In the following, all algebraic equations are polynomials in the ring $\mathbf{K}[\underline{x}] = \mathbf{K}[x_1, \ldots, x_n]$ where $\mathbf{K}$ is a field like $\mathbb{Q}$ (the rational numbers) or $\mathbb{C}$ (the complex numbers).

4.1 *Ideals and Affine Varieties*

At first, polynomial ideals are defined which are the basic objects for everything else.

Definition 1 A set $I \subseteq \mathbf{K}[\underline{x}]$ is called an **ideal** if the following conditions are fulfilled:

- $\forall f, g \in I : f + g \in I$,
- $\forall f \in I$ and $\forall h \in \mathbf{K}[\underline{x}] : hf \in I$.

It follows that almost all ideals are infinite sets of polynomials and cannot be written down as a whole. The sole exception is the ideal $\{0\}$ which is also a proper subset resp. subideal of all other ideals because 0 is contained in every ideal. There is also an ideal which is a proper superset resp. superideal of them, namely the ideal which contains the constant polynomial 1 and with it all polynomials of $\mathbf{K}[\underline{x}]$.

Using Definition 1, it is possible to define the ideal generated by a set of given polynomials $f_1, \ldots, f_s$.

Definition 2 Let $f_1, \ldots, f_s$ be polynomials in $\mathbf{K}[\underline{x}]$. Then the set

$$\langle f_1, \ldots, f_s \rangle = \{g \in \mathbf{K}[\underline{x}] : g = \sum_{i=1}^{s} h_i f_i \ \ and \ \ h_1, \ldots, h_s \in \mathbf{K}[\underline{x}]\}$$

is the **ideal generated by** $f_1, \ldots, f_s$.

The ideal generated be the given polynomials is the set of all combinations of these polynomials using coefficients from $\mathbf{K}[\underline{x}]$. The polynomials $f_1, \ldots, f_s$ form a so-called *basis* of the ideal. Due to the fact that the same ideal can be generated by another set of polynomials, such a basis is not unique. Furthermore, the ideal is obviously *finitely generated* and it can be shown that every ideal of $\mathbf{K}[\underline{x}]$ can be generated by a finite set of polynomials (Hilbert's basis theorem). So the two special ideals mentioned above can be written as $\langle 0\rangle$ and $\langle 1\rangle$.

Example 14 A circle shall be intersected with an ellipse where the corresponding algebraic equations are $f_1 = (x_1 - 1)^2 + (x_2 - 2)^2 - 4 = 0$ and $f_2 = x_1^2 + 3\,x_2^2 - 5 = 0$. To formulate this problem using Maple, one has to type in the following to define the corresponding ideal:

```
with(PolynomialIdeals);

I1:=<(x[1]-1)^2+(x[2]-2)^2-4,x[1]^2+3*x[2]^2-5>;
```

All commands in the package `Groebner` allow to use the notation [...] instead of $< \cdots >$.

Remark 3 In this newly introduced language, the three constraint equations h_1, h_2, h_3 of Example 3 (Eq. 21) constitute an ideal. Definition 1 guarantees that the operations that were used to derive the univariate polynomial (e.g., taking the differences of two of the given equations) result in polynomials that are contained in the ideal.

As it was mentioned above, such a basis is not unique. For example, the ideals $\langle f_1, f_2\rangle$ and $\langle f_1 - f_2, f_2\rangle$ are completely the same. But how can this be found out for two given ideals if they are equal or not?

Two ideals I and J are equal if each element of I is contained in J and vice versa. It is sufficient to test if the basis of one ideal is contained in the other, and vice versa. To find out if a given polynomial is a member of an ideal it is necessary to test if the polynomial can be written as a combination of the ideal's basis. How this can be done in a systematic way will be explained later.

Before that affine varieties are introduced.

Definition 3 For a given ideal $I = \langle f_1, \ldots, f_s\rangle \subseteq \mathbf{K}[\underline{x}]$, the set

$$\mathbf{V}(I) = \{(a_1, \ldots, a_n) \in \mathbf{K}^n : f_i(a_1, \ldots, a_n) = 0 \;\; for\ all \;\; 1 \le i \le s\} \subseteq \mathbf{K}^n$$

is called the **affine variety** of the ideal I.

For each ideal $I = \langle f_1, \ldots, f_s\rangle$, there exists a unique variety $\mathbf{V}(I)$ which is the set of all solutions of the polynomials equations $f_1 = 0, \ldots, f_s = 0$, the so-called *vanishing set*. It follows immediately that all bases of the ideal describe the same variety. In general, the variety of an ideal is the more interesting thing, not the ideal itself, because the variety is exactly the set of solutions of the input equations $f_1, \ldots, f_s$.

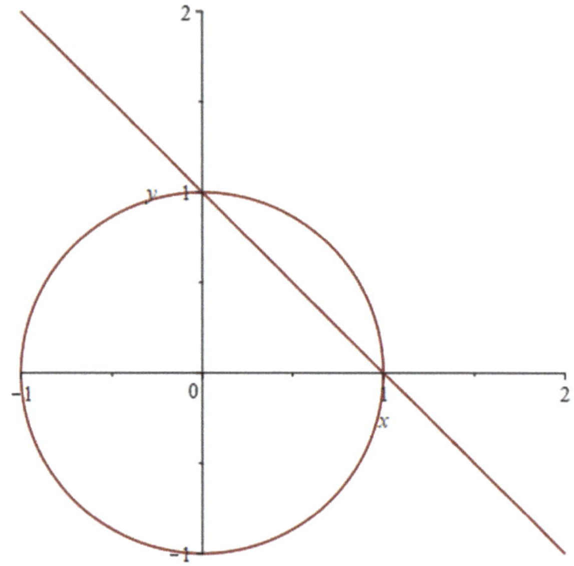

Fig. 15 Affine varieties belonging to f_1, f_2

It has to be mentioned explicitly that the variety does not contain information about the multiplicity of solutions. It is just a set of points in $\mathbf{K}[\underline{x}]$, nothing more.

Two special varieties are $\emptyset$ and $\mathbf{K}^n$ which are the vanishing sets of the ideals $\langle 1 \rangle$ and $\langle 0 \rangle$ which appeared earlier.

Example 15 A circle with center $(0, 0)$ and a line are given by $x_1^2 + x_2^2 - 1 = 0$ and $x_1 + x_2 - 1 = 0$. Then the ideal generated by these two equations is given by $I = \langle x_1^2 + x_2^2 - 1, x_1 + x_2 - 1 \rangle = \langle f_1, f_2 \rangle$, and the corresponding variety is $\{(1, 0), (0, 1)\}$.

The affine variety belonging to the first equation is the set of points that are on the circle, and those of the second equation is the set of points that are on the line. The affine variety belonging to the generated ideal I is the set of points that belong to both varieties, i.e., the set of intersection points (Fig. 15). A third polynomial of the ideal is constructed $f_3 = 3f_1 - 10f_2^3 = -10x_1^3 - 30x_1^2x_2 - 30x_1x_2^2 - 10x_2^3 + 33x_1^2 + 60x_1x_2 + 33x_2^2 - 30x_1 - 30x_2 + 7$ (Fig. 16). In (Fig. 17), one can see that all three polynomials have the same solution set, i.e., they constitute the same affine variety.

It is also possible that different ideals describe the same variety. This is related to the fact that solutions can appear with higher multiplicities (Fig. 17).

Example 16 The following polynomial ideals I and I' are given, each by two possible bases.

$$
\begin{aligned}
I &= \langle x_1^2 + x_2^2 - 1, x_1 + x_2 - 1 \rangle = \langle x_2^2 - x_2, x_1 + x_2 - 1 \rangle, \\
I' &= \langle (x_1^2 + x_2^2 - 1)^2, x_1 + x_2 - 1 \rangle = \langle x_2^4 - 2\,x_2^3 + x_2^2, x_1 + x_2 - 1 \rangle.
\end{aligned}
$$

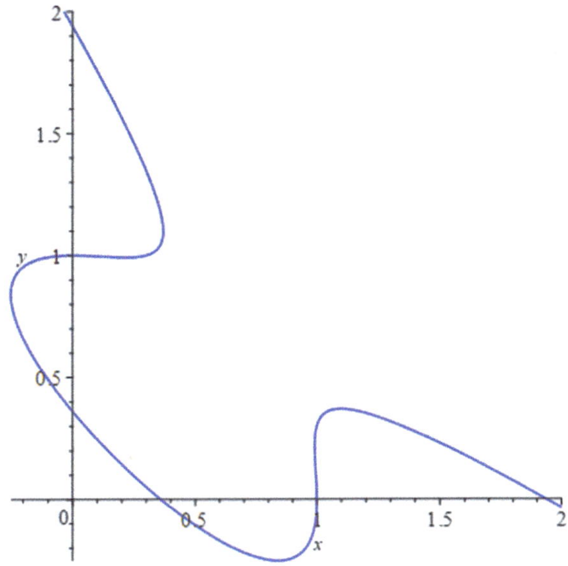

Fig. 16 Affine variety belonging to f_3

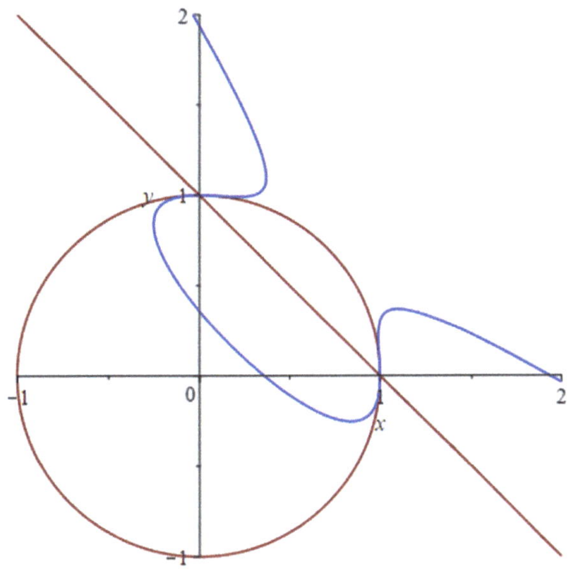

Fig. 17 Affine varieties belonging to f_1, f_2, f_3

It can easily be seen that $\mathbf{V}(I) = \mathbf{V}(I') = \{(1, 0), (0, 1)\}$, but the ideals are not the same because $x_2^2 - x_2$ cannot be written as a combination of $x_2^4 - 2\,x_2^3 + x_2^2$ and $x_1 + x_2 - 1$.

To test if two ideals describe the same variety, *radicals* are introduced.

Definition 4 Let $I \subseteq \mathbf{K}[\underline{x}]$ be an ideal. The set

$$\sqrt{I} := \{f \in \mathbf{K}[\underline{x}] : \exists\ m \in \mathbb{N}, m \geq 1 \ \ with \ \ f^m \in I\}$$

is called the **radical** of I.

The computation of the radical of an ideal I can be seen as reducing I down to the most important things, relevant for its vanishing set by taking out multiplicities. In the example above, the radical is the same: $\sqrt{I} = \sqrt{I'} = I$.

Example 17 An ideal $I = \langle (2\,x_1 - x_2 - 2)x_1, (2\,x_1 - x_2 - 2)x_2^2 \rangle$ is defined. Its vanishing set $\mathbf{V}(I)$ is the line described by $2\,x_1 - x_2 - 2$ and the isolated point $(0, 0)$. To be exact, there are two copies of the point $(0, 0)$ when multiplicities are taken into account.

Computation of the radical using Maple leads to the slightly simpler ideal $\sqrt{I} = \langle -(2\,x_1 + x_2)(2\,x_1 - x_2 - 2), -(2\,x_1 - x_2 - 2)x_1 \rangle$ which has the same vanishing set, but now the point $(0, 0)$ does not appear with higher multiplicity. The code for Maple is the following:

```
with(PolynomialIdeals);

I1:=<(2*x[1]-x[2]-2)*x[1], (2*x[1]-x[2]-2)*x[2]^2>;

rad:=Radical(I1);
```

Next some operations are given which can be applied to varieties.

Definition 5 Let $I = \langle f_1, \ldots, f_s \rangle$ and $J = \langle g_1, \ldots, g_t \rangle$ be ideals with corresponding varieties $V = \mathbf{V}(I)$ and $W = \mathbf{V}(J)$. Then the **union** and **intersection** of V and W can be described as follows:

$$\begin{aligned} V \cap W &= \mathbf{V}(\langle f_1, \ldots, f_s, g_1, \ldots, g_t \rangle), \\ V \cup W &= \mathbf{V}(\langle f_i g_j : 1 \leq i \leq s, 1 \leq j \leq t \rangle). \end{aligned}$$

The first equality is quite clear if one is searching for the solutions two systems have in common, the equations are combined and the resulting system is examined. The second equality can be used to construct varieties which are a composition of simpler varieties.

Example 18 Two very simple varieties are given to show what happens, when varieties are intersected or joined.

$$V_1 = \mathbf{V}(\langle x_1 \rangle), \quad V_2 = \mathbf{V}(\langle x_2 \rangle),$$

$$\begin{aligned} V_1 \cap V_2 &= \mathbf{V}(\langle x_1, x_2 \rangle) = \{(0, 0)\}, \\ V_1 \cup V_2 &= \mathbf{V}(\langle x_1 x_2 \rangle). \end{aligned}$$

What happens when two varieties are joined where one is a subset of the other variety?

$$V_1 \cup \mathbf{V}(\langle x_1, x_2 \rangle) = \mathbf{V}(\langle x_1^2, x_1 x_2 \rangle).$$

It can easily be seen that the vanishing set is the same, but when multiplicities are taken into account the point (0, 0) appears twice.

The code for Maple and the corresponding commands reads:

```
with(PolynomialIdeals);

J1:=<x[1]>;  J2:=<x[2]>;

J_union:=Add(J1,J2);

J_inters:=Multiply(J1,J2);

Multiply(J1,J_union);
```

Now we go back to the different ways to generate an ideal. As already mentioned, there are lots of different bases which describe the same ideal, e.g., the ideals

$$I = \langle f_1, f_2, f_3 \rangle \subseteq \mathbf{K}[x_1, x_2, x_3, x_4],$$
$$I' = \langle f_1, x_2^2 f_1 - f_2, 2 f_3 + (x_3 + 5) f_2 - f_1 \rangle.$$

It can easily be verified that each combination of the generators of I' can be written as a combination of generators of I and vice versa. But what to do when the ideals are more complex? It is necessary to have a systematic way for testing if a polynomial is a combination of some other polynomials. Therefore, the concept of *multivariate division with remainder* is introduced, which is similar to the well-known *univariate division with remainder* which will be discussed first.

Theorem 1 *Let $f, g \in \mathbf{K}[x_1]$ be univariate polynomials with $g \neq 0$. Then there exist unique polynomials q and r such that*

$$f = qg + r$$

with either $r = 0$ or $deg(r) < deg(g)$.

In the corresponding division algorithm, an appropriate multiple of g is subtracted from f such that the monomial with highest degree is canceled from f. This procedure is repeated until the remainder is either 0 or has degree less than $deg(g)$.

Example 19 The division algorithm is demonstrated using the two polynomials $f = 2x^2 - 3x - 7$ and $g = x + 5$:

$$f - \mathbf{2x}\, g = 2x^2 - 3x - 7 - 2x(x + 5) = -13x - 7,$$
$$(-13x - 7) - (\mathbf{-13})g = -13x - 7 + 13x + 65 = \mathbf{58}.$$

It follows that $q = 2x - 13$ and $r = 58$. To get these results with Maple, the commands would be

```
quo(f,g,x); rem(f,g,x);
```

The process stops when the highest monomial of the remainder is not divisible by the highest monomial of g. So in the univariate case, the degree of monomials can be seen as a natural *order* on the set of monomials, which guides the user through the algorithm.

In the following different *termorders* are introduced which allow ordering of the monomials of a multivariate polynomial. With these termorders, an analog algorithm to the univariate case can be defined for multivariate polynomials.

Definition 6 Let $\underline{x}^{\alpha} = x_1^{\alpha_1} \dots x_n^{\alpha_n}$ and $\underline{x}^{\beta} = x_1^{\beta_1} \dots x_n^{\beta_n}$ be monomials in $\mathbf{K}[\underline{x}] = \mathbf{K}[x_1, \dots, x_n]$. To order these monomials, a **monomial ordering** or **termorder** $>_{\underline{x}}$ on the set of monomials in $\mathbf{K}[\underline{x}]$ is defined by an ordering $>$ on the n-tuples $\alpha, \beta \in \mathbb{Z}^n_{\geq 0}$ which has to fulfill the following conditions:

- $>$ is a total ordering on $\mathbb{Z}^n_{\geq 0}$,
- if $\alpha > \beta$ and $\gamma \in \mathbb{Z}^n_{\geq 0}$, then $\alpha + \gamma > \beta + \gamma$,
- every nonempty subset of $\mathbb{Z}^n_{\geq 0}$ has a smallest element under $>$.

If such an ordering $>$ on $\mathbb{Z}_{\geq 0}$ is given, the monomials are ordered using the following equivalence:

$$\underline{x}^{\alpha} >_{\underline{x}} \underline{x}^{\beta} \Longleftrightarrow \alpha > \beta.$$

So monomials are ordered by comparing the ordered n-tuples constructed from the powers of each variable. Next the most important termorderings are given.

Definition 7 (*Lexicographic Order*) Let $\alpha = (\alpha_1, \dots, \alpha_n)$ and $\beta = (\beta_1, \dots, \beta_n)$ be elements of $\mathbb{Z}^n_{\geq 0}$. We define $\alpha >_{lex} \beta$ if the leftmost nonzero entry of the vector-difference $\alpha - \beta \in \mathbb{Z}^n$ is positive.

The Maple keyword for this ordering is `plex`, e.g., a possible lexicographic termorder for polynomials containing the unknowns $\{x_1, x_2, x_3\}$ could be `plex(x[3], x[1],x[2])`.

Example 20 How this ordering looks like for monomials in $\mathbf{K}[x_1, x_2]$ can be seen in the following. It is a sketch how the set of all monomials is ordered; first the 2-tuples are given and then the corresponding monomials starting with the smallest.

$$(0,0) <_{lex} (0,1) <_{lex} (0,2) \dots <_{lex} (1,0) <_{lex} (1,1) <_{lex} (1,2) \dots$$

$$1 <_{lex} x_2 <_{lex} x_2^2 \dots <_{lex} x_1 <_{lex} x_1x_2 <_{lex} x_1x_2^2 \dots$$

Maple has the command `TestOrder` to test if a monomial is smaller than another one.

```
with(Groebner);

TestOrder(x[1]*x[2],x[2]^5,plex(x[1],x[2]));

TestOrder(x[2]^4,x[1]^2*x[2],plex(x[1],x[2]));
```

The result of the first test will be *false*, and the second result will be *true*.

For the next termorderings, the total degree of monomials is needed. For a monomial $\underline{x}^\alpha$, the total degree is denoted by $|\alpha|$ where $|\alpha| = \alpha_1 + \cdots + \alpha_n$.

Definition 8 (*Graded Lex Order*) Let $\alpha = (\alpha_1, \ldots, \alpha_n)$ and $\beta = (\beta_1, \ldots, \beta_n)$ be elements of $\mathbb{Z}^n_{\geq 0}$. We say $\alpha >_{grlex} \beta$ if $|\alpha| > |\beta|$ or, in case of $|\alpha| = |\beta|$, if $\alpha >_{lex} \beta$.

In this termorder monomials are first ordered by the total degree and then ties are broken using $>_{lex}$. The keyword for Maple is `grlex`, and a possible termorder could be `grlex(x[2],x[1],x[3])`.

Example 21 Here is a sketch how the monomials in $\mathbf{K}[x_1, x_2]$ are ordered in graded lex order.

$$(0,0) <_{grlex} (0,1) <_{grlex} (1,0) <_{grlex} (0,2) <_{grlex} (1,1) <_{grlex} (2,0) \ldots$$

$$1 <_{grlex} x_2 <_{grlex} x_1 <_{grlex} x_2^2 <_{grlex} x_1x_2 <_{grlex} x_1^2 \ldots$$

The corresponding Maple command is the same as in Example 20; only `plex` has to be replaced with `grlex`.

Next the most commonly used termorder is discussed. Computations using this order tend to be faster than computations w.r.t. other orders.

Definition 9 (*Graded Reverse Lex Order*) Let $\alpha = (\alpha_1, \ldots, \alpha_n)$ and $\beta = (\beta_1, \ldots, \beta_n)$ be elements of $\mathbb{Z}^n_{\geq 0}$. We define $\alpha >_{grevlex} \beta$ if $|\alpha| > |\beta|$ or, in case of $|\alpha| = |\beta|$, if the rightmost nonzero entry of $\alpha - \beta \in \mathbb{Z}^n$ is negative.

Again monomials are first ordered by the total degree. But ties are now broken using a different condition. In Maple, this termorder is used with the keyword `tdeg`.

Example 22 Here is a sketch how the monomials in $\mathbf{K}[x_1, x_2]$ are ordered when $<_{grevlex}$ is used.

$$(0,0) <_{grevlex} (0,1) <_{grevlex} (1,0) <_{grevlex} (0,2) <_{grevlex} (1,1) <_{grevlex} (2,0) \ldots$$

$$1 <_{grevlex} x_2 <_{grevlex} x_1 <_{grevlex} x_2^2 <_{grevlex} x_1x_2 <_{grevlex} x_1^2 \ldots$$

As it can easily be seen, monomials in this example are sorted the same way as with the graded lex order. This is a special property of $>_{grevlex}$, namely if there are only two variables it leads to the same ordering as $>_{grlex}$. For three or more variables, the ordering is really different, e.g.,

$$x_1^2x_3^2 >_{grlex} x_1x_2^2x_3 \quad but \quad x_1^2x_3^2 <_{grevlex} x_1x_2^2x_3.$$

All these orderings can also be combined which leads to the so called *Product Orders*. An example for a product order on $\mathbf{K}[x_1, x_2, x_3, x_4]$ in Maple would be `prod(plex(x[1],x[2]),tdeg(x[3],x[4]))`. This means that monomials are first compared using the `plex` order and ties are broken using the `tdeg` order. Even more than two partial orders are allowed.

It has to be noted explicitly that also the ordering of the variables can be varied, not only the type. All in all there are lots of different ways to define an order for ordering monomials.

With these preparations, it is possible to order the monomials of a polynomial. Before the multivariate polynomial division can be defined another definition is necessary.

Definition 10 Let $f \in \mathbf{K}[\underline{x}]$ be a polynomial with $f = \sum_\alpha a_\alpha x^\alpha$ and let $>_{\underline{x}}$ be a monomial ordering on $\mathbf{K}[\underline{x}]$. We define the **leading monomial** $LM(f)$ as the highest monomial of f with respect to $>_{\underline{x}}$, the **leading coefficient** $LC(f)$ as the coefficient of the highest monomial, and the **leading term** as $LT(f) = LC(f) \cdot LM(f)$.

The most important issue in this definition is the leading monomial. It will appear quite often in the following definitions and theorems.

Example 23 Let $f = x_1^2x_2^3 - 5x_1x_2x_3 + 4\,x_1^3x_3^2$ be a polynomial where the termorder $>_{grlex}$ on $\mathbf{K}[\underline{x}]$ is used:

$$deg(f) = 5, \quad LM(f) = x_1^3x_3^2, \quad LC(f) = 4, \quad LT(f) = 4\,x_1^3x_3^2.$$

To obtain these results with Maple, the following commands can be used:

```
with(Groebner);

f:=x[1]^2*x[2]^3-5*x[1]*x[2]*x[3]+4*x[1]^3*x[3]^2;

degree(f,[x[1],x[2],x[3]]);

LeadingMonomial(f,grlex(x[1],x[2],x[3]));

LeadingCoefficient(f,grlex(x[1],x[2],x[3]));

LeadingTerm(f,grlex(x[1],x[2],x[3]));
```

For the leading term, Maple does not return the product, but the pair $LC(f)$, $LM(f)$. If the termorder *plex(x[2],x[3],x[1])* is used instead, the results are as follows:

$$deg(f) = 5, \quad LM(f) = x_1^2x_2^3, \quad LC(f) = 1, \quad LT(f) = x_1^2x_2^3.$$

Now all the ingredients are defined, and it is possible to introduce the multivariate division.

Definition 11 Let $F = [f_1, \ldots, f_s]$ be an ordered list of polynomials in $\mathbf{K}[\underline{x}]$ and $>_{\underline{x}}$ a monomial order. Then every polynomial $f \in \mathbf{K}[\underline{x}]$ can be written in the form

$$f = a_1 f_1 + \cdots + a_s f_s + r$$

where all a_i and r are elements of $\mathbf{K}[\underline{x}]$ and r is either 0 or a polynomial where no monomial is divisible by any of $LM(f_1), \ldots, LM(f_s)$. r is called **a remainder** of f on division by F.

The algorithm which produces the a_i and the remainder r acts in the following way: It is tested if $LM(f)$ is divisible by $LM(f_1)$, then by $LM(f_2)$ and so on. If the result of such a test is true for a leading monomial $LM(f_i)$, testing is stopped and an appropriate multiple of f_i is subtracted from f such that $LT(f)$ is canceled. The result is again named f, and the testing starts again. In the other case that $LM(f)$ is not divisible by $LM(f_1), \ldots, LM(f_s)$, the term $LT(f)$ is added to the remainder. It follows that in each step $LT(f)$ is removed, either by cancellation or by moving it to the remainder. The process is finished when f is 0.

Example 24 As an example, the polynomial $f = x_1^2 x_2 + x_1 x_2^2 + x_2^2$ is divided by the polynomials $f_1 = x_1 x_2 - 1$ and $f_2 = x_2^2 - 1$, where the termorder $>_{lex}$ is used in $\mathbf{K}[x_1, x_2]$.

$$\begin{aligned}
&f = x_1^2 x_2 + x_1 x_2^2 + x_2^2, \quad r = 0,\ a_1 = 0,\ a_2 = 0,\\
&f \leftarrow x_1 x_2^2 + x_1 + x_2^2, \quad r = 0,\ a_1 = x_1,\ a_2 = 0,\\
&f \leftarrow x_1 + x_2^2 + x_2, \quad r = 0,\ a_1 = x_1 + x_2,\ a_2 = 0,\\
&f \leftarrow x_2^2 + x_2, \quad r = x_1,\ a_1 = x_1 + x_2,\ a_2 = 0,\\
&f \leftarrow x_2 + 1, \quad r = x_1,\ a_1 = x_1 + x_2,\ a_2 = 1,\\
&f \leftarrow 1, \quad r = x_1 + x_2,\ a_1 = x_1 + x_2,\ a_2 = 1,\\
&f \leftarrow 0, \quad r = x_1 + x_2 + 1,\ a_1 = x_1 + x_2,\ a_2 = 1.
\end{aligned}$$

It follows that f can be written as

$$f = a_1 f_1 + a_2 f_2 + r = (x_1 + x_2)(x_1 x_2 - 1) + (1)(x_2^2 - 1) + (x_1 + x_2 + 1),$$

where no monomial in the remainder $x_1 + x_2 + 1$ is divisible by $LM(f_1)$ or $LM(f_2)$.

It has to be said clearly that the result r is *a remainder* of f on division by the ordered list F. If f_1 is exchanged with f_2, the following result is obtained:

$$f = a_1' f_1 + a_2' f_2 + r' = (x_1 + 1)(x_2^2 - 1) + (x_1)(x_1 x_2 - 1) + (2\,x_1 + 1).$$

So the remainder r depends on the order of the polynomials in the list F and of course, on the monomial order which has to be chosen first of all.

Using the multivariate division, the notion of *reduction* can be defined.

Definition 12 Let $F = [f_1, \ldots, f_s]$ be an ordered list of polynomials in $\mathbf{K}[\underline{x}]$, $f \in \mathbf{K}[\underline{x}]$ and $>_{\underline{x}}$ a monomial order. Then we call the process of dividing f by F **reduction** and denote the remainder with $\overline{f}^F$. The choice of a monomial order is required to be able to compute the remainder.

This reduction will be used quite often in the following definitions and theorems. But before that the corresponding Maple command is given.

Example 25 The polynomial $f = x_1^2x_2 + x_1x_2^2 + x_2^2$ is to be divided by the polynomials $f_1 = x_1x_2 - 1$ and $f_2 = x_2^2 - 1$, where the termorder $>_{lex}$ is used in $\mathbf{K}[x_1, x_2]$.

```
with(Groebner);

f:=x[1]^2*x[2]+x[1]*x[2]^2+x[2]^2;

f1:=x[1]*x[2]-1;

f2:=x[2]^2-1;

r:=Reduce(f,[f1,f2],plex(x[1],x[2]),'s','a');

s;  a;
```

The result is a remainder r, a list of quotients a and a number s such that

$$f = \sum_{i=1}^{2} a_i f_i + \frac{r}{s}.$$

The result for this example is as follows:

$$r = x_1 + x_2 + 1, \qquad a = [x_1 + x_2, 1,] \qquad s = 1.$$

The necessity of a monomial order will not be mentioned explicitly from now on. Next we define *interreduction*.

Definition 13 Let $F = [f_1, \ldots, f_s]$ be an ordered list of polynomials in $\mathbf{K}[\underline{x}]$. The process of replacing each polynomial f_i by $\overline{f_i}^{F\setminus\{f_i\}}$ is called **interreduction** of the list F.

This means that every polynomial in F is reduced with respect to all other elements of the list. An important property of an interreduction is that the original set of polynomials and the result of the interreduction generate the same ideal. Interreduction can sometimes be used to simplify generating sets (bases) of an ideal, to obtain shorter polynomials. The right choice of the monomial order is important.

Example 26 The ideal $I = \langle (x_1 - 1)^2 + (x_2 + 5)^2 - 6, 2\,x_1^2 + 2\,x_2^2 - 4\rangle$ is interreduced. The appropriate commands are the following:

```
with(Groebner);

I1:=[(x[1]-1)^2+(x[2]+5)^2-6,2*x[1]^2+2*x[2]^2-5];

ir:=InterReduce(I1,plex(x[1],x[2]));
```

And the result is

$$ir = [416\,x_2^2 + 1800\,x_2 + 1985, 4\,x_1 - 20\,x_2 - 45].$$

Example 27 Interreduction is very efficient for the constraint polynomials h_1, h_2, h_3 of Example 3:

```
Groebner[InterReduce]([h_1,h_2,h_3],tdeg(y_1,y_2,x_3));
```

yields the three simplified polynomials

$$\begin{aligned}
g_1 &: (-5239\sqrt{3} - 1876)x_3^2 + (624\sqrt{3}y_2 + 4550\sqrt{3} + 174y_2 + 11684)x_3 + 104\sqrt{3}y_1 \\
&- 1755\sqrt{3} + 218y_1 - 720y_2 - 774 = 0, \\
g_2 &: 547x_3^2 + (18y_1 - 48y_2 - 476)x_3 - 8y_1 + 18y_2 + 99 = 0, \\
g_3 &: -8527x_3^2 + (552y_2 + 7616)x_3 + 36y_1^2 + 36y_2^2 + 344y_1 - 1359 = 0,
\end{aligned}$$

which have the property that y_1 and y_2 appear linear.

Now back to ideals generated by a set of polynomials. There was already the question how one could decide if a given polynomial f is an element of $I = \langle f_1, \ldots, f_s\rangle$. The concept of reduction would be a good method to answer that question. First a result is given which is not the perfect solution.

Theorem 2 *Let $I = \langle f_1, \ldots, f_s\rangle \subseteq \mathbf{K}[\underline{x}]$ be an ideal and $f \in \mathbf{K}[\underline{x}]$. If $\overline{f}^F = 0$ then f is an element of I.*

The disadvantage of this result is that if $\overline{f}^F \neq 0$ the question is open as before. The crucial point is that the leading terms $LT(f_1), \ldots, LT(f_s)$ which are used in the division algorithm are in general "bad" representatives for the set of all leading terms which are possible in the ideal I.

Example 28 Let $F = [f_1, f_2] = [x_1x_2 + 1, x_2^2 - 1]$ be an ordered list of polynomials in $\mathbf{K}[\underline{x}]$, and $p_1 = x_1 f_1 + x_2 f_2$ and $p_2 = x_2 f_1 + x_1 f_2$ be combinations of these polynomials. It is clear that $p_1, p_2 \in \langle f_1, f_2\rangle$. But if the reductions are computed w.r.t. `plex(x[1],x[2])`, the following results are obtained:

$$\overline{p_1}^F = 0, \qquad \overline{p_2}^F = -x_1 - x_2.$$

In the first case Theorem 2 works quite fine, it does not in the second.

Now *standard bases* are defined which are "much better" representatives (generators) for an ideal. They are still dependent from a chosen monomial order but nevertheless very useful to deduce information about the ideal and the corresponding variety.

4.2 Standard Bases

Definition 14 For a fixed monomial order and an ideal $I \in \mathbf{K}[\underline{x}]$, a finite subset $G = \{g_1, \ldots, g_t\}$ of I is called a **Groebner basis** or **standard basis** if

$$\langle LM(g_1), \ldots, LM(g_t)\rangle = \langle LM(I)\rangle,$$

where $LM(I)$ is the ideal generated by all the leading terms of the elements of I.

With other words, a basis is also a Groebner basis if the leading monomials of the generators generate the same ideal as the leading monomials of all ideal elements.

A Groebner basis is a very special generating set of the ideal I with some nice properties. Such a basis is not unique due to the fact that adding another polynomial to the generators does not change this property. Another reason for nonuniqueness is the fact that a monomial order has to be chosen first.

In this introduction, it will not be explained in detail how such a Groebner basis is computed explicitly, starting with a set of generators and a termorder. Just a short appetizer: For all possible pairs of generators, the so-called *S-polynomials* are computed and reduced with the list of generators. All remainders which are not 0 are added to the list, and the procedure starts again. Such rounds are repeated until all remainders of the S-polynomials are all 0 after reduction. See the book "Ideals, Varieties, and Algorithms" by Cox, Little and O'Shea for a detailed description of the algorithm and enhancements of it.

With a Groebner basis, we get a better result for the ideal membership question.

Theorem 3 *Let* $G = \{g_1, \ldots, g_t\} \subseteq \mathbf{K}[\underline{x}]$ *be a Groebner basis of an ideal I and* $f \in \mathbf{K}[\underline{x}]$. *The polynomial f is an element of I if and only if* $\overline{f}^G = 0$.

All in all the procedure for the ideal membership question is the following: First a termorder is fixed, then a Groebner basis of the ideal is computed and after that the polynomial f is reduced with respect to this basis. f is only then an element of I if the result of the reduction is 0.

It has to be mentioned that the result of a reduction with respect to a Groebner basis is independent of the order of the basis elements. This was not the case when the reduction was done with respect to a normal basis.

Example 29 As an example we take the polynomials from Example 28. Let $I = \langle f_1, f_2 \rangle = \langle x_1x_2 + 1, x_2^2 - 1 \rangle$ be an ideal in $\mathbf{K}[\underline{x}]$ and $p_1 = x_1 f_1 + x_2 f_2$ and $p_2 = x_2 f_1 + x_1 f_2$ polynomials which are definitely elements of I.

Now a Groebner basis is computed w.r.t. *plex(x[1],x[2])*, and p_1, p_2 are reduced with that basis.

```
with(Groebner);

with(PolynomialIdeals):

f1:=x[1]*x[2]+1;   f2:=x[2]^2-1;

G:=Basis(<f1,f2>,plex(x[1],x[2]));

p1:=x[1]*f1+x[2]*f2;

p2:=x[2]*f1+x[1]*f2;

Reduce(p1,G,plex(x[1],x[2]));

Reduce(p2,G,plex(x[1],x[2]));
```

The Groebner basis is

$$G = \langle x_2^2 - 1, x_1 + x_2 \rangle$$

and now the expected results are obtained:

$$\overline{p_1}^F = 0, \qquad \overline{p_2}^F = 0.$$

The package *PolynomialIdeals* contains also two commands for testing if a polynomial or an ideal is contained in another ideal. They are used as follows:

```
IdealMembership(p1,G);

IdealContainment(<p1,p2>,G);
```

The result of both commands will be *true* for this example.

As mentioned above, adding another polynomial of the ideal to the generators does not change the property of being a Groebner basis. To get a unique version of such a basis, the so-called *reduced Groebner bases* are defined.

Definition 15 Let $G = \{g_1, \ldots, g_t\} \subseteq \mathbf{K}[\underline{x}]$ be a Groebner basis of a given ideal I. The generators G are a **reduced Groebner basis** if G is interreduced and all leading coefficients are 1.

Fortunately, all bases returned by Maple are interreduced. So at least the more important condition is fulfilled and the bases are quasi unique. This is not the case when *Singular* is used. But there exists a command to switch on basis interreduction.

Now some remarks about Groebner bases:

- The number of polynomials in a Groebner basis can be very large, and it depends on the complexity of the ideal and, of course, on the chosen monomial order. It is quite possible that only a part of the Groebner basis would be enough to generate the ideal I, but then the relevant condition $\langle LM(g_1), \ldots, LM(g_t)\rangle = \langle LM(I)\rangle$ is no more fulfilled.
- There are different methods to compute Groebner bases. The sketch of the algorithm above describes only the main concept. Lots of improvements were made to accelerate basis computations. Depending on the termorder, the computations can be very expensive with respect to time and memory. Sometimes it is better to compute the basis for an "easy" termorder and then to convert it to the desired termorder using one of the conversion algorithms.
- The monomial order $>_{lex}$ (in Maple `plex`) tends to be expensive in general, whereas the order $>_{grevlex}$ (in Maple `tdeg`) tends to be "relatively" cheap. If the type of order is chosen, the computational costs can again be influenced by the order of unknowns.
- The typical way to order the polynomials in a basis is by sorting them w.r.t. the leading monomials, starting with the smallest one.

Example 30 Now as an example, reduced Groebner bases are computed for the ideal

$$I = \langle x_1^2x_2 + x_1x_2^2 + x_2^2x_3, x_1x_2 - x_3, x_2^2 - 1\rangle$$

using different termorders:

$>_{lex}$ *in* $\mathbf{K}[x_1, x_2, x_3]$:

$$G = \{x_3^3 + 2\,x_3^2, x_2x_3 + x_3^2 + x_3, x_2^2 - 1, x_1 + x_3^2 + x_3\}$$

$>_{lex}$ *in* $\mathbf{K}[x_3, x_1, x_2]$:

$$G = \{x_2^2 - 1, x_1^2 + x_1x_2 + x_1, x_3 - x_1x_2\}$$

$>_{grlex}$ *in* $\mathbf{K}[x_1, x_2, x_3]$:

$$G = \{x_3^2 + x_1 + x_3, x_2x_3 - x_1, x_2^2 - 1, x_1x_3 + x_1 + x_3, x_1x_2 - x_3, x_1^2 + x_1 + x_3\}$$

The code for Maple looks as follows:

```
with(Groebner);

with(PolynomialIdeals):

f1:=x[1]^2*x[2]+x[1]*x[2]^2+x[2]^2*x[3];

f2:=x[1]*x[2]-x[3];   f3:=x[2]^2-1;

Basis(<f1,f2,f3>,plex(x[1],x[2],x[3]));
```

```
Basis(<f1,f2,f3>,plex(x[3],x[1],x[2]));

Basis(<f1,f2,f3>,grlex(x[1],x[2],x[3]));
```

It is no coincidence that in all the bases w.r.t. to $>_{lex}$, the first polynomial is univariate. This is because the lex-orders have the so-called *elimination property*.

Example 31 Taking the three polynomial equations g_1, g_2, g_3 from Example 27 and computing a Groebner basis using the command

```
Groebner[Basis]([g_1,g_2,g_3],plex(y1,y2,x3));
```

yield immediately a univariate polynomial of degree six in x_3 and two more polynomials in which y_1 and y_2 appear linearly.

What is elimination resp. an elimination ideal?

4.3 Elimination

First of all, the so-called *elimination ideals* are defined.

Definition 16 Let I be an ideal in $\mathbf{K}[\underline{x}] = \mathbf{K}[x_1, \ldots, x_n]$ and $1 \leq l < n$. Then the ideal $I_l = I \cap \mathbf{K}[x_{l+1}, \ldots, x_n]$ is called the **l-th elimination ideal** of I. I_l contains all elements of I which do not contain the variables $x_1, \ldots, x_l$.

If a termorder has the elimination property, then a basis w.r.t. this order can be used to extract bases for the elimination ideals.

Theorem 4 *Let $G = \{g_1, \ldots, g_t\} \subseteq \mathbf{K}[x_1, \ldots, x_n]$ be a Groebner basis of an ideal I with respect to $>_{lex}$. Furthermore, let $H = \{h_1, \ldots, h_k\}$ be the first k polynomials of G which do not contain the unknowns $x_1, \ldots, x_l$. Then H is a Groebner basis of the l-th elimination ideal I_l.*

It follows an example where it can be seen more clearly what this means.

Example 32 The ideal $I = \langle x_1^2 x_2 + x_1 x_2^2 + x_2^2 x_3, x_1 x_2 - x_3, x_2^2 - 1\rangle$ is given with the Groebner basis w.r.t. *plex(x[1],x[2],x[3])*

$$G = \langle x_3^3 + 2\,x_3^2, x_2 x_3 + x_3^2 + x_3, x_2^2 - 1, x_1 + x_3^2 + x_3\rangle.$$

Then the following generators are all Groebner bases of the corresponding elimination ideal:

$$\begin{aligned} I_2 &= \langle x_3^3 + 2\,x_3^2\rangle, \\ I_1 &= \langle x_3^3 + 2\,x_3^2, x_2 x_3 + x_3^2 + x_3, x_2^2 - 1\rangle. \end{aligned}$$

There is also another command to do the elimination directly. It is contained in the package *PolynomialIdeals*, and to compute I_1 directly from the input polynomials, it is used as follows:

```
EliminationIdeal(<f1,f2,f3>,{x[2],x[3]});
```

The second argument gives the unknowns which shall not be eliminated. Furthermore, no termorder has to be chosen, and this is done by Maple internally. Unfortunately, this choice is session-dependent.

It can easily be seen that with these elimination ideals, it is quite easy to solve a system of equations with only finitely many solutions.

Example 33 Here once again the Groebner basis G from above is discussed including I_2 and I_1.

$$\begin{aligned} I_2 &= \langle x_3^3 + 2\,x_3^2 \rangle, \\ I_1 &= \langle x_3^3 + 2\,x_3^2, x_2x_3 + x_3^2 + x_3, x_2^2 - 1 \rangle, \\ G &= \langle x_3^3 + 2\,x_3^2, x_2x_3 + x_3^2 + x_3, x_2^2 - 1, x_1 + x_3^2 + x_3 \rangle. \end{aligned}$$

Now the system can be solved step by step by solving the partial systems and extending the solutions.

```
with(Groebner);

with(PolynomialIdeals):

f1:=x[1]^2*x[2]+x[1]*x[2]^2+x[2]^2*x[3];

f2:=x[1]*x[2]-x[3];   f3:=x[2]^2-1;

I0:=<f1,f2,f3>;

G:=<op(Basis(I0,plex(x[1],x[2],x[3])))>;

NumberOfSolutions(I0);

I2:=EliminationIdeal(<f1,f2,f3>,{x[3]});

I1:=EliminationIdeal(<f1,f2,f3>,{x[2],x[3]});

L1:={solve(Generators(I2),{x[3]})};

L2:=map(y->op(map(z->z union y,{solve(eval(Generators(I1),y),
{x[2]})})),L1);

L3:=map(y->op(map(z->z union y,{solve(eval(Generators(G),y),
{x[1]})})),L2);
```

The result for the vanishing set is $\mathbf{V}(I) = \{(-2, 1, -2), (0, -1, 0), (0, 1, 0)\}$, where the second solution has multiplicity 2.

How does the variety of the l-th elimination ideal $\mathbf{V}(I_l)$ correspond to the original variety $\mathbf{V}(I)$?

Theorem 5 *The variety* $\mathbf{V}(I_l) \in \mathbf{K}^{n-l}$ *is the smallest variety (w.r.t. inclusion) which contains* $\pi(\mathbf{V}(I))$, *where* $\pi(\mathbf{V}(I))$ *is the orthonormal projection of* $\mathbf{V}(I)$ *onto the subspace generated by equations* $x_1 = \cdots = x_l = 0$.

An important fact is that the equality $\mathbf{V}(I_l) = \pi(\mathbf{V}(I))$ only holds when I is a *projective variety* and I_l a *projective elimination ideal*. This means that here it can happen that a solution of I_l cannot be extended to a solution of I_{l-1}.

Another application of elimination is implicitization. The following example shows how a variety can be deduced from a parameterization.

Example 34 A parameterization of an ellipse is given.

$$x[1] = \frac{5\,(1-t^2)}{1+t^2} \qquad x[2] = \frac{3\,(2\,t)}{1+t^2}$$

For all values of $t \in \mathbb{R}$, a point of the ellipse is obtained. The only point which cannot be reached is the left apex $(-5, 0)$. To compute the smallest variety which contains all these points (should be the ellipse), the following code could be used:

```
with(Groebner);

with(PolynomialIdeals):

par1:=x[1]-(5*(1-t^2))/(1+t^2);

par2:=x[2]-(3*(2*t))/(1+t^2);

J:=<numer(par1),numer(par2)>;

G:=Basis(J,plex(t,x[1],x[2]));

J1:=<G[1]>;
```

The first elimination ideal $J_1 = \langle 9\,x_1^2 + 25\,x_2^2 - 225\rangle$ has exactly the ellipse as vanishing set.

In the last subsection, two very important concepts are discussed that play major roles in the analysis of more complicated kinematic structures. The first one has to do with the degree of freedom of a mechanism, and the second one can help to find different operation modes of a kinematic structure.

4.4 Dimension, Primary Decomposition

It happens quite often that one is first of all interested in the *dimension* of a variety. Here is an example where the variety contains parts with different dimensions, and it is shown how the dimension is computed.

Example 35 The ideal $I = \langle (2x_1 - x_2 - 2)x_1, (2x_1 - x_2 - 2)x_2^2 \rangle$ is given. $\mathbf{V}(I)$ consists of a line and an isolated point. Using the Hilbert polynomial, the dimension is computed (which should be 1).

```
with(PolynomialIdeals):

I1:=<(2*x[1]-x[2]-2)*x[1], (2*x[1]-x[2]-2)*x[2]^2>;

HilbertDimension(I1,{x[1],x[2]});
```

If a variety contains parts with different dimensions, then the dimension of the whole variety is defined to be the largest of these numbers.

In kinematics, the dimension of the ideal generated by the constraint equations yields the degrees of freedom of the mechanical structure. But one has to be very careful with the obtained number. As mentioned above, the Hilbert dimension yields the largest dimension. At special points of the corresponding variety, this number can be larger or smaller.

Example 36 Taking the three constraint equations h_1, h_2, h_3 of Example 3 and computing the Hilbert dimension of the generated ideal, one can use the Maple command

```
Groebner[HilbertDimension]([h1,h2,h3],tdeg(x3,y1,y2));
```

The answer will be 0, and this means that the solution of the corresponding system of equations is a set of points. The command

```
Groebner[Basis]([h1,h2,h3],plex(y1,y2,x3));
```

yields immediately, without further manipulations, a univariate polynomial of degree 6 in x_3 and two more polynomials which are linear in y_1 and y_2.

Substituting on the other hand the following set of design and input parameters

$$A_2 = 16, A_3 = 9, B_3 = 6, a_2 = 16, a_3 = 9, b_3 = 6, l_1 = 10, l_2 = 10, l_3 = 10,$$
$$k_1 = 5, k_2 = 5, k_3 = 5, u = 1, v = 1, w = 1,$$

into the general system of constraint equations Eq. (20), one obtains a new system of constraint equation which reads:

$$\begin{aligned}
g_1 &: 75x_3^2 + 40x_3y_1 + 4y_1^2 + 4y_2^2 + 40y_2 + 75 = 0,\\
g_2 &: 1099x_3^2 + 40x_3y_1 - 128x_3y_2 + 4y_1^2 + 4y_2^2 - 640x_3 + 40y_2 + 75 = 0,\\
g_3 &: 783x_3^2 + 88x_3y1 - 72x_3y_2 + 4y_1^2 + 4y_2^2 - 360x_3 + 40y_2 + 75 = 0.
\end{aligned}$$

Invoking again the Maple command

```
Groebner[HilbertDimension]([g1,g2,g3],tdeg(x3,y1,y2));
```

yields now the answer 1 for the Hilbert dimension. This means that the manipulator allows a one-DOF motion with all inputs locked. In this case, the inputs were all set to 1, which corresponds to 90°. The Groebner base for this set of equations is:

```
Groebner[Basis]([h1,h2,h3],plex(y1,y2,x3));
[x3*(201*x3^2-20), -x3*(8*x3-y2-5), x3*(11*x3+4*y1),
-35*x3^2+4*y1^2+4*y2^2+40*y2+75].
```

The base consists of four equations, and the first equation is $x_3(201x_3^2 - 20) = 0$. One can see immediately that the solution $x_3 = 0$ causes the second and the third equations to vanish. Only the fourth equation remains. $x_3 = 0$ and $4y_1^2 + 4y_2^2 + 40y_2 + 75 = 0$ determine the self-motion, which is the well-known parallel bar motion. But when $x_3 \neq 0$ then the four equations have two distinct solutions, one of them reads

$$x_3 = \frac{2}{201}\sqrt{1005},\ y_1 = -\frac{11}{402}\sqrt{1005},\ y_2 = \frac{16}{201}\sqrt{1005} - 5.$$

The two solutions yield two distinct assemblies of the manipulators. From point of view of algebraic geometry, the ideal has two components which have different dimensions. The corresponding affine varieties consist of a curve (Fig. 18, left) and two points (Fig. 18, right). In this example, it is not possible to move directly from one component to the other. One has to disassemble the mechanism and to reassemble to obtain a solution of the other component. Each component corresponds to an operation mode of the mechanism or robot. As it will be shown in the next section, it can happen that different components can have intersections and this would mean that a mechanism is able to move at certain poses from one operation mode into an other operation mode.

As already mentioned, it is possible that an ideal describes a variety which is made up of some simpler varieties, e.g., the variety from the Examples 35 and 36. The question is how such a decomposition can be found if there is one.

Definition 17 A **primary decomposition** of a given ideal I is an expression of I as an intersection of primary ideals, namely $I = \bigcap_{i=1}^{r} Q_i$. Such a decomposition is called **minimal** if the radicals $\sqrt{Q_i}$ are all different and $Q_i \not\supseteq \bigcap_{i \neq j} Q_j$. Furthermore, if no radical $\sqrt{Q_i}$ is strictly contained in another radical $\sqrt{Q_j}$, then the primary components Q_i are uniquely determined.

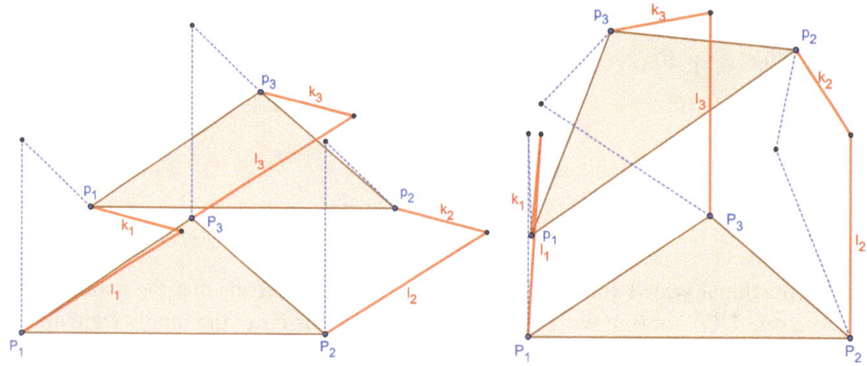

Fig. 18 Different Hilbert dimensions of the constraint equations g_1, g_2, g_3 left: self-motion, right: rigid assembly

The radicals $\sqrt{Q_i} =: P_i$ are the corresponding prime ideals.

It follows an example where the variety can be decomposed into three different parts.

Example 37 A rather large ideal J is given.

$$
\begin{aligned}
J = \langle & 4\,{x_3}^2{x_1}^3 + 4\,{x_3}^2 x_1 {x_2}^2 + 4\,{x_3}^4 x_1 - 16\,x_1{x_3}^2 + 5\,{x_3}^2 x_2 {x_1}^2 + 5\,{x_3}^2{x_2}^3 + \\
& +5\,{x_3}^4 x_2 - 20\,x_2{x_3}^2 - 6\,{x_3}^2{x_1}^2 - 6\,{x_3}^2{x_2}^2 - 6\,{x_3}^4 + 24\,{x_3}^2, \\
& 4\,{x_1}^4 + 4\,{x_1}^2{x_2}^2 + 4\,{x_3}^2{x_1}^2 + 2\,{x_1}^2 + 5\,{x_1}^3 x_2 + 5\,x_1{x_2}^3 + 5\,x_1x_2{x_3}^2 - \\
& -20\,x_1x_2 - 18\,{x_1}^3 - 18\,x_1{x_2}^2 - 18\,x_1{x_3}^2 + 72\,x_1 - 15\,x_2{x_1}^2 - 15\,{x_2}^3 - \\
& -15\,x_2{x_3}^2 + 60\,x_2 + 18\,{x_2}^2 + 18\,{x_3}^2 - 72\rangle
\end{aligned}
$$

Using the command

```
PrimaryDecomposition(J);
```

the primary decomposition is computed and the result is

$$
Q_1 = \langle 4\,x_1 + 5\,x_2 - 6\rangle, \qquad Q_2 = \langle {x_1}^2 + {x_2}^2 + {x_3}^2 - 4\rangle, \qquad Q_3 = \langle {x_3}^2, x_1 - 3\rangle,
$$

which means that $\mathbf{V}(I)$ can be decomposed into a line, a sphere and an isolated point which appears with multiplicity 2.

Such decompositions are quite convenient if the variety has to be intersected with other varieties, because then each of the components can be treated separately. The worst case is that one has to deal with an ideal which is primary or even prime. In this case, one has to take the ideal as a whole.

Examples of the application of primary decomposition in the analysis of parallel manipulators are given in Sect. 5.

As a final comment, it has to be mentioned that it is possible to extend the concept of ideals and varieties to projective spaces. Due to the fact that lots of subtleties appear there, the reader is once more referred to the book "*Ideals, Varieties, and Algorithms*" by *David Cox, John Little* and *Donal O'Shea*.

5 Complete Kinematic Analysis of the TSAI Parallel Manipulator

Within this section, the algebraic methods derived in the previous sections are applied to the complete kinematic analysis of a parallel manipulator introduced by L. W. Tsai in 1996. It will be shown how the direct kinematics can be solved. Moreover, this manipulator exhibits different interesting operation modes, which can be detected using primary decomposition. The analysis continues with a singularity analysis of each operation mode and the discussion of all poses in which the manipulator can change from one operation mode into another.

5.1 Introduction

In Tsai (1996) designed three-DOF parallel manipulators with the aim to generate pure translational motion. As a result, he presented 3-UPU parallel manipulators where the axes of the U-joints have to fulfill some special conditions in order to allow the mechanism a translational motion. A generalization followed in 1998 by Di Gregorio and Parenti-Castelli by decomposing the U-joints into two simple R-joints without any condition of intersection or perpendicularity. There exist many publications about singularities of these mechanisms, especially because the SNU 3-UPU, which was built in about 2001 at the Seoul National University, showed unexpected motion in its home position, leading to many different attempts to explain this behavior, see, e.g., (Bonev and Zlatanov 2001), (Wolf et al. 2002), (Han et al. 2002). A complete analysis using algebraic methods can be found in Walter et al. (2009), where also the strange behavior of the mechanism at the home pose is explained. The crucial point in the design of the SNU 3-UPU is the fact that all axes in the base (resp. platform) intersect in a single point. The type of manipulator which will be discussed in detail in this section is the classic 3-UPU mechanism, where the axes of the U-joints are in a more general position. An analysis of the manipulator's translational kinematics was published by (Di Gregorio and Parenti-Castelli 1999) where also singular configurations are discussed. Concerning such configurations, there is also an interesting thesis by Chebbi (2011), where 3-UPU mechanisms are discussed in detail regarding singularities and strategies for selecting appropriate architectures (see also Chebbi and Parenti-Castelli 2010). The topic of workspace optimization is covered by several publications, e.g., (Parenti-Castelli et al. 2000), (Badescu et al. 2002), (Tsai and Joshi 2000) and also (Dwarakanath and Bhutani 2012) where the capabilities of the translational 3-UPU manipulator are tested with a high-precision prototype model. Another part of publications about 3-UPU mechanisms discusses generalizations and slightly exotic varieties of the manipulator, e.g., (Ji and Wu 2003) and (Varedi et al. 2009) where the U-joints are decomposed into R-joints with offsets, and (Hu and Lu 2011) where the first axes of the lower U-joints are perpendicular to the base.

The topic of this section is the classical fully symmetric TSAI 3-UPU mechanism, where the anchor points in the base (resp. platform) constitute an equilateral triangle, and the axes embedded in base and platform are tangent to the circumcircle of the corresponding triangle. Goal of this section is to analyze the motion capabilities of this manipulator by a purely algebraic approach, which was developed in the previous sections of this chapter in terms of a description by algebraic constraint equations. Due to the fact that the manipulator's translational motion was the main issue in most publications, the question about other types of motion that are allowed by this design arises immediately. The present description via algebraic equations gives a complete analysis about absolutely all poses the mechanism can attain and allows therefore a more general view. Some results of other authors appear again naturally as a part of this analysis, e.g., the singular poses of the translational operation mode.

5.2 *Manipulator Design*

First of all, an exact description of the manipulator's design is necessary. Figure 19 depicts a sketch of the manipulator including all necessary notations. Due to the fact that the TSAI 3-UPU can be obtained from the SNU 3-UPU by simply rotating each limb by 90° about the axis of the prismatic joint, the following is almost identical to the description of the SNU 3-UPU in Walter et al. (2009). In the base, there are three points $\mathbf{A}_1$, $\mathbf{A}_2$ and $\mathbf{A}_3$ forming an equilateral triangle with circumradius h_1. The frame Σ_0 is fixed in the base such that its origin lies in the circumcenter of the triangle, its yz-plane coincides with the plane containing the triangle, and its z-axis passes through $\mathbf{A}_3$. The same situation is established in the platform. There we have an equilateral triangle with vertices $\mathbf{B}_1$, $\mathbf{B}_2$, $\mathbf{B}_3$, circumradius h_2, and the platform frame is denoted by Σ_1. The parameters h_1 and h_2 are the first two design parameters, which are always different, if not mentioned explicitly. Each pair of corresponding points $\mathbf{A}_i$, $\mathbf{B}_i$ is connected by a limb L_i with U-joints at each end. The length of each limb is denoted by r_i and adjusted by a prismatic joint. The first and the fourth axes are embedded in the base resp. platform such that each of them is tangent to the corresponding circumcircle. Second and third axes of this link combination are parallel to each other and perpendicular to the axis of the limb and its first and fourth axes. With this setup, the conditions for translational motion as stated in Di Gregorio and Parenti-Castelli (1998) are fulfilled. All together there are five parameters to

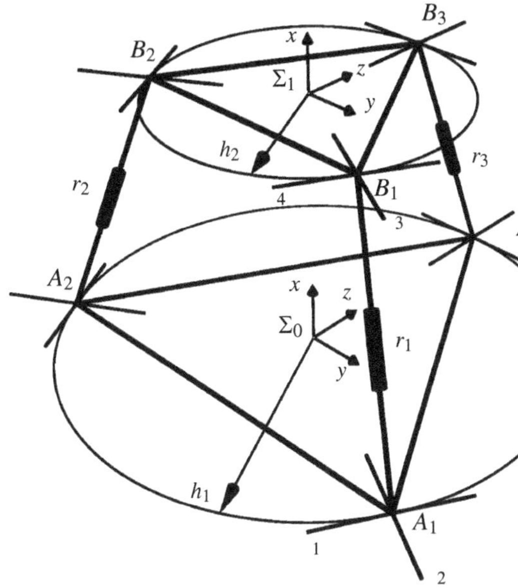

Fig. 19 TSAI 3-UPU manipulator

describe the TSAI 3-UPU mechanism: r_1, r_2, r_3, h_1 and h_2, where the first three of these are used to actuate the manipulator, the latter are fixed design parameters.

5.3 Deduction of Constraint Equations

In the following, algebraic equations are derived whose solutions describe the pose of Σ_1 with respect to Σ_0, and with it the pose of the platform with respect to the fixed base. For each limb L_i, the kinematic chain can be described by a sequence of transformations which was already explained in Eq. (40) on page 125. The pose of frame Σ_1 with respect to Σ_0 via limb L_i can be described by the matrix product

$$\mathbf{T}_i = \mathbf{F}_{i1}\mathbf{M}_{i1}\mathbf{G}_{i1}\mathbf{M}_{i2}\mathbf{G}_{i2}\mathbf{M}_{i3}\mathbf{G}_{i3}\mathbf{M}_{i4}\mathbf{F}_{i2}. \tag{57}$$

The matrices $\mathbf{F}_{ij}$ at the beginning and at the end are fixed transformations, only depending on the design parameters h_1 and h_2 and the location of the anchor points:

$$\mathbf{F}_{11} = \begin{pmatrix} 1 & 0 & 0 & 0 \\ 0 & 1 & 0 & 0 \\ \frac{\sqrt{3}}{2}h_1 & 0 & \frac{\sqrt{3}}{2} & \frac{1}{2} \\ -\frac{1}{2}h_1 & 0 & -\frac{1}{2} & \frac{\sqrt{3}}{2} \end{pmatrix}, \quad \mathbf{F}_{12} = \begin{pmatrix} 1 & 0 & 0 & 0 \\ 0 & 1 & 0 & 0 \\ -h_2 & 0 & \frac{\sqrt{3}}{2} & -\frac{1}{2} \\ 0 & 0 & \frac{1}{2} & \frac{\sqrt{3}}{2} \end{pmatrix},$$

$$\mathbf{F}_{21} = \begin{pmatrix} 1 & 0 & 0 & 0 \\ 0 & 1 & 0 & 0 \\ -\frac{\sqrt{3}}{2}h_1 & 0 & \frac{\sqrt{3}}{2} & -\frac{1}{2} \\ -\frac{1}{2}h_1 & 0 & \frac{1}{2} & \frac{\sqrt{3}}{2} \end{pmatrix}, \quad \mathbf{F}_{22} = \begin{pmatrix} 1 & 0 & 0 & 0 \\ 0 & 1 & 0 & 0 \\ h_2 & 0 & \frac{\sqrt{3}}{2} & \frac{1}{2} \\ 0 & 0 & -\frac{1}{2} & \frac{\sqrt{3}}{2} \end{pmatrix},$$

$$\mathbf{F}_{31} = \begin{pmatrix} 1 & 0 & 0 & 0 \\ 0 & 1 & 0 & 0 \\ 0 & 0 & 0 & -1 \\ h_1 & 0 & 1 & 0 \end{pmatrix}, \quad \mathbf{F}_{32} = \begin{pmatrix} 1 & 0 & 0 & 0 \\ 0 & 1 & 0 & 0 \\ -h_2 & 0 & 0 & 1 \\ 0 & 0 & -1 & 0 \end{pmatrix}.$$

Matrices $\mathbf{M}_{ij}$ are responsible for the passive rotations about the axes of the U-joints depending on rotation angles u_{ij}, and the matrices $\mathbf{G_{ij}}$ manage the transformations from one rotational axis to the next one:

$$\mathbf{M}_{ij} = \begin{pmatrix} 1 & 0 & 0 & 0 \\ 0 & \cos(u_{ij}) & -\sin(u_{ij}) & 0 \\ 0 & \sin(u_{ij}) & \cos(u_{ij}) & 0 \\ 0 & 0 & 0 & 1 \end{pmatrix}, \quad \mathbf{G}_{ij} = \begin{pmatrix} 1 & 0 & 0 & 0 \\ a_j & 1 & 0 & 0 \\ 0 & 0 & \cos(\alpha_j) & -\sin(\alpha_j) \\ d_j & 0 & \sin(\alpha_j) & \cos(\alpha_j) \end{pmatrix}.$$

The corresponding Denavit–Hartenberg parameters are given in Table 2.

Due to the special arrangement of the axes in each limb, no separate transformation matrix was introduced for the active prismatic joints, and their parameter r_i appears in the matrix $\mathbf{G}_{i2}$. A clear and brief introduction of the Denavit–Hartenberg convention and the systematic deduction of the forward transformation matrices $\mathbf{T}_i$ can be found, e.g., in Tsai (1999), Husty et al. (1997) and Pfurner (2006).

In the next step, half-tangent substitutions for all u_{ij} are performed to get rid of the trigonometric functions:

$$\cos(u_{ij}) = \frac{1 - t_{ij}^2}{1 + t_{ij}^2}, \quad \sin(u_{ij}) = \frac{2t_{ij}}{1 + t_{ij}^2} \tag{58}$$

This results in displacement matrices $\overline{\mathbf{T}}_i$ containing new parameters t_{ij} for the rotation angles of the passive joints. Then from each matrix $\overline{\mathbf{T}}_i$, the Study parameters are computed using Eqs. (7) and (8), leading to expressions

Table 2 Denavit–Hartenberg parameters for limb L_i

j	a_j	d_j	α_j
1	0	0	$\pi/2$
2	r_i	0	0
3	0	0	$-\pi/2$

$$
\begin{aligned}
x_0 &= f_{i1}(t_{i1}, t_{i2}, t_{i3}, t_{i4}) \\
&\vdots \\
y_3 &= f_{8i}(t_{i1}, t_{i2}, t_{i3}, t_{i4}) \quad i = 1, \ldots, 3.
\end{aligned} \tag{59}
$$

From each matrix $\overline{\mathbf{T}}_i$, we obtain a parameterization of a subset of S_6^2 depending on four parameters. This subset essentially describes the motion capability of one limb with fixed active prismatic joint. The final step is then to eliminate from each parameterization the corresponding four joint parameters $t_{i1}, t_{i2}, t_{i3}, t_{i4}$ to obtain algebraic equations which contain only the Study parameters and, of course, the design parameters h_1 and h_2 and the active motion parameters r_1, r_2, r_3. This can be achieved easily by using the linear implicitization algorithm (LIA).

To check the result, another method was used to deduce these equations. In contrast to the systematic way from before, this method is based on geometric insight and for more complex link combinations, it is quite likely that it does not work. Aside from Eq. (5), two other equations have to be deduced for each leg. As it can be seen in Fig. 19, the distance of point $\mathbf{A}_i$ and $\mathbf{B}_i$ always remains constant r_i. By describing both points with respect to Σ_0 using matrices Eqs. (38) and (39) and transformations $\mathbf{F}_{ij}$, the Euclidean distance $|\mathbf{A}_i\mathbf{B}_i|^2$ can be computed and equated with r_i^2. After removal of denominators, an equation of degree four in $x_0, x_1, x_2, x_3, y_0, y_1, y_2, y_3$ remains. After applying a trick which was used by Husty (1996), the degree can be lowered to two. The trick is to add a multiple of the square of Eq. (5) to obtain an expression with the factor Eq. (6), which can immediately be removed because it will never vanish. All equations generated this way appear also in the result of the previous method. Note that these equations were also derived in Example 11 Eq. (53) on page 131. The other condition is that the first and last axes in each leg are constrained to stay coplanar. By using points $\mathbf{A}_i$ and $\mathbf{B}_i$ from above and another two points lying on the axes, a 4×4 matrix and its determinant can be computed. This determinant is the condition for coplanarity and has again degree four. It is possible to lower the degree to two by subtracting a proper multiple of Eq. (5) and removing the emerging factor Eq. (6). The resulting equations are again identical to those derived by the LIA. After some simplifications of all seven equations to remove $\sqrt{3}$ from at least some equations, the following equations are obtained:

$$
\begin{aligned}
g_1 :& x_0\, y_0 + x_1\, y_1 + x_2\, y_2 + x_3\, y_3 = 0 \\
g_2 :& (h_1 - h_2)\, x_0\, x_2 + (h_1 + h_2)\, x_1\, x_3 - x_2\, y_3 - x_3\, y_2 = 0 \\
g_3 :& (h_1 - h_2)\, x_0\, x_3 - (h_1 + h_2)\, x_1\, x_2 - 4\, x_1\, y_1 - 3\, x_2\, y_2 - x_3\, y_3 = 0 \\
g_4 :& (h_1 - h_2)\, x_0\, x_3 - (h_1 + h_2)\, x_1\, x_2 + 2\, x_1\, y_1 + 2\, x_3\, y_3 = 0 \\
g_5 :& (h_1^2 - 2\, h_1\, h_2 + h_2^2 - r_1^2)\, x_0^2 + 2\sqrt{3}\,(h_1 - h_2)\, x_0\, y_2 - 2\,(h_1 - h_2)\, x_0\, y_3 + \\
& (h_1^2 + 2\, h_1\, h_2 + h_2^2 - r_1^2)\, x_1^2 - 2\,(h_1 + h_2)\, x_1\, y_2 - 2\sqrt{3}\,(h_1 + h_2)\, x_1\, y_3 + \\
& (h_1^2 - h_1\, h_2 + h_2^2 - r_1^2)\, x_2^2 + 2\sqrt{3}\, h_1\, h_2\, x_2\, x_3 - 2\sqrt{3}\,(h_1 - h_2)\, x_2\, y_0 + \\
& 2\,(h_1 + h_2)\, x_2\, y_1 + (h_1^2 + h_1\, h_2 + h_2^2 - r_1^2)\, x_3^2 + 2\,(h_1 - h_2)\, x_3\, y_0 + \\
& 2\sqrt{3}\,(h_1 + h_2)\, x_3\, y_1 + 4\,(y_0^2 + y_1^2 + y_2^2 + y_3^2) = 0
\end{aligned} \tag{60}
$$

$$
\begin{aligned}
g_6 :& (h_1^2 - 2\,h_1\,h_2 + h_2^2 - r_2^2)\,x_0^2 - 2\sqrt{3}\,(h_1 - h_2)\,x_0\,y_2 - 2\,(h_1 - h_2)\,x_0\,y_3 + \\
& (h_1^2 + 2\,h_1\,h_2 + h_2^2 - r_2^2)\,x_1^2 - 2\,(h_1 + h_2)\,x_1\,y_2 + 2\sqrt{3}\,(h_1 + h_2)\,x_1\,y_3 + \\
& (h_1^2 - h_1\,h_2 + h_2^2 - r_2^2)\,x_2^2 - 2\sqrt{3}\,h_1\,h_2\,x_2\,x_3 + 2\sqrt{3}\,(h_1 - h_2)\,x_2\,y_0 + \\
& 2\,(h_1 + h_2)\,x_2\,y_1 + (h_1^2 + h_1\,h_2 + h_2^2 - r_2^2)\,x_3^2 + 2\,(h_1 - h_2)\,x_3\,y_0 - \\
& 2\sqrt{3}\,(h_1 + h_2)\,x_3\,y_1 + 4\,(y_0^2 + y_1^2 + y_2^2 + y_3^2) = 0 \\
g_7 :& (h_1^2 - 2\,h_1\,h_2 + h_2^2 - r_3^2)\,x_0^2 + 4\,(h_1 - h_2)\,x_0\,y_3 + (h_1^2 + 2\,h_1\,h_2 + h_2^2 - r_3^2)\,x_1^2 + \\
& 4\,(h_1 + h_2)\,x_1\,y_2 + (h_1^2 + 2\,h_1\,h_2 + h_2^2 - r_3^2)\,x_2^2 - 4\,(h_1 + h_2)\,x_2\,y_1 + \\
& (h_1^2 - 2\,h_1\,h_2 + h_2^2 - r_3^2)\,x_3^2 - 4\,(h_1 - h_2)\,x_3\,y_0 + 4\,(y_0^2 + y_1^2 + y_2^2 + y_3^2) = 0
\end{aligned}
$$

This system of algebraic equations describes the mechanism. Fixing the motion parameters r_i, one can ask for all projective points in $\mathbb{P}^7$ which fulfill all these seven equations, under the condition that $x_0^2 + x_1^2 + x_2^2 + x_3^2 \neq 0$. These points represent then all possible poses of the platform (assemblies of the manipulator) which are the solution of the direct kinematics of the TSAI 3-UPU manipulator for given fixed inputs. Because it is more convenient to do all computations in affine space, without loss of generality, the following normalization equation is added:

$$
g_8 :\ x_0^2 + x_1^2 + x_2^2 + x_3^2 - 1 = 0 \tag{61}
$$

Now it is guaranteed that no solution of this final system lies in the forbidden variety described by $x_0^2 + x_1^2 + x_2^2 + x_3^2 = 0$ (see Eq. 6). The downside of the normalization is that for each projective solution point, two affine representatives as solutions for Eqs. (60)–(61) are obtained. This has to be taken in account when different solutions are counted.

5.4 *Solving the System*

In the previous subsection, a set of eight algebraic equations was deduced whose solutions describe all poses of the TSAI 3-UPU mechanism, depending on fixed parameters h_1 and h_2 and of course the active joint parameters r_1, r_2 and r_3. In the following, this system of equations is always written as a polynomial ideal in the variables $x_0, x_1, x_2, x_3, y_0, y_1, y_2, y_3$ over the coefficient ring $\mathbb{C}[h_1, h_2, r_1, r_2, r_3]$. The ideal that has to be dealt with is

$$
\mathscr{I} = \langle g_1, g_2, g_3, g_4, g_5, g_6, g_7, g_8 \rangle,
$$

where each g_i here denotes the polynomial on the left-hand side of the corresponding constraint equation. To begin with the following ideal $\mathscr{J}$ is inspected, which is obviously independent of the joint parameters r_1, r_2 and r_3

$$\mathscr{J} = \langle g_1, g_2, g_3, g_4 \rangle.$$

Although that ideal is not that complicated, it is checked if it is possible to split $\mathscr{J}$ into smaller ideals such that $\mathscr{J}$ can be written as

$$\mathscr{J} = \bigcap_i \mathscr{J}_i.$$

The benefit of such a decomposition would be that the vanishing set $\mathscr{V}(J)$ of $\mathscr{J}$ would be the union of the vanishing sets (i.e., varieties) of the ideals $\mathscr{J}_i$. As a simple example, one could imagine a polynomial equation which can be written as product of several factors, i.e., the variety of the equation is a union of simpler varieties. This concept can be extended to arbitrary ideals and is called *primary (prime) decomposition*, where the difference between *prime* and *primary* lies only in the way how the resulting ideals are returned. *Prime decomposition* is the more strict version and returns all ideals reduced down to the bare essentials (the results are *radicals*). Relating to the example mentioned above, *primary decomposition* would return the factors with possible powers, whereas *prime decomposition* only the factors with multiplicity one (see Sect. 4).

The computation of the prime decomposition of $\mathscr{J}$ shows that it can indeed be written in a very simple way:

$$\mathscr{J} = \bigcap_{i=1}^{6} \mathscr{J}_i$$

with

$$\mathscr{J}_1 = \langle y_0, x_1, x_2, x_3 \rangle, \quad \mathscr{J}_2 = \langle x_0, y_1, x_2, x_3 \rangle,$$

$$\mathscr{J}_3 = \langle y_0, y_1, x_2, x_3 \rangle, \quad \mathscr{J}_4 = \langle x_0, x_1, y_2, y_3 \rangle,$$

$$\begin{aligned}\mathscr{J}_5 = \langle &(h_1 - h_2)\, x_0\, x_2 + (h_1 + h_2)\, x_1\, x_3 - x_2\, y_3 - x_3\, y_2, (h_1 - h_2)\, x_0\, x_3 - \\ &(h_1 + h_2)\, x_1\, x_2 - x_2\, y_2 + x_3\, y_3, 2\, x_1\, y_1 + x_2\, y_2 + x_3\, y_3, x_0\, y_0 - x_1\, y_1, \\ &(h_1 - h_2)^2\, x_0^2 + (h_1 + h_2)^2\, x_1^2 - y_2^2 - y_3^2, (h_1 + h_2)\, x_2^3\, y_0 - 3\,(h_1 - h_2)\, x_2^2\, x_3\, y_1 - \\ &2\, x_2^2\, y_0\, y_1 - 3\,(h_1 - h_2)\, x_2\, x_3^2\, y_0 + (h_1 - h_2)\, x_3^3\, y_1 - 2\, x_3^2\, y_0\, y_1 \rangle\end{aligned}$$

$$\mathscr{J}_6 = \langle x_0, x_1, x_2, x_3 \rangle.$$

This primary decomposition was computed over the ring $\mathbb{Q}[x_0, x_1, x_2, x_3, y_0, y_1, y_2, y_3, h_1, h_2]$ to find possible other decompositions or changes of dimension, which could occur for special values of h_1 and h_2. Fortunately, over this ring ideal $\mathscr{J}_5$ does not split and has always the same dimension, i.e., dimension 3. Although ideal $\mathscr{J}_6$ is a valid component of $\mathscr{J}$ it can be removed, because in connection with Eq. (6)

it leads to an inconsistent system. For the zero set or vanishing set $\mathscr{V}(J)$ of $\mathscr{J}$, it follows that

$$\mathscr{V}(J) = \bigcup_{i=1}^{5} \mathscr{V}(\mathscr{J}_i).$$

Now the remaining equations are added and by generating the subsystems

$$\mathscr{K}_i := \mathscr{J}_i \cup \langle g_5, g_6, g_7, g_8 \rangle$$

the vanishing set of the whole system $\mathscr{I}$ can be written as

$$\mathscr{V}(I) = \bigcup_{i=1}^{5} \mathscr{V}(\mathscr{K}_i).$$

So, instead of studying the system as a whole, each of the smaller subsystems $\mathscr{K}_i$ can be inspected separately. Then the solution of system $\mathscr{I}$ is the union of the solutions of the subsystems.

Furthermore, it has to be noted that for each assembly mode described by a solution of direct kinematics, there exists another solution where the manipulator is assembled mirrored with respect to the base.

Solutions for arbitrary design parameters In the following subsection, all computations for direct kinematics are performed under the assumption that the five parameters are arbitrary, i.e., generic. To find out the Hilbert dimension of each ideal $\mathscr{K}_i$, the necessary Groebner bases are computed for general parameters, except the basis for $\mathscr{K}_5$, where for reasons of simplification randomly chosen parameters h_1, h_2 are substituted before the computation of the basis. So, for arbitrary design parameters, the result is that

$$dim(\mathscr{K}_i) = 0, \ \ i = 1, \ldots, 5$$

which means that all subsystems have finitely many solutions. Reusing the computed bases from above the number of solutions can be determined easily for each system $\mathscr{K}_i$. Due to the fact that always two solutions of a system describe the same position of the platform, each of these numbers has to be halved (see paragraph below Eq. 61). In the following, only these essentially different solutions are considered. The number of solutions for each system $\mathscr{K}_i$ in the generic case is listed in Table 3. So all together there are 78 essentially different solutions for given generic parameters h_1, h_2 and joint parameters r_1, r_2, and r_3, i.e., 78 possible poses of the platform, theoretically. It is clear that for arbitrarily chosen parameters, all these solutions will be complex. Mechanically, this means that the manipulator cannot be assembled because of, e.g., too different limb lengths. But of course, when design and joint parameters are chosen thoughtfully, many of the solution poses also can be real. And, as already mentioned, if there are real solutions at all, they appear in pairs where one pose

Table 3 Essentially different solutions for each ideal $\mathcal{K}_1, \ldots, \mathcal{K}_5$

$\mathcal{K}_1$	$\mathcal{K}_2$	$\mathcal{K}_3$	$\mathcal{K}_4$	$\mathcal{K}_5$	Total
2	2	4	6	64	78

Table 4 Essentially different solutions, limb lengths are equal

$\mathcal{K}_1$	$\mathcal{K}_2$	$\mathcal{K}_3$	$\mathcal{K}_4$	$\mathcal{K}_5$	Total
2	2	2	6	60	72

can be assembled "upward," the other one downward. In Sect. 6, design parameters will be given for almost all $\mathcal{K}_i$ such that all solutions of the corresponding system are real. Due to the fact that systems $\mathcal{K}_1, \mathcal{K}_2, \mathcal{K}_3$ allow a closed-form solution, it can be shown that it is not possible that all systems can reach their maximum of real solutions simultaneously. To get an idea what the total number of real solutions for the manipulator can be, some attempts were made using reasonably chosen parameters. Surprisingly, this number never exceeded 28 in our computations. According to private correspondence with Damien Chablat, he found a set of parameters where the number of essentially different solutions is 32. But it is absolutely not clear if this is the upper bound for the number of real solutions.

Solutions for equal limb lengths Here it is assumed that all limbs are of equal length. Because of the structure of the equations, this is a nongeneric case and has to be treated separately.

$$r_1 := r \quad r_2 := r \quad r_3 := r.$$

Now the same computations can be performed which were done in the previous subsection to obtain the Hilbert dimension of each ideal. Due to the fact that there are less parameters, all Groebner bases can be computed without specifying any parameters. For the dimension, the result is the same as in the previous case.

$$dim(\mathcal{K}_i) = 0, \;\; i = 1, \ldots, 5.$$

The number of solutions is computed for each system and halved afterward. The results are listed in Table 4.

Here there exist theoretically 72 solutions for the platform's position, six less than before. Concerning the question where they could have gone, one should not forget that there is this forbidden subvariety on the Study quadric. It is possible that for special design parameters, a solution lies on this subvariety. In the previous paragraph, it was mentioned that for special design parameters 28 real solutions can be obtained. Actually, it was a set of parameters with equal limb lengths. The exact values for the parameters were

$$h_1 = 12,\ h_2 = 7,\ r_1 = r_2 = r_3 = \frac{181}{13} \approx 13.923.$$

A very important difference to the SNU 3-UPU manipulator which was discussed in Walter et al. (2009) is the fact that here in general all 72 solutions have multiplicity 1, i.e., particularly the solution which corresponds to the so-called home position has multiplicity 1, and not 4, which is the case for the SNU 3-UPU. One could conclude that the TSAI 3-UPU should show a better behavior in the home position, not being that shaky as the SNU 3-UPU. According to the results from Dwarakanath and Bhutani (2012), this is absolutely the case provided that the mechanism is built with minimal clearances in all joints. The home pose can be seen in Fig. 19 and is described by the following values

$$x_0 = 1,\ \ x_1 = 0,\ \ x_2 = 0,\ \ x_3 = 0,$$

$$y_0 = 0,\ \ y_1 = \sqrt{r^2 - (h_1 - h_2)^2}/2,\ \ y_2 = 0,\ \ y_3 = 0.$$

6 The Manipulator's Operation Modes

Until now r_1, r_2 and r_3 were treated as fixed design parameters for the direct kinematics. In this section, they will be treated as parameters which are allowed to change, i.e., the behavior of this mechanism will be studied, when the prismatic joints are actuated. Computation of the Hilbert dimension of each ideal $\mathscr{K}_i$ with r_1, r_2, r_3 used as unknowns shows that

$$\overline{dim}(\mathscr{K}_i) = 3,\ \ i = 1, \dots, 5$$

where $\overline{dim}$ denotes the dimension over $\mathbb{C}[h_1, h_2]$, in contrast to dim which denotes the dimension over $\mathbb{C}[h_1, h_2, r_1, r_2, r_3]$ as in the previous sections. It follows that for generic design parameters, the 3-UPU manipulator has three-DOFs.

As it was shown in Walter et al. (2009) for the SNU 3-UPU manipulator, each subsystem $\mathscr{K}_i$ of a mechanism's set of equations corresponds to a specific operation mode of the manipulator. In the following each system, $\mathscr{K}_i$ will be discussed separately. Possible singularities which can emerge either as singular poses of a mode itself, or as poses where a change of modes is feasible will be discussed in the next sections.

The following algorithm is applied in the next paragraphs: Each system $\mathscr{J}_i$ is solved, and the solution is substituted into the transformation matrix **M** Eq. (3). From the obtained results, properties of the solutions of the subsystems $\mathscr{K}_i$ can be deduced and from these properties follow properties of the motion of the platform. It is absolutely not necessary to use equations $g_5 - g_7$ from Eq. (60) for this inspection, because they describe only the limb lengths which are now free parameters. Equation (61) is used to simplify the matrix entries, if possible, i.e., if three unknowns of Eq. (61) are 0, the remaining unknown is set to 1. Furthermore, the position of the platform is described by a series of simpler transformations, starting from the "planar home position."

System $\mathcal{K}_1$: Translational Mode

$$\{y_0 = 0,\ x_1 = 0,\ x_2 = 0,\ x_3 = 0\}$$

$$\mathbf{M} = \begin{pmatrix} 1 & 0 & 0 & 0 \\ -2\,y_1 & 1 & 0 & 0 \\ -2\,y_2 & 0 & 1 & 0 \\ -2\,y_3 & 0 & 0 & 1 \end{pmatrix}.$$

This is the simple translational operation mode which was discussed in many papers about this manipulator. The transformation matrix was simplified by substituting $x_0 = 1$. From the matrix, it can easily be seen that the translational motion can be parameterized using y_1, y_2, y_3. If r_1, r_2, r_3 are given, this mode leads to two assembly modes (see Table 3) and both are real, e.g., for the parameters

$$h_1 = 12,\ h_2 = 7,\ r_1 = \frac{670}{21},\ r_2 = \frac{243}{7},\ r_3 = \frac{611}{21}.$$

System $\mathcal{K}_2$: Twisted Translational Mode

$$\{x_0 = 0,\ y_1 = 0,\ x_2 = 0,\ x_3 = 0\}$$

$$\mathbf{M} = \begin{pmatrix} 1 & 0 & 0 & 0 \\ 2\,y_0 & 1 & 0 & 0 \\ 2\,y_3 & 0 & -1 & 0 \\ -2\,y_2 & 0 & 0 & -1 \end{pmatrix}.$$

This is a mode which was already mentioned in Chebbi and Parenti-Castelli (2010), where singularities of different geometries of the 3-UPU translational manipulator are investigated. The twisted translational mode appears in this paper as geometry B.1. Here $x_1 = 1$ was used to simplify the result. Each solution of system $\mathcal{K}_2$ corresponds to a rotation of the platform about its normal axis N by 180° and a subsequent translation. It follows that the described operation mode is basically a pure translation. To parameterize this motion, one could use y_0, y_2, y_3 as parameters. All solutions for direct kinematics are real for, e.g., the parameters

$$h_1 = 12,\ h_2 = 7,\ r_1 = \frac{121}{3},\ r_2 = \frac{965}{21},\ r_3 = \frac{62}{3}.$$

System $\mathcal{K}_3$: Planar Mode

$$\{y_0 = 0,\ y_1 = 0,\ x_2 = 0,\ x_3 = 0\}$$

$$\mathbf{M} = \begin{pmatrix} 1 & 0 & 0 & 0 \\ 0 & 1 & 0 & 0 \\ 2\,(-x_0\,y_2 + x_1\,y_3) & 0 & x_0^2 - x_1^2 & -2\,x_0\,x_1 \\ 2\,(-x_0\,y_3 - x_1\,y_2) & 0 & 2\,x_0\,x_1 & x_0^2 - x_1^2 \end{pmatrix}.$$

Solutions of $\mathscr{K}_3$ correspond to poses of the platform where it is coplanar to the base. To parameterize this planar operation mode, one could use x_1, y_2, y_3 in connection with $x_0^2 + x_1^2 = 1$, where x_1 is responsible for the rotation of the platform about its normal axis N and y_2, y_3 for the translation in the base plane, i.e free planar motion is available in this mode. Concerning assembly modes belonging to this mode, all four solutions of direct kinematics are real, for e.g., the parameters

$$h_1 = 12,\ h_2 = 7,\ r_1 = \frac{125}{7},\ r_2 = \frac{493}{21},\ r_3 = \frac{66}{7}.$$

System $\mathscr{K}_4$: Upside-Down Planar Mode

$$\{x_0 = 0,\ x_1 = 0,\ y_2 = 0,\ y_3 = 0\}$$

$$\mathbf{M} = \begin{pmatrix} 1 & 0 & 0 & 0 \\ 0 & -1 & 0 & 0 \\ 2\,(x_2\,y_0 - x_3\,y_1) & 0 & x_2^2 - x_3^2 & 2\,x_2\,x_3 \\ 2\,(x_2\,y_1 + x_3\,y_0) & 0 & 2\,x_2\,x_3 & -x_2^2 + x_3^2 \end{pmatrix}.$$

Solutions of $\mathscr{K}_4$ correspond to positions of the platform where it is turned upside down and coplanar to the base. This can be achieved by starting in the planar home position with disconnected legs, turning the platform about its y-axis by 180°, and attaching the legs again. To parameterize the upside-down planar operation mode, one could use x_3, y_0, y_1 in connection with $x_2^2 + x_3^2 = 1$, where x_3 is responsible for the rotation of the platform about its normal axis N and y_0, y_1 for the translation in the base plane. It was possible to find parameters where all six solutions of direct kinematics are real, which are, e.g.,

$$h_1 = 12,\ h_2 = 7,\ r_1 = \frac{125}{7},\ r_2 = \frac{493}{21},\ r_3 = \frac{66}{7}.$$

System $\mathscr{K}_5$: General Mode

This mode is the most difficult one, because of the complexity of the equations in ideal $\mathscr{J}_5$. What definitely can be said is that it has in general for given limb length 64 solutions and that the system has 28 real solutions if the parameters are

$$h_1 = 12,\ h_2 = 7,\ r_1 = r_2 = r_3 = \frac{181}{13}.$$

Although the equations of $\mathcal{J}_5$ can be solved linearly for x_0, x_1,y_2 and y_3, due to the complexity of the equations there is no possibility to find a neat description of this mode, neither the condition for singularities nor a description of them.

Summing up there are five essentially different operation modes of the TSAI 3-UPU manipulator. It is quite obvious that this mechanism is more complex than the SNU 3-UPU whose essentially seven operation modes were quite simple.

7 Conditions for Singular Poses

In this section, singular poses of the TSAI 3-UPU manipulator are discussed. A singular pose is a pose where the Jacobian matrix of Eqs. (60)–(61) with respect to the unknowns $x_0, x_1, x_2, x_3, y_0, y_1, y_2, y_3$ has vanishing determinant. If this determinant is computed, a complete description of all possible forward singular poses is available. Because the manipulator is actuated via prismatic joints, there are no input singularities in the kinematic chains of the legs. Due to the fact that the manipulator has more than one operation mode, singular poses can be divided into two different kinds. Either such a pose belongs to a single operation mode, or it is a pose which becomes singular because it lies in more than one operation mode. The latter is then a pose where the manipulator can change its operation mode. Seen from the point of ideals, the ideal $\mathcal{I}$ describes a three-dimensional variety in $\mathbb{C}[h_1, h_2][x_0, x_1, x_2, x_3, y_0, y_1, y_2, y_3, r_1, r_2, r_3]$, which is composed of five simpler varieties. Singular poses lie now either on one of these components or in an intersection, which can be at most two dimensional.

In this subsection, necessary conditions on the lengths r_1, r_2, r_2 are deduced for obtaining singular poses of the first kind. Because these poses always correspond to a single operation mode, it is possible to use ideals $\mathcal{K}_i$ instead of $\mathcal{I}$. For almost all ideals, the procedure is the same:

The 8×8 Jacobian matrix and its determinant are computed. This equation is added to the ideal $\mathcal{K}_i$, and from the resulting ideal, all variables $x_0, x_1, x_2, x_3, y_0, y_1, y_2, y_3$ are eliminated. The result is an equation in $\mathbb{Q}[h_1, h_2][r_1, r_2, r_3]$ which is the necessary condition that the direct kinematics of the manipulator leads to a singular pose in this mode. The ideal for the *General Mode* needs a special treatment because it is too complex for the standard elimination methods.

Due to the fact that leg lengths r_1, r_2, r_2 appear squared in the equations, the result of the elimination describes a hypersurface in the space of these parameters, which has full octahedral symmetry. The surface looks the same in all octants, and therefore, only the first octant is shown in the following figures.

Singular poses: Translational Mode The condition is an equation of degree 4 in r_1, r_2, r_3 and looks as follows:

$$r_1^4 + r_2^4 + r_3^4 - r_1^2 r_2^2 - r_1^2 r_3^2 - r_2^2 r_3^2 - 3\,(h_1 - h_2)^2\,(r_1^2 + r_2^2 + r_3^2) + 9\,(h_1 - h_2)^4 = 0. \tag{62}$$

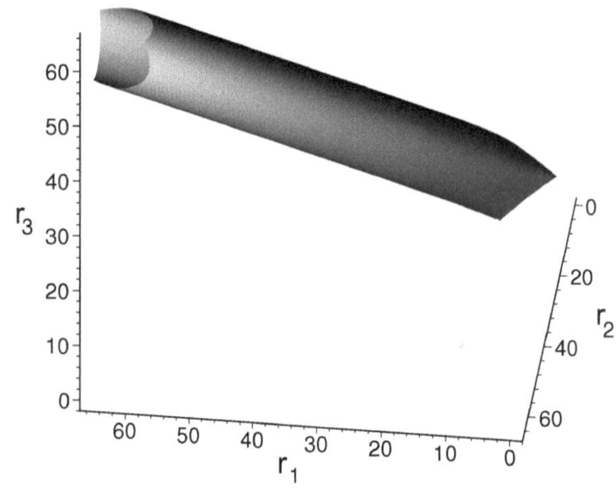

Fig. 20 The singularity surface of the translational mode

Further inspection shows that for given values for h_1 and h_2, Eq. (62) describes a Kummer surface in the joint parameter space r_1, r_2, r_3. The Kummer surface is a very famous algebraic degree four surface, possessing the maximum number of 16 double points (see, e.g., Hudson 1905). If the joint parameters r_1, r_2, r_3 are chosen inside resp. outside the "pipe" (Fig. 20), both solutions of the direct kinematics are real resp. complex. On the other hand, if the parameters of the limbs are chosen such that the condition Eq. (62) is fulfilled, it is not difficult to show that both solutions coincide and the corresponding positions are described by

$$\mathbf{M} = \begin{pmatrix} 1 & 0 & 0 & 0 \\ 0 & 1 & 0 & 0 \\ p_2 & 0 & 1 & 0 \\ p_3 & 0 & 0 & 1 \end{pmatrix},$$

i.e., the singular positions of the translational mode are those, where base and platform lie in the same plane (Fig. 21).

Singular poses: Twisted Translational Mode The necessary condition for singular positions:

$$r_1^4 + r_2^4 + r_3^4 - r_1^2 r_2^2 - r_1^2 r_3^2 - r_2^2 r_3^2 - 3\,(h_1 + h_2)^2\,(r_1^2 + r_2^2 + r_3^2) + 9\,(h_1 + h_2)^4 = 0 \tag{63}$$

It is again a Kummer surface with the difference that the "pipe" (Fig. 22) has a larger diameter. Again, if the limb parameters are chosen inside resp. outside the "pipe," both solutions of the direct kinematics are real resp. complex. If the condition Eq. (63) is fulfilled, both solutions coincide and the corresponding positions are described by

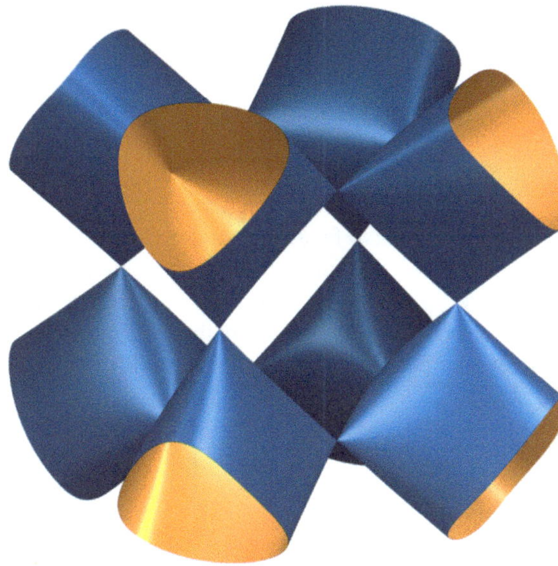

Fig. 21 Kummer singularity surface in all eight octants

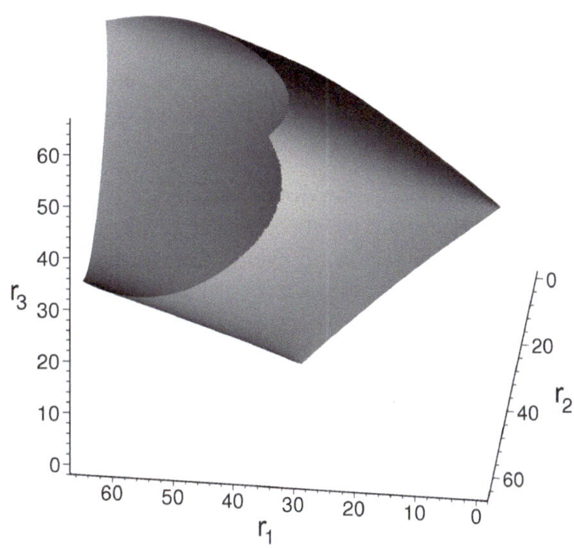

Fig. 22 The singularity surface of the twisted translational mode

$$\mathbf{M} = \begin{pmatrix} 1 & 0 & 0 & 0 \\ 0 & 1 & 0 & 0 \\ p_2 & 0 & -1 & 0 \\ p_3 & 0 & 0 & -1 \end{pmatrix},$$

i.e., the singular positions of the twisted translational mode are those, where base and platform lie in the same plane and the platform is rotated by 180° in this plane.

Singular poses: Planar Mode The condition is an equation of degree 12 which can be factorized into

$$F_1\, F_2\,(r_1 + r_2 - r_3)\,(r_1 + r_3 - r_2)\,(r_2 + r_3 - r_1)\,F_3 = 0 \tag{64}$$

where F_1 and F_2 are exactly the singularity conditions from the previous modes and F_3 is the factor $r_1 + r_2 + r_3$ which does never vanish. The union of the corresponding varieties of these five factors separates the joint parameter space into several zones where zero, two, or four real solutions are obtained. Due to the fact that the condition is a product of essentially five factors, there are more than one representatives for a singular position. First of all, both matrices from the previous modes, describing their singular positions belong to this condition. Furthermore, from the small factors one obtains

$$\mathbf{M} = \begin{pmatrix} 1 & 0 & 0 & 0 \\ 0 & 1 & 0 & 0 \\ p_2 & 0 & v_{33} & v_{32} \\ p_3 & 0 & -v_{32} & v_{33} \end{pmatrix}$$

with

$$v_{32}^2 + v_{33}^2 = 1, \quad p_2^2 + p_3^2 = h_1^2 + h_2^2 - 2\,h_1\,h_2\,v_{33},$$

i.e., among others a singular position is obtained when the origin of the platform frame lies on a circle in the base plane whose radius is determined by the rotation of the frame about its x-axis.

Singular poses: Upside-down Planar Mode Computation of the singularity condition was rather hard in this case. The result is a very lengthy equation of degree 24 which cannot be factorized over $\mathbb{Q}$. Due to space limitations, it is not displayed here. Several plots of it were made for given h_1, h_2. The variety is again tube-like but it has also self-intersections which again separate the parameter space in different zones. It is not clear how the singular poses arising from this condition can be described in a short and concise way (Fig. 23).

Singular poses: General Mode Although the equations of $\mathscr{J}_5$ can be solved linearly for x_0, x_1,y_2 and y_3, due to the complexity of the equations, there is no possibility to find a neat description of this mode. Surprisingly, the condition for singularities could be computed for given values of h_1 and h_2, but in a very different way. First of all, the two design parameters are set to

$$h_1 = 12,\ h_2 = 7.$$

The ideal describing all singular poses of this mode is computed in an indirect way. The reason for this is the fact that the number of describing equations is not eight as in the previous cases. To begin with, the Jacobian determinant is derived from Eqs. (60)–(61) and it is added to the ideal $\mathscr{I}$. This ideal has dimension 2 and describes on one hand all singular poses of all modes and on the other hand all singular poses arising

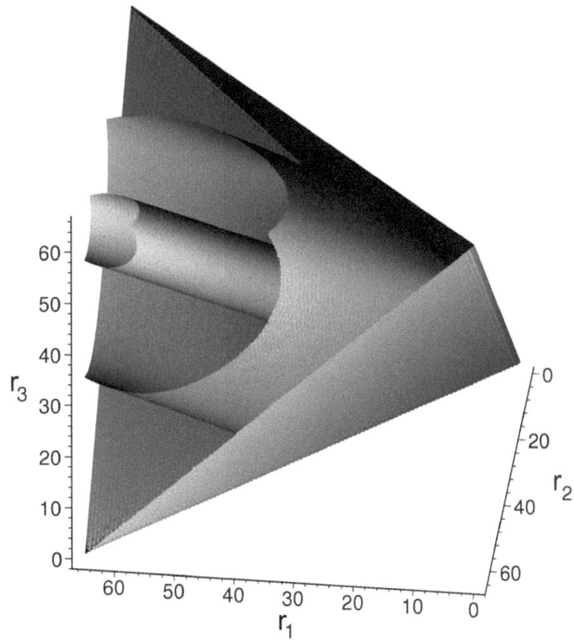

Fig. 23 The singularity surface of the planar mode. It contains the surfaces from Figs. 20 and 22

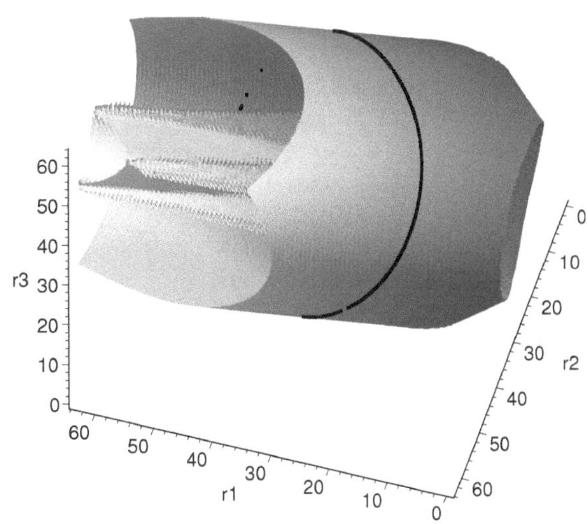

Fig. 24 The singularity surface of the upside-down planar mode

of intersections of components. But especially it is again a composition of various varieties. All components of this variety which are not of interest are described by ideals which contain an ideal $\mathscr{J}_i$ with $i = 1, \ldots, 4$. To obtain the desired component, all superfluous components have to be removed. Essentially, this can be achieved by the technique of *saturation* with respect to each of the polynomials x_0, x_1, x_2, x_3, y_0 and y_1, which are the generating elements of ideals $\mathscr{J}_1, \ldots, \mathscr{J}_4$. By inspection

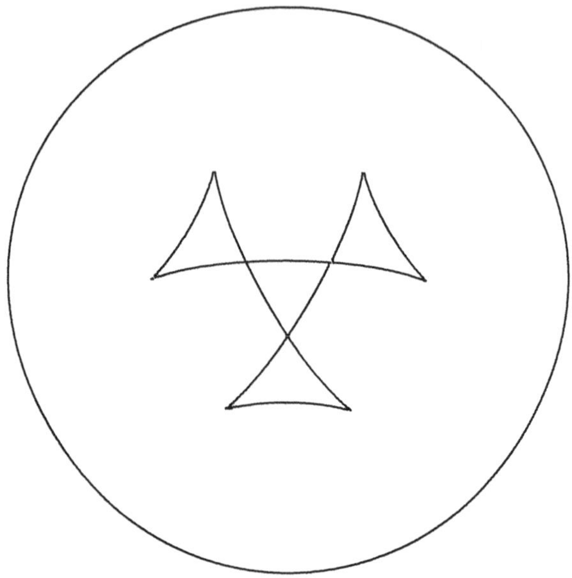

Fig. 25 Intersection curve of the "pipe" of Fig. 24

of the intersection of the variety with each of these hyperplanes, it is checked that the only two-dimensional components where one of these variables is zero are the superfluous ones. Saturation of an ideal with respect to a polynomial (which can be seen as a principal ideal) is nothing else then the computation of their quotient for multiple times until the result does not change anymore. A possible analogy would be the removal of a power of a prime number from a composite number by repeated division. The result is an ideal $\mathscr{L}$ which describes all singular poses of the general mode.

The tricky part is the elimination of the Study parameters. Due to the fact that all present methods and algorithms fail for reasons of memory and time, another way has to be found. First of all, randomly chosen parameters are substituted for r_1 and r_2. Then the elimination can be done, and the result is a univariate polynomial in r_3 of degree 208 (only even powers). From this, it is concluded that the equation wanted has in general degree 208 in each unknown. For the total degree, one has to guess and the following computations show that a guess of also 208 is correct. Then a general ansatz for a polynomial in r_1, r_2, r_3 of degree 208 is made, introducing 34497 unknown coefficients $C_1, \ldots, C_{34497}$ which have to be determined. Now the following steps are repeated for several pairs of values for r_1 and r_2 (for 1413 pairs to be exact): At first, the values of the pair are substituted into ideal $\mathscr{L}$ and the univariate polynomial in r_3 is obtained. Then the substitution is made in the general ansatz, leading to a polynomial in r_3 where the coefficients are linear expressions in the coefficients C_i. After multiplication of the univariate with an unknown scaling factor, comparison of coefficients leads to a system of 105 linear equations.

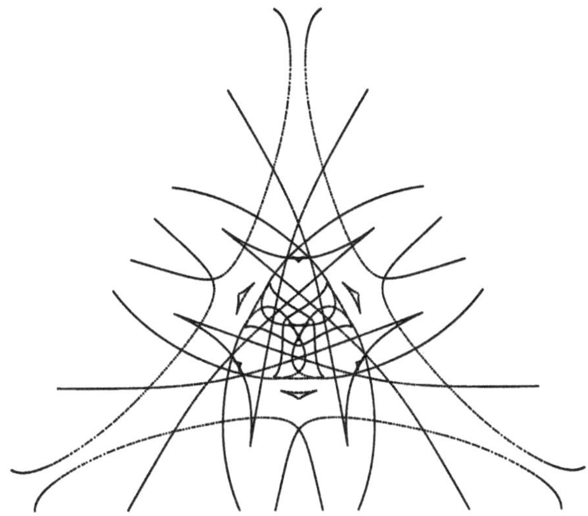

Fig. 26 Intersection curve of the surface from the general mode, first octant, cut out by a plane orthogonal to $(1, 1, 1)^T$

For each pair, 105 such equations are obtained which means that in the end there are 148365 equations in 35910 unknowns. As a precaution and for reasons of linear dependencies, the number of pairs of values is set so high. In the end, all coefficients $C_1, \ldots, C_{34497}$ can be computed up to a scaling factor and after substitution into the general ansatz the condition for singular poses of the general mode is available. Due to the extreme size of this equation, a 3D plot is not possible, but in a roundabout way a plot of a planar intersection of the surface in the first octant can be generated, like for the surface in Fig. 24 resp. Fig. 25. See Fig. 26 for an impression of the complexity of this surface. Due to the fact that the surface has several connections to the parts in neighboring octants, there is no "pipe" shape anymore.

7.1 Changing Operation Modes

As already mentioned, there exist poses where the mechanism can change from one mode into another mode. One of them is, e.g., the "planar home position" where the mechanism can bifurcate into the planar mode or into the translational mode. In the following, an overview is given of those poses where a mode change is possible. This is done by inspecting each pair of ideals $\{\mathscr{K}_i, \mathscr{K}_j\}$ with respect to common solutions. For each pair, the dimension of the intersection of the corresponding varieties is computed and the following results are obtained.

The numbers in Table 5 correspond to the dimension of the intersection variety. As it can easily be seen there are four possible combinations of operation modes which have no pose in common, to change from one to the other a detour has to be made, which is always possible via the general mode, corresponding to $\mathscr{K}_5$. Once more it has to be noted that mode changing poses of the manipulator are also singular poses,

Table 5 All values of $\overline{dim}(\mathscr{K}_i \cup \mathscr{K}_j)$

	$\mathscr{K}_1$	$\mathscr{K}_2$	$\mathscr{K}_3$	$\mathscr{K}_4$	$\mathscr{K}_5$
$\mathscr{K}_1$	...	−1	2	−1	2
$\mathscr{K}_2$	−1	...	2	−1	2
$\mathscr{K}_3$	2	2	...	−1	2
$\mathscr{K}_4$	−1	−1	−1	...	2
$\mathscr{K}_5$	2	2	2	2	...

because each of them lies in the intersection of two varieties. In the following, all possible changes are discussed with respect to necessary conditions and a description of related poses.

Translational mode ($\mathscr{K}_1$) $\longleftrightarrow$ **planar mode** ($\mathscr{K}_3$) The condition for this change is exactly the singularity condition from the translational mode, i.e., all singularities of this mode coincide with the intersection singularities with the planar mode. Hence, the possible mode change poses have already been given in Sect. 7 and the corresponding singularity surface can be seen in Fig. 20.

Translational mode ($\mathscr{K}_1$) $\longleftrightarrow$ **general mode** ($\mathscr{K}_5$) For this change, a new condition appears, that is

$$r_1^4 + r_2^4 + r_3^4 - r_1^2 r_2^2 - r_1^2 r_3^2 - r_2^2 r_3^2 - 36\,(h_1 - h_2)^4 = 0 \qquad (65)$$

and for the corresponding positions of the platform the following description can be deduced:

$$\mathbf{M} = \begin{pmatrix} 1 & 0 & 0 & 0 \\ p_1 & 1 & 0 & 0 \\ p_2 & 0 & 1 & 0 \\ p_3 & 0 & 0 & 1 \end{pmatrix}$$

with

$$p_2^2 + p_3^2 = 4\,(h_1 - h_2)^2,$$

i.e., platform and base have the same orientation and the origin of the platform's frame has to lie on a cylinder with radius $2\,(h_1 - h_2)$. This is a result which verifies statements about singularity loci from Chebbi and Parenti-Castelli (2010).

Twisted translational mode ($\mathscr{K}_2$) $\longleftrightarrow$ **planar mode** ($\mathscr{K}_3$) The condition for this change is exactly the singularity condition of the twisted translational mode, i.e., all singularities of this mode coincide with the intersection singularities with the planar mode. Hence the possible mode change poses have already been mentioned and the corresponding singularity surface can be seen in Fig. 22.

Twisted translational mode ($\mathscr{K}_2$) $\longleftrightarrow$ **general mode** ($\mathscr{K}_5$) The condition for this mode change is as follows:

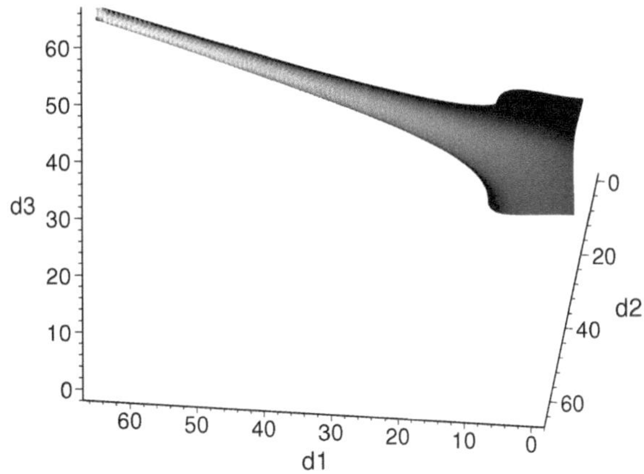

Fig. 27 Surface describing the condition for changes from translational to general mode

$$r_1^4 + r_2^4 + r_3^4 - r_1^2 r_2^2 - r_1^2 r_3^2 - r_2^2 r_3^2 - 36\,(h_1 + h_2)^4 = 0. \tag{66}$$

For the corresponding poses of the platform, the following description can be deduced:

$$\mathbf{M} = \begin{pmatrix} 1 & 0 & 0 & 0 \\ p_1 & 1 & 0 & 0 \\ p_2 & 0 & -1 & 0 \\ p_3 & 0 & 0 & -1 \end{pmatrix}$$

with

$$p_2^2 + p_3^2 = 4\,(h_1 + h_2)^2,$$

i.e., once more platform and base have the same orientation and the origin of the platform's frame has to lie on a cylinder, but this time with radius $2\,(h_1 + h_2)$. The same result can again be found in Chebbi and Parenti-Castelli (2010), p. 598 (Fig. 27).

Planar mode ($\mathscr{K}_3$) $\longleftrightarrow$ **general mode** ($\mathscr{K}_5$) The condition which has to be fulfilled for this case reads

$$7\,(r_1^4 + r_2^4 + r_3^4) - 11\,(r_1^2 r_2^2 - r_1^2 r_3^2 - r_2^2 r_3^2) = 0. \tag{67}$$

The corresponding poses of the platform are given by

$$\mathbf{M} = \begin{pmatrix} 1 & 0 & 0 & 0 \\ 0 & 1 & 0 & 0 \\ p_2 & 0 & v_{33} & v_{32} \\ p_3 & 0 & -v_{32} & v_{33} \end{pmatrix}$$

with

$$v_{32}^2 + v_{33}^2 = 1, \quad p_2^2 + p_3^2 = 4\,(h_1^2 + h_2^2 - 2\,h_1\,h_2\,v_{33}).$$

Fig. 28 Surface describing the condition for changes from twisted translational mode to general mode

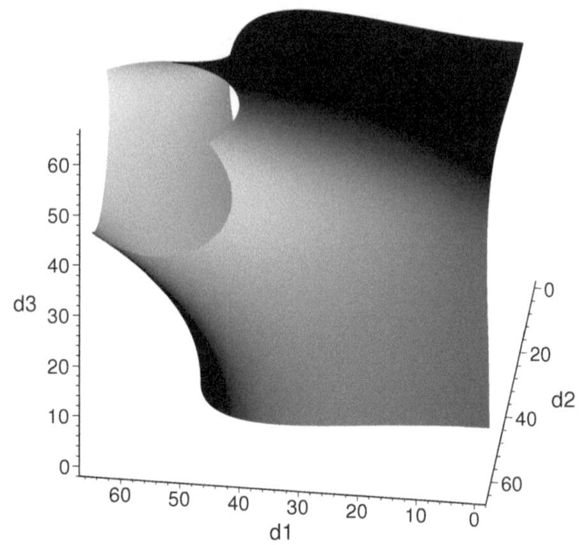

Fig. 29 Surface describing the condition for changes from planar mode to general mode

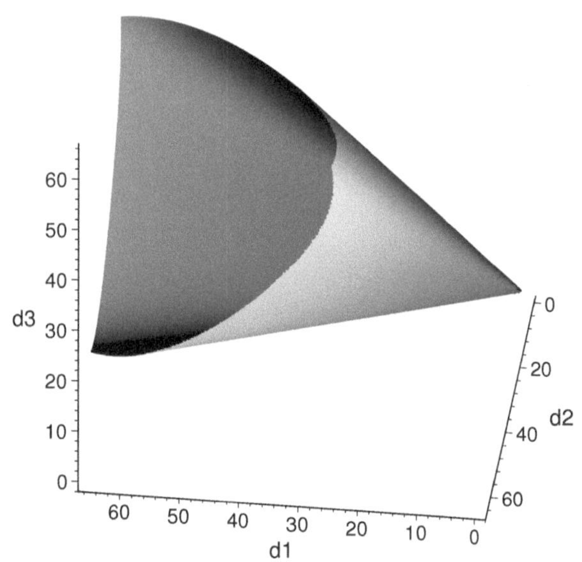

It is noticeable that these positions are very similar to the singular poses of the planar mode itself (Figs. 28 and 29).

Upside-down planar mode ($\mathscr{K}_4$) $\longleftrightarrow$ **general mode** ($\mathscr{K}_5$) The condition for this case can be computed but it is rather complicated. It is an equation of degree 24, and it is not equal to the singularity condition of the upside-down planar mode itself. As a result of the complexity, up to now also no description of the platform's poses could be found.

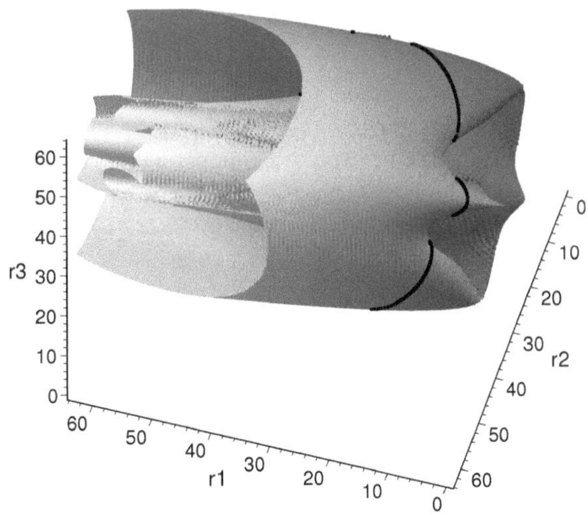

Fig. 30 Surface describing the condition for changes from upside-down planar mode to general mode

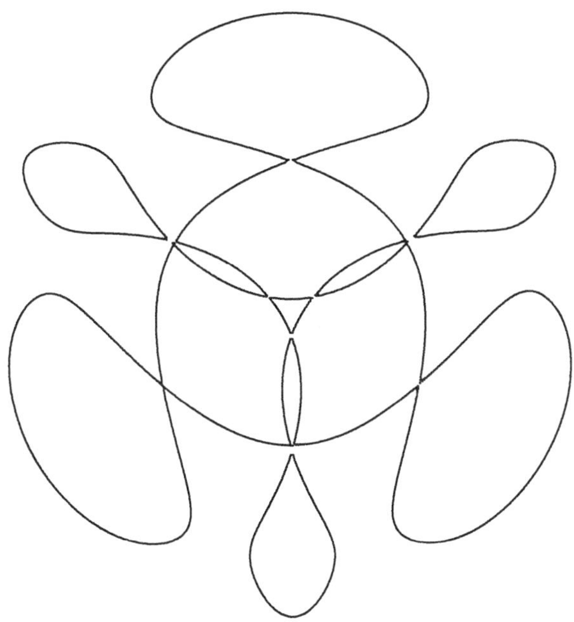

Fig. 31 Intersection curve of the "pipe" of Fig. 30

All the other combinations which are $\{\mathcal{K}_1, \mathcal{K}_2\}$, $\{\mathcal{K}_1, \mathcal{K}_4\}$, $\{\mathcal{K}_2, \mathcal{K}_4\}$ and $\{\mathcal{K}_3, \mathcal{K}_4\}$ do not allow any operation mode change (Fig. 31).

8 Conclusion

In this chapter, an algebraic approach to the analysis of mechanisms was presented. To establish this methodology, the Study representation of spatial displacements was introduced. Starting from the kinematic chain as basic element, different methods to derive the algebraic constraint equations for kinematic chains were demonstrated. After that, in a primer on methods from algebraic geometry, the most important and powerful algebraic tools for the analysis and solution of these equations were introduced. In the last section, the application of all these methods was demonstrated in the complete global analysis of the 3-UPU-TSAI parallel manipulator.

References

A. Angerer, Rechentechnische Realisierung des inversen Roboterkinematik Problems. Ph.D. thesis, UMIT Hall, Tyrol, Austria, 2016

M. Badescu, J. Morman, C. Mavroidis, Workspace optimization of 3-UPU parallel platforms with joint constraints. In *ICRA*, (2002), pp. 3678–3683

H. Bo, L. Yi, Solving stiffness and deformation of a 3-UPU parallel manipulator with one translation and two rotations. Robotica **29**(06), 815–822 (2011)

I. Bonev, D. Zlatanov, The mystery of the singular SNU translational parallel robot (2001). http://www.parallemic.org/Re-views/Review004.html

I.A. Bonev, Geometric analysis of parallel mechanisms. Ph.D. thesis, Université Laval, Québec, 2002

O. Bottema, B. Roth, *Theoretical Kinematics*. (North-Holland Publishing Company, 1979). ISBN 0-486-66346-9

A.H. Chebbi. The potential of the 3-UPU translational parallel manipulator and a procedure to select the best architecture. Ph.D. thesis, Università di Bologna, 2011

A.H. Chebbi, V. Parenti-Castelli, Geometric and manufacturing issues of the 3-UPU pure translational manipulator, in *New Trends in Mechanism Science: Analysis and Design*, ed. by D. Pisla, M. Ceccarelli, M. Husty, B. Corves (Springer, 2010), pp. 595–603

D. Cox, J. Little, D. O'Shea, in *Ideals, Varieties, and Algorithms*, 3th edn. (Springer, 2007)

R. Di Gregorio, V. Parenti-Castelli, A translational 3-DOF parallel manipulator, in *Advances in Robot Kinematics: Analysis and Control*, ed. by J. Lenarčič, M. Husty (Kluwer Academic Publishers, 1998), pp. 49–58

R. Di Gregorio, V. Parenti-Castelli, Mobility analysis of the 3-UPU parallel mechanism assembled for a pure translational motion, in *Proceedings of IEEE/ASME International Conference on Advanced Intelligent Mechatronics, Atlanta* (1999), pp. 520–525

T. Dwarakanath, G. Bhutani, Novel design solution to high precision 3 axes translational parallel mechanism, in *Machines and Mechanisms*, ed. by S. Bandyopadhyay, G. Kumar, P. Ramu (Narosa Publishing House, New Delhi, 2012), pp. 373–381

C. Gosselin, J. Angeles, Singularity analysis of closed-loop kinematic chains. IEEE Trans. Robot. Autom. **6**(3), 281–290 (1990). https://doi.org/10.1109/70.56660, ISSN 1042-296X

C. Han, J. Kim, J. Kim, F.C. Park, Kinematic sensitivity analysis of the 3-UPU parallel mechanism. Mech. Mach. Theory **37**(8), 787–798 (2002)

R.W.H.T Hudson, *Kummer's Quartic Surface* (Cambridge University Press, 1905). (reprint 1990 by Cambridge University Press)

M.L. Husty, An algorithm for solving the direct kinematics of general Stewart-Gough platforms. Mech. Mach. Theory **31**(4), 365–380 (1996)

M.L. Husty, H.-P. Schröcker, Kinematics and Algebraic Geometry, in *21st Century Kinematics* (Springer, 2012), pp. 85–123
M.L. Husty, A. Karger, H. Sachs, W. Steinhilper, *Kinematik und Robotik* (Springer, 1997)
M.L. Husty, M. Pfurner, H.-P. Schröcker, A new and efficient algorithm for the inverse kinematics of a general serial 6R manipulator. Mech. Mach. Theory **42**(1), 66–81 (2007a). ISSN 0094-114X
M.L. Husty, M. Pfurner, H.-P. Schröcker, K. Brunnthaler, Algebraic methods in mechanism analysis and synthesis. Robotica **25**, 661–675 (2007b)
P. Ji, W. Hongtao, Kinematics analysis of an offset 3-UPU translational parallel robotic manipulator. Robot. Auton. Syst. **42**(2), 117–123 (2003)
A. Nayak, Th. Stigger, M. Husty, Ph. Wenger, St. Caro, Operation mode analysis of 3-RPS parallel manipulators based on their design parameters. Comput. Aided Geom. Des. **63**, 122–134 (2018). https://doi.org/10.1016/j.cagd.2018.05.003, https://hal.archives-ouvertes.fr/hal-01860789
L. Nurahmi, J. Schadlbauer, M.L. Husty, Ph. Wenger, St. Caro, Kinematic analysis of the 3-RPS cube parallel manipulator. ASME J. Mech. Robot. **7**(1), 1–11 (2015)
L. Nurahmi, St. Caro, Ph. Wenger, J. Schadlbauer, M.L. Husty, Reconfiguration analysis of a 4-RUU parallel manipulator. Mech. Mach. Theory **96**, 269–289 (2016)
V. Parenti-Castelli, R. Di Gregorio, F. Bubani, Workspace and Optimal Design of a Pure Translation Parallel Manipulator. Meccanica **35**(3), 203–214 (2000)
M. Pfurner, Analysis of spatial serial manipulators using kinematic mapping. Ph.D. thesis, University of Innsbruck, 2006. http://repository.uibk.ac.at
M. Pfurner, H.-P. Schröcker, M.L. Husty, Path planning in kinematic image space without the study condition, in *Advances in Robot Kinematics*, ed. by J.-P. Merlet, J. Lenarcic (2016). https://hal.archives--ouvertes.fr/hal--01339423
J. Schadlbauer, Algebraic methods in kinematics and line geometry. Doctoral thesis, University Innsbruck, 2014
J. Schadlbauer, D.R Walter, M.L. Husty, The 3-RPS parallel manipulator from an algebraic viewpoint. Mech. Mach. Theory **75**, 161–176 (2014). ISSN 0094114X
J. Schadlbauer, L. Nurahmi, M.L. Husty, P. Wenger, St. Caro, Operation modes in lower-mobility parallel manipulators, in, *IAK: Proceedings of the International Conference, Lima, Peru, 9–11 Sept 2013* (Springer International Publishing, 2015), pp. 1–9
J.M. Selig, *Geometric Fundamentals of Robotics*, 2nd edn. (Springer Science+Business Media Inc., 2005), ISBN 0-387-20874-7
A.J. Sommese, C.W. Wampler, The numerical solution of systems of polynomials arising in engineering and science. World Sci. (2005), ISBN 981-256-184-6
Th. Stigger, A. Nayak, St. Caro, Ph. Wenger, M. Pfurner, M.L. Husty, Algebraic analysis of a 3-RUU parallel manipulator, in *16th International Symposium on Advances in Robot Kinematics* (2018a), pp. 141–149
Th. Stigger, M. Pfurner, M.L. Husty, Workspace and Singularity Analysis of a 3-RUU Parallel Manipulator, in *Proceedings of the 7th European Conference on Mechanism Science, Mechanism and Machine Science*, vol. 59 (Springer, Berlin, 2018b), pp. 325–332
E. Study, *Geometrie der Dynamen* (Teubner, Leipzig, 1903)
L.-W. Tsai, Kinematics of a three-DOF platform with three extensible limbs, in *Recent Advances in Robot Kinematics*, ed. by J. Lenarčič, V. Parenti-Castelli (Kluwer Academic Publishers, 1996), pp. 401–410
L.-W. Tsai, *Robot Analysis: The Mechanics of Serial and Parallel Manipulators* (Wiley, 1999)
L.-W. Tsai, S. Joshi, Kinematics and optimization of a spatial 3-UPU parallel manipulator. J. Mech. Des. **122**, 439–446 (2000)
S.-M. Varedi, H.M. Daniali, D.D. Ganji, Kinematics of an offset 3-UPU translational parallel manipulator by the homotopy continuation method. Nonlinear Anal. Real World Appl. **10**(3), 1767–1774 (2009)
D.R. Walter, M.L. Husty, On implicitization of kinematic constraint equations, in *Machine Design & Research (CCMMS 2010)*, vol. 26 (Shanghai, 2010), pp. 218–226

D.R. Walter, M.L. Husty, M. Pfurner, A complete kinematic analysis of the SNU 3-UPU parallel robot, in *Interactions of Classical and Numerical Algebraic Geometry of Contemporary Mathematics*, vol. 496, ed. by D. Bates, G.M. Besana, S. Di Rocco, C. Wampler (American Mathematical Society, 2009), pp. 331–346

A. Wolf, M. Shoham, F. Chongwoo Park, An investigation of the singularities and self-motions of the 3-UPU robot, in *Advances in Robot Kinematics* (Kluwer Academic Publishers, 2002), pp. 165–174

D. Zlatanov, R.G. Fenton, B. Benhabib, Analysis of the instantaneous kinematics and singular configurations of hybrid-chain manipulators, in *Proceedings of the ASME 23rd Biennial Mechanisms Conference*, vol. DE-Vol. 72, (Minneapolis, MN, USA, 11–14 Sept 1994) pp. 467–476

Local Investigation of Mobility and Singularities of Linkages

Andreas Müller

Abstract Kinematic singularities affect the mobility and thus the stability of a mechanism. Understanding the possible motions a mechanism can perform when passing through a singularity is hence crucial. In this chapter, the basic definitions of singularities are recalled, where configuration space (c-space) singularities, kinematic singularities, and constraint singularities are distinguished. It is shown that these types of singularities can occur independently from each other. The focus of this chapter is the local higher-order analysis in order to understand how a mechanism can behave when passing through such singularities. An exhaustive investigation involves separate analysis of the c-space geometry as well as finite curves in c-space, but also of the set of configurations of certain rank and finite curves within this set. Computational methods and necessary formulations are presented to this end. All formulations are given in terms of joint screw coordinates so that the analysis requires simple algebraic operations only.

1 Introduction

A mechanism is a mechanical system relating input (actuator) motion to output (end-effector) motion. Its kinematic performance is thus determined by the mechanical linkage as well as by its actuation and interaction with the environment. Critical situations can thus occur regarding the relation of the mechanism's motion to the input as well as the output motion, as described by Zlatanov in Chap. 1, but also regarding the internal motion of the mechanism. This chapter deals with the mechanical linkage as such, i.e., independently of input and output.

Kinematic singularities of linkages are configurations where the differential DOF, and thus the rank of the constraint Jacobian $\mathbf{J}$, changes. These are critical configurations since they consequently affect the mobility of a linkage possibly rendering it unstable or uncontrollable. When dealing with singularities of the mechanical linkage

A. Müller (✉)
Institute of Robotics, Johannes Kepler University, Linz, Austria
e-mail: a.mueller@jku.at

A. Müller and D. Zlatanov (eds.), *Singular Configurations of Mechanisms and Manipulators*, CISM International Centre for Mechanical Sciences 589,
https://doi.org/10.1007/978-3-030-05219-5_5

(but not with the input–output characteristics of a mechanism), the central objects are the loop closure constraints and the configuration space (c-space) V defined by them. Most of the past research has therefore addressed constraint singularities (Agrawal 1990; Bonev and Gosselin 2003; Gosselin and Angeles 1990; Zlatanov et al. 1998, 1995). It was even concluded by Zlatanov et al. (2002) that constraint singularities are singularities of the c-space, which is not always true, however. An assumption implicitly underlying most of these approaches is that an *instantaneous* first-order analysis is sufficient to identify singularities and to determine the finite mobility, which boils down to check the rank of $\mathbf{J}$. Since the latter is determined by the number of linearly independent joint screws, the classification of screw systems of a certain rank (Gibson and Hunt 1990a, b; Martinez and Duffy 1992a, b) provides geometric conditions on the instantaneous screws and thus on the linkage configuration. This was also used to identify input singularities of parallel manipulators (Merlet 1991).

C-space singularities are points of V where the latter ceases to be a manifold. Constraint singularities are points $\mathbf{q} \in V$ where the constraint Jacobian $\mathbf{J}$ does not have full rank. Kinematic singularities are points where rank $\mathbf{J}$ is not constant in any neighborhood of $\mathbf{q}$ in V. Clearly, a constraint singularity may neither be a kinematic nor be a c-space singularity. First-order considerations are thus not sufficient to analyze either of these singularities in general; rather, a *local* higher-order analysis is necessary, which reveals possible finite motions at a singularity. Unfortunately, there is no established approach to such a higher-order analysis, and various attempts have been reported in the literature that differ regarding the approximation order but more importantly regarding what is actually approximated. Existing approaches can be classified as those that use a higher-order approximation of solution curves of the constraints and those that use a Taylor series expansion of the constraints. A second-order approximation of the curves satisfying the constraints was used by Bandyopadhyay and Ghosal (2004) and de Bustos et al. (2012), and a third-order approximation by Kieffer (1994). A formulation for the third-order (jerk) analysis was presented by Martinez et al. (1999), Gallardo-Alvarado and Rico-Martinez (2001), and Lerbet (1999) using screw and Lie group theory. Extensions to arbitrary higher orders were presented in (Müller 2002; Müller and Rico 2018; Müller 2016b). Since these approximations are derived from derivatives of the constraints w.r.t. a path parameter (e.g., time), they reveal possible motions of a certain order and eventually tangents to *smooth* solution curves in V. There are linkages, however, possessing c-space singularities that do not allow for smooth curves (e.g., cusps), and it is known that an approximation of smooth curves through such a singularity may not reveal the local geometry of V (Connelly and Servatius 1994). An approach to the second-order mobility analysis using a Taylor expansion of the loop constraints was presented by Chen (2011). Because this approach makes use of a series expansion of the constraints, it does in fact provide a second-order approximation of V.

It was further observed that the c-space can be a smooth manifold even if the linkage exhibits kinematic singularities (Müller and Zlatanov 2015; Müller 2016b). A local analysis must thus additionally take into account the rank of $\mathbf{J}$ in an open neighborhood.

In this chapter, a comprehensive definition of singularities of linkages and a computational framework for the higher-order local analysis will be introduced considering the differential geometric properties of the c-space V (an approach to the global analysis using algebraic methods is presented by Husty in Chap. 4). The kinematics is formulated in terms of joint screw coordinates and joint variables so that the loop closure constraints are analytic and the c-space is hence an analytic variety. This gives rise to closed-form relations for the required higher-order constraints and differentials that will be used in the computational framework.

2 Kinematic Model of Single-Loop Linkages

A linkage is a mechanical system composed of lower pair joints connecting links, which will be assumed rigid. Lower pair joints are mechanical generators of subgroups of $SE(3)$, the group of rigid body motions. Their motions can hence be expressed by an exponential mapping in terms of screw coordinates (Bottema and Roth 1979; Selig 2005). This leads to a central result in kinematics (Brockett 1984; Herve 1982) stating that the motion of the members of a kinematic chain is described by a product of exponentials in terms of joint screws and joint variables. Throughout this chapter, it is assumed that all members of a linkage are part of a kinematic loop. The key elements in the kinematic modeling are thus the loop closure constraints.

In this section, a single kinematic loop comprising n joints is considered. Without losing generality, all joints are assumed to be one-DOF lower pairs; i.e., they are either revolute, prismatic, or helical joints. The joints are numbered consecutively with $i = 1, \ldots, n$ within the loop. Figure 1 shows the numbering for a planar four-bar linkage. The arrangement of joints and links (i.e., the kinematic topology) is represented by a linear graph (the topological graph Γ). This is trivial for a single loop but will be important for multi-loop linkages (Sect. 3).

2.1 *Geometric Loop Closure Constraints and the Configuration Space*

To joint i is assigned a joint variable (angle, displacement), denoted with q_i. The set of all joint variables of the linkage is summarized in the vector $\mathbf{q} = (q_1, \ldots, q_n) \in \mathbb{V}^n$, where $\mathbb{V}^n := \mathbb{R}^{n_P} \times \mathbb{T}^{n_R}$ is called the joint space of a linkage with n_P prismatic or helical and n_R revolute joints ($n := n_P + n_R$). Since the joint coordinates represent a local parameterization of the configuration of the linkage, $\mathbf{q}$ will interchangeably be referred to as a *point* (of $\mathbb{V}^n$) or as a *configuration* (of the linkage).

Remark 2.1 It is important to remember that a linkage is a set of bodies in E^3 geometrically interconnected by lower pair joints. That is, the pose of each body is coordinate-free represented by a point in $SE(3)$, and a configuration of the linkage is

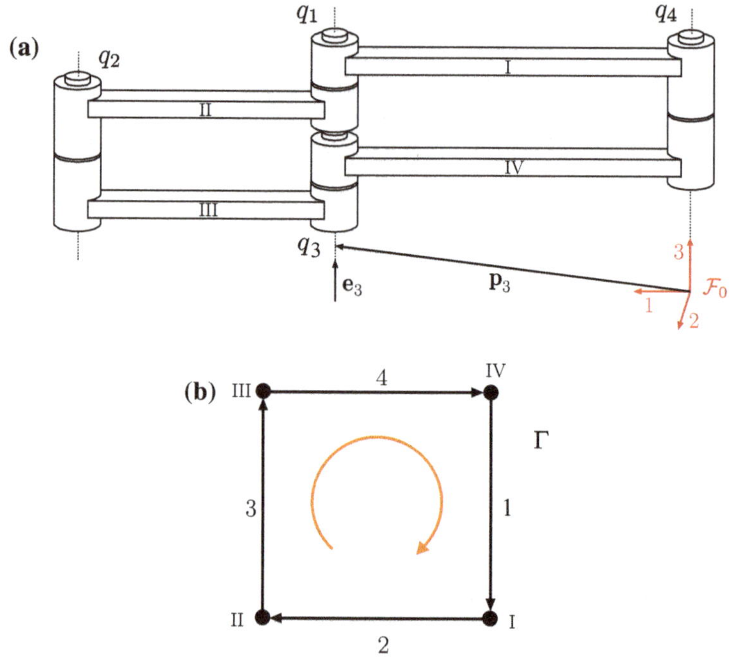

Fig. 1 **a** Planar four-bar linkage. **b** Topological graph Γ for the planar four-bar linkage

an element of $SE(3)^n$. The set of feasible configurations is the subspace of $SE(3)^n$ defined by the joint constraints. If all joints are one-DOF lower pairs, joint coordinates provide a global chart on $SE(3)^n$. If there are spherical joints, and if they are replaced by three consecutive revolute joints, the corresponding joint variables are local coordinates only, due to the singularity of such parameterization of rotations.

The loop is regarded as a kinematic chain where the body connected to joint 1 and n is used as reference body (ground) at which a global reference frame $\mathcal{F}_0$ is defined.

A screw coordinate vector is introduced for each joint. In the reference configuration $\mathbf{q} = \mathbf{0}$, denote with $\mathbf{e}_i$ a unit vector along the axis of joint i, and with $\mathbf{p}_i$ the position vector to a point on that axis. Figure 1a shows this for joint 3 of the four-bar linkage. All vectors are resolved in $\mathcal{F}_0$. The *joint screw coordinate vector* for a revolute/helical joint and a prismatic joint, respectively, is (Murray et al. 1994)

$$\mathbf{Y}_i = \begin{pmatrix} \mathbf{e}_i \\ \mathbf{p}_i \times \mathbf{e}_i + h_i \mathbf{e}_i \end{pmatrix}, \quad \mathbf{Y}_i = \begin{pmatrix} \mathbf{0} \\ \mathbf{e}_i \end{pmatrix} \tag{1}$$

The relative motion of the two bodies connected by a joint with screw coordinates $\mathbf{Y} \in \mathbb{R}^6$ is given by the exponential mapping that reads explicitly for a revolute/helical joint (finite pitch h)

$$\exp(\varphi\mathbf{Y}) = \begin{pmatrix} \exp(\varphi\widetilde{\mathbf{e}}) & (\mathbf{I}-\exp(\varphi\widetilde{\mathbf{e}}))\mathbf{p}+\varphi h\mathbf{e} \\ \mathbf{0} & 1 \end{pmatrix}, \text{ for} h \neq \infty \quad (2)$$

where φ is the rotation angle, and for a prismatic joint (infinite pitch h)

$$\exp(\varphi\mathbf{Y}) = \begin{pmatrix} \mathbf{I} & \varphi\mathbf{e} \\ \mathbf{0} & 1 \end{pmatrix}, \text{ for} h = \infty \quad (3)$$

where now φ is the translation variable. The exp mapping for the rotation part is given by the Euler–Rodrigues formula, which can be expressed as $\exp(\varphi\widetilde{\mathbf{e}}) = \mathbf{I} + \sin\varphi\widetilde{\mathbf{e}} + (1-\cos\varphi)\,\widetilde{\mathbf{e}}^2$ (Müller 2016a; Murray et al. 1994).

The combination of successive relative motions of all joints in the loop is described by the mapping $f : \mathbb{V}^n \to SE\,(3)$ defined by the product of exponentials (Brockett 1984; Müller 2016b)

$$f\,(\mathbf{q}) = \exp\left(\mathbf{Y}_1 q_1\right)\exp\left(\mathbf{Y}_2 q_2\right)\cdot\ldots\cdot\exp\left(\mathbf{Y}_n q_n\right). \quad (4)$$

Any possible configuration of a closed kinematic loop must satisfy the *geometric loop constraint* $f\,(\mathbf{q}) = \mathbf{I}$. Therefore, f is referred to as the constraint mapping. The solution set

$$V := \left\{\mathbf{q} \in \mathbb{V}^n | f\,(\mathbf{q}) = \mathbf{I}\right\} = f^{-1}\,(\mathbf{I}) \cap \mathbb{V}^n \quad (5)$$

is referred to as the *configuration space* (c-space) of the single-loop linkage and $\mathbf{q} \in V$ as the *configuration* (or representing point). This is a real analytic variety since the exp mapping is analytic. In general, V is only locally a smooth manifold, but may possess singular points. Throughout this chapter, the c-space always refers to the real point set defined in (5).

A motion of the linkage is a curve in V. It may happen that the configuration space V is not connected so that there are configurations $\mathbf{q} \in V$ which cannot be reached by finite motions. Well known is the example of a 6R linkage in Fig. 2 that can be assembled in two different assembly modes (Chai and Chen 2010). The linkage must be disassembled in order to switch between the two modes. Yet, the same constraint mapping, i.e., the same joint screws $\mathbf{Y}_i$, describes the kinematics. The c-space is the disjoint union of an isolated point and a smooth curve, corresponding to the rigid structure in Fig. 2a and the mobile Bricard linkage in Fig. 2b, respectively.

It is assumed in the following that the *reference configuration* $\mathbf{q} = \mathbf{0}$ is admissible, so that $\mathbf{q} = \mathbf{0} \in V$. This can always be achieved by appropriate definition of joint variables.

2.2 Velocity Loop Closure Constraints

In configuration $\mathbf{q} \in \mathbb{V}^n$, the relative twist of the two bodies connected by joint i (represented in $\mathcal{F}_0$) is $\mathbf{S}_i(\mathbf{q})\,\dot{q}_i$, where

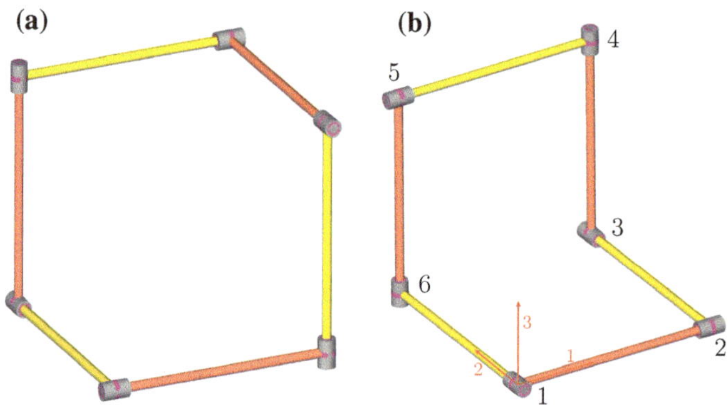

Fig. 2 Two assembly modes of a 6R chain. The c-space V consists of an isolated point and a smooth curve. The point corresponds to the immobile 6R linkage in (**a**), and the curve to the line-symmetric Bricard linkage in (**b**)

$$\mathbf{S}_i(\mathbf{q}) := \mathbf{Ad}_{g_i(\mathbf{q})}\mathbf{Y}_i \tag{6}$$

is the instantaneous screw coordinate vector of joint i (which is indeed the instantaneous screw axis (ISA) of the relative motion). Relation (6) shows that $\mathbf{S}_i$ is obtained by transformation of $\mathbf{Y}_i$ to the current configuration according to $g_i(\mathbf{q}) := \exp(\mathbf{Y}_1 q_1) \cdot \ldots \cdot \exp(\mathbf{Y}_i q_i)$ of body i given by the frame transformation $\mathbf{Ad}_{g_i}$ of screw coordinates (Selig 2005). The instantaneous screw coordinate vector of joint i represented in world frame $\mathcal{F}_0$ can readily be constructed as $\mathbf{S}_i(\mathbf{q}) = (\mathbf{e}_i(\mathbf{q}), \mathbf{s}_i(\mathbf{q}) \times \mathbf{e}_i(\mathbf{q}) + h_i\mathbf{e}_i(\mathbf{q}))^T$, where $\mathbf{e}_i(\mathbf{q})$ and $\mathbf{s}_i(\mathbf{q})$ are the respective vectors in the current configuration resolved in $\mathcal{F}_0$. However, the relation (6) is a key to the systematic derivation of higher-order constraints in the next section.

The net velocity of the closed loop due to the motion of all joints must be zero. This velocity constraint can be written in the form

$$\mathbf{J}(\mathbf{q})\dot{\mathbf{q}} = \mathbf{0}, \tag{7}$$

where the $6 \times n$ *constraint Jacobian* $\mathbf{J}(\mathbf{q}) : \mathbb{R}^n \to se(3)$

$$\mathbf{J}(\mathbf{q}) = \Big(\mathbf{S}_1(\mathbf{q}), \mathbf{S}_2(\mathbf{q}), \cdots, \mathbf{S}_n(\mathbf{q})\Big). \tag{8}$$

Remark 2.2 (Algebraic formulation) The constraint mapping f is analytic (simply because it is defined by an exponential mapping). Therefore, the c-space V is an analytic variety (hence the symbol). The use of the exponential mapping in terms of screw coordinates gives rise to explicit closed-form relations for its derivatives and differentials, as will be seen in the following. Alternatively, the kinematics and constraints can be formulated algebraically, i.e., by a system of polynomials (see Chap. 4 by Husty and Husty et al. 2007). This approach is frequently used when aiming at a global analysis of mechanisms. The c-space is then defined by a polynomial system and is thus an algebraic variety. As for the analytic formulation, there are in principle two ways to model the kinematics of a rigid body mechanism: using absolute or relative coordinate constraints. In the absolute coordinate approach, the pose of each body is described with an algebraic parameterization, such as study parameters (Husty et al. 2007) or unit quaternions and position vectors, and for each joint a set of constraints is introduced. The relative coordinate approach treats the loop as described above with the difference that the relative joint motion is described algebraically, instead of using joint angles. This can be achieved by treating $s_i := \sin q_i$ and $c_i := \cos q_i$ as variables subjected to $c_i^2 + s_i^2 = 1$ or by using the tangent half substitution. Both approaches work for so-called algebraic joints, i.e., for joints whose motion can be parameterized algebraically. All lower pairs except the screw joint are algebraic. There is powerful machinery from computational algebraic geometry that shall allow to deal with the resulting polynomial systems and thus to analyze the algebraic variety (Cox et al. 2007). It is know, however, that there is no straightforward procedure and moreover, since the involved algorithms are NP-hard, it is not guarantied that such computation is even feasible. A result that can ease the computation is that the constraint variety of a revolute joint can be decomposed a priori, which was reported by Arponen et al. (2004). Also a strategic rearrangement of the equations can help, as shown in by Manfred Husty.

3 Kinematic Model of Multi-loop Linkages

Linkages in general possess several kinematic loops. A system of topologically independent loops is given by the fundamental cycles (FCs) of the topological graph Γ representing the kinematic topology (Müller 2018c). The number of FCs (the Euler characteristic) is denoted with γ. The latter is unique, whereas the actual choice of FCs is not. A general multi-loop linkage can thus be considered as an interconnection of γ single-loop linkages corresponding to the FCs.

Figure 3 shows the topological graph Γ for a planar two-loop linkage obtained from a Watt linkage by adding a joint. For a selected spanning tree $\mathcal{G}$ of Γ, the co-tree $\mathcal{H}$ is the complement of $\mathcal{G}$ in Γ. A FC is a closed path in Γ containing exactly one co-tree edge. The FCs are denoted with $\Lambda_l, l = 1, ..., \gamma$. The Watt linkage in Fig. 3 possesses $\gamma = 2$ FCs Λ_1 and Λ_2. Since the same joint can be traversed in different directions within different FCs, an orientation of the joints and the FCs is introduced, represented by the directed spanning tree $\vec{\mathcal{G}}$ and co-tree $\vec{\mathcal{H}}$. An indicator function $\sigma_l\,(i)$ is defined for Λ_l, so that $\sigma_{l,i} = 1$, if edge i has the same orientation as Λ_l, and $\sigma_{l,i} = -1$, if it is oriented opposite to Λ_l. Finally, an orientation on Λ_l is introduced that allows identifying predecessors of joints within the FC. A Joint $\bar{l}$ regarded as the terminal joint in the chain. Joint i is a predecessor of j if it is visited after j when traversing Λ_l according to its orientation, which is denoted as $i <_l j$. The loop is thus regarded as a kinematic chain with its first joint indexed with $\underline{l}$ and terminal joint $\bar{l}$, i.e., $\underline{l} <_l i <_l j <_l \ldots <_l \bar{l}$, for $\bar{l}, i, j, \ldots, \underline{l} \in \Lambda_l$. If joint j comes immediately

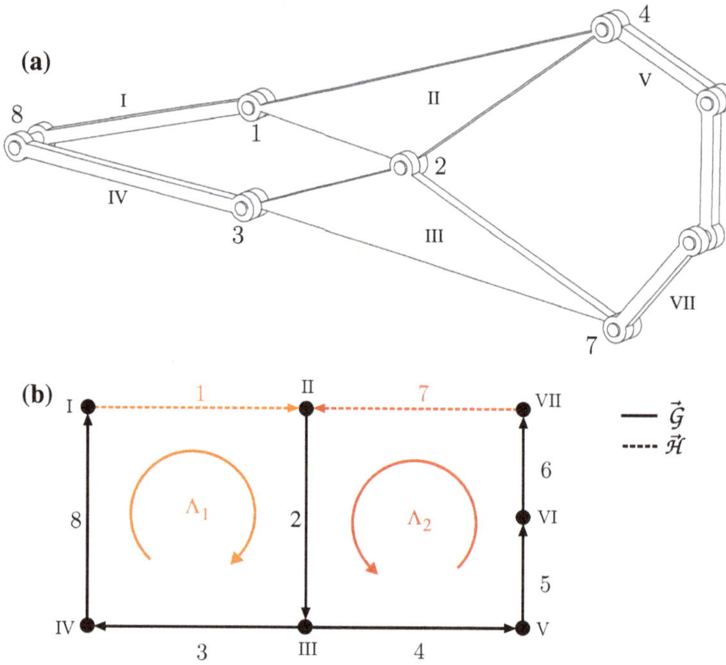

Fig. 3 **a** Extended Watt linkage in a two-DOF motion mode and **b** its topological graph

after joint i, it is its direct predecessor denoted by $j = i -_l 1$. If there is no danger of confusion, the subscript l will be omitted.

The constraint mapping for FC Λ_l is thus

$$f_l\,(\mathbf{q}) := \exp(\sigma_{l,\underline{l}}\mathbf{Y}_{\underline{l}}q_{\underline{l}})\exp(\sigma_{l,i}\mathbf{Y}_i q_i)\exp(\sigma_{l,j}\mathbf{Y}_j q_j)\cdot\ldots\cdot\exp(\sigma_{l,\overline{l}}\mathbf{Y}_{\overline{l}}q_{\overline{l}}),$$
$$\underline{l} <_l i <_l j <_l \ldots <_l \overline{l} \in \Lambda_l \qquad (9)$$

which resembles (4) except that joint $\underline{l}$ serves as first and joint $\overline{l}$ as the last joint in the chain. The system of loop closure constraints defines the c-space of the linkage as

$$V = f_1^{-1}\,(\mathbf{I}) \cap \ldots \cap f_\gamma^{-1}\,(\mathbf{I}) \cap \mathbb{V}^n \qquad (10)$$

with FCs $\Lambda_1, \ldots, \Lambda_\gamma$.

The corresponding velocity constraints for the multi-loop linkage are of the form (7) with constraint Jacobian

$$\mathbf{J} = \begin{pmatrix} \mathbf{J}_1 \\ \vdots \\ \mathbf{J}_\gamma \end{pmatrix} \qquad (11)$$

where $\mathbf{J}_l$ is the constraint Jacobian for the FC l composed of the relevant joint screw coordinates

$$\mathbf{S}_i^l(\mathbf{q}) := \sigma_{l,i}\mathbf{Ad}_{g_{l,i}(\mathbf{q})}\mathbf{Y}_i \tag{12}$$

with

$$g_{l,i}(\mathbf{q}) := \exp(\sigma_{l,\underline{l}}q_{\underline{l}}\mathbf{Y}_{\underline{l}}) \cdot \ldots \cdot \exp(\sigma_{l,i-1}q_{i-1}\mathbf{Y}_{i-1})\exp(\sigma_{l,i}q_i\mathbf{Y}_i). \tag{13}$$

For the linkage in Fig. 3a, for example, using the FCs in Fig. 3b, the constraint Jacobians $\mathbf{J}_l, l = 1, 2$ are

$$\begin{aligned} \mathbf{J}_1 &= \left(\mathbf{S}_1, \mathbf{S}_2, \mathbf{S}_3, \mathbf{0}, \mathbf{0}, \mathbf{0}, \mathbf{0}, \mathbf{S}_8\right) \\ \mathbf{J}_2 &= \left(\mathbf{0}, \mathbf{S}_2, \mathbf{0}, \mathbf{S}_4, \mathbf{S}_5, \mathbf{S}_6, \mathbf{S}_7, \mathbf{0}\right). \end{aligned} \tag{14}$$

4 Phenomenology and Definitions

A finite motion of a linkage[1] corresponds to a curve in its c-space. The basic kinematic characteristic of a linkage is its mobility, i.e., how it can move, whereas its DOF is the dimension of possible motions. It is further necessary to distinguish between finite and infinitesimal motions. The important point (in fact the constitutive feature for the occurrence of singularities) is that both are configuration dependent.

Figure 4 shows a three-dimensional section of the c-space V of the planar four-bar linkage in Fig. 1. The c-space of this one-DOF linkage is the union of curves. Except at the intersections of these curves, the c-space is a smooth manifold, and the linkage moves continuously. At any of these 'smooth' points, there is a well-defined tangent space, which in this one-dimensional case is simply the tangent to the curve. The tangent space at a given point seems to reveal possible infinitesimal (instantaneous) motions at that point. Apparently, at smooth points, the dimension of the c-space and its tangent space, denoted $T_{\mathbf{q}}V$, is equal, while at the bifurcation point $\mathbf{q}_0$ the tangent space $T_{\mathbf{q}_0}V$ is a two-dimensional plane as it accounts for tangents to both of the intersecting c-space branches. The tangent space does clearly not provide information about finite motions through a general point. Such a separation of tangent vectors according to the motion branches of the four bars is provided by the tangent cone, denoted $C_{\mathbf{q}_0}V$ (Whitney 1965), which can provisionally be regarded as the set of tangents to curves through $\mathbf{q}_0 \in V$. Clearly, when V is smooth at $\mathbf{q} \in V$, then $C_{\mathbf{q}}V = T_{\mathbf{q}}V$. For this example, the tangent space is given at any point by $T_{\mathbf{q}}V = \ker \mathbf{J}(\mathbf{q})$.

Now, consider the pathological four-bar linkage in Fig. 5. This linkage is immobile since the sum of the lengths of the three shorter links equals that of the longest link.

[1] The term 'linkage' will be used in the following, although the definitions apply to general mechanisms.

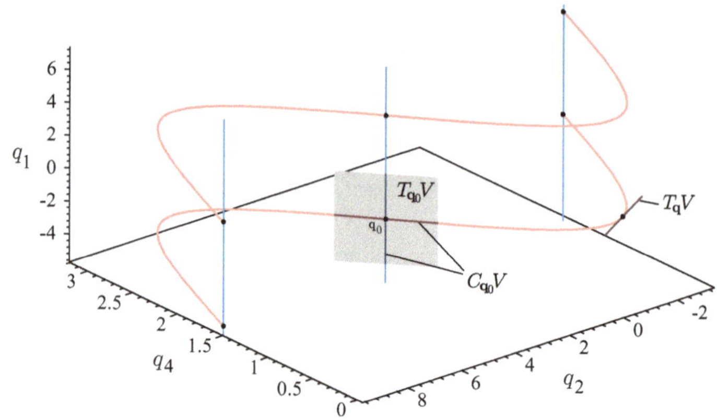

Fig. 4 q_1-q_2-q_4 section of the c-space of the planar four-bar linkage

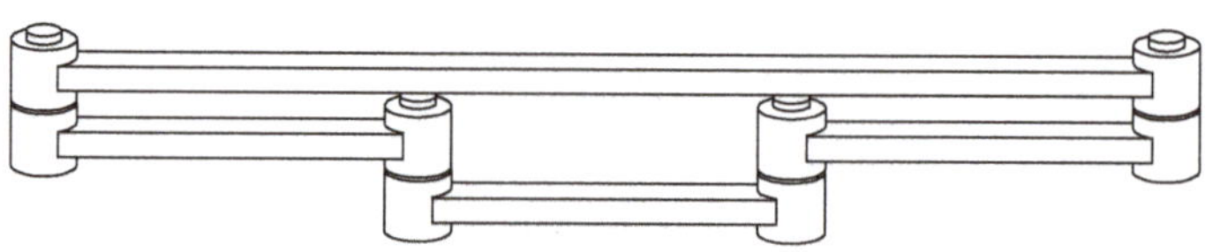

Fig. 5 Immobile planar four-bar linkage

The c-space is hence a single point, i.e., the zero-dimensional smooth manifold $V = \{\mathbf{q}_0\}$. There is indeed no curve through $\mathbf{q}_0$. Yet, since the Jacobian is rank deficient at $\mathbf{q}_0$, the linkage possesses infinitesimal two-dimensional motions defined by the null-space of the constraint Jacobian. Since the tangent space to the zero-dimensional manifold is $T_{\mathbf{q}_0}V = \{\mathbf{0}\}$, it is clearly not defined by $\ker \mathbf{J}(\mathbf{q}_0)$. Consequently, the tangent space does not reveal information about the instantaneous mobility.

As a final introductory example, consider the 'extended Watt' linkage in Fig. 3. This linkage possesses motion modes with a different finite DOF (Figs. 3 and 6). The linkage transits between the different motion modes via the singular configuration in Fig. 7. Linkages that can enter motion modes with different finite mobilities, without the need for reassembling, are called kinematotropic (clearly different from the 6R example in Fig. 2). The c-space is thus the union of one-dimensional (curves) and two-dimensional (surfaces) components, as depicted in Fig. 8. Similar to the immobile four bars, this linkage exhibits higher-dimensional infinitesimal than finite motions in its one-DOF motion mode in Fig. 6. That is, in parts of the c-space the tangent space does not reflect the infinitesimal motions even when the c-space is smooth. This leads to the notion of shakiness (Wohlhart 1999, 2010).

These examples show that the geometry of V reveals the finite motion characteristic of a linkage, whereas the infinitesimal aspects are determined by the particular system of loop constraints rather than by the differential aspect of V.

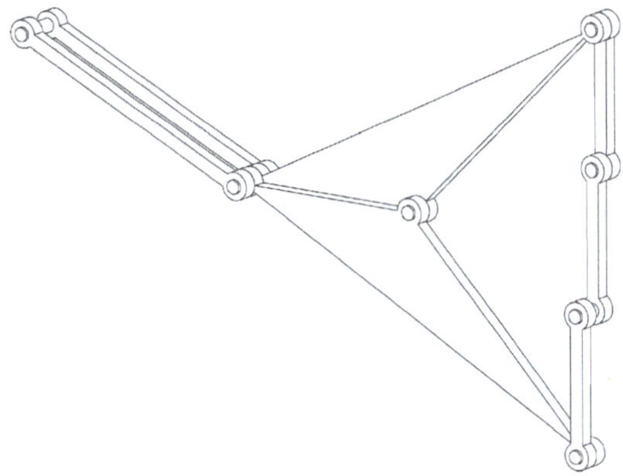

Fig. 6 Extended two-loop Watt linkage in its one-DOF motion mode

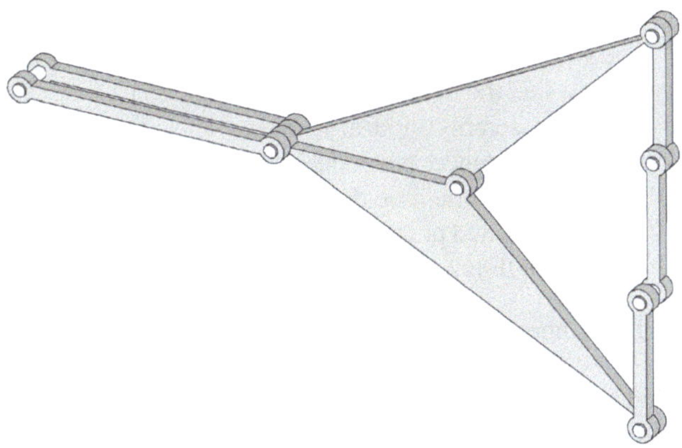

Fig. 7 Extended two-loop Watt linkage in a singular configuration where it can bifurcate to its one-DOF and two-DOF motion modes

Fig. 8 Schematic representation of the local geometry of the c-space of the extended Watt linkage

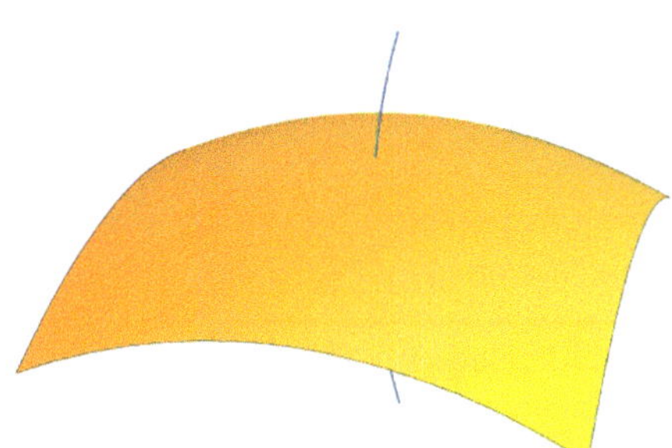

Definition 4.1 The *differential DOF* (also called instantaneous DOF) of the linkage at $\mathbf{q} \in V$ is $\delta_{\text{diff}}(\mathbf{q}) := n - \operatorname{rank} \mathbf{J}(\mathbf{q})$.

Definition 4.2 The *local (finite) DOF* (or simply *DOF*) at $\mathbf{q} \in V$ is the local dimension of the variety V, denoted $\delta(\mathbf{q}) := \dim_{\mathbf{q}} V$. This is the highest dimension of manifolds passing through $\mathbf{q}$.

Here, 'local' indicates that this is the (highest) finite DOF in a sufficiently small neighborhood of $\mathbf{q}$, i.e., the local dimension of V. This is sometimes indicated by the notation $\delta_{\text{loc}}(\mathbf{q})$ (Müller 2016b, 2018b), which is omitted here. The terminology must not be confused with the term 'local mobility' occasionally used to refer to the mobility of a subsystem of a mechanism (see, e.g., Takeda 2006). In order to simplify notation, the argument in $\delta(\mathbf{q})$ is often omitted if the point $\mathbf{q}$ is clear from the context. Further, V consists of manifolds of regular points and the linkage usually moves along a curve of regular configurations. When referring to the corresponding motion, then δ is understood as the finite DOF at these configurations.

Definition 4.3 The *global DOF* $\bar{\delta}$ is the highest local DOF the linkage possesses at feasible configurations, i.e., the highest local dimension of V.

If there are different assembly modes, it makes sense to restrict this definition to the one of interest. For instance, when dealing with the rigid structure in Fig. 2a, the global DOF is $\bar{\delta} = 0$ and should be deduced from the corresponding connected component of V, i.e., the isolated point, whereas when the Bricard linkage in Fig. 2b) is considered the global DOF is $\bar{\delta} = 1$.

Definition 4.4 The connected components of V are the *assembly modes* of the linkage.

This definition must not be confused with the use of the term 'assembly mode' that is frequently used to refer to the inverse kinematic solution of serial and parallel manipulators. A connection to the latter can be made observing that for fixed end-effector a serial manipulator forms a kinematic loop so that the inverse kinematic solution determines how the manipulator must be assembled in between the ground and the end-effector (analogously for parallel manipulators).

Definition 4.5 A linkage is said to be *kinematotropic* if an assembly mode comprises manifolds of different dimensions (a different local DOF).

In other words, a linkage is kinematotropic if it can change its finite DOF via a finite motion (Wohlhart 1996).

Before defining 'singularities,' some more preparation is in order. Consider again the four-bar linkage. The configuration $\mathbf{q}_0$ is a bifurcation point of the c-space V. Such points, where V ceases to be a manifold, are called singular points of V as a topological set. While this being in obvious agreement with the differential geometric notions of non-smooth points, there are more subtle points to take into account.

The first point is that the differential DOF can change even if the c-space is a smooth manifold. This may, at first sight, seem to contradict the experience of most readers, which is supported by publications such as Zlatanov et al. (2002).

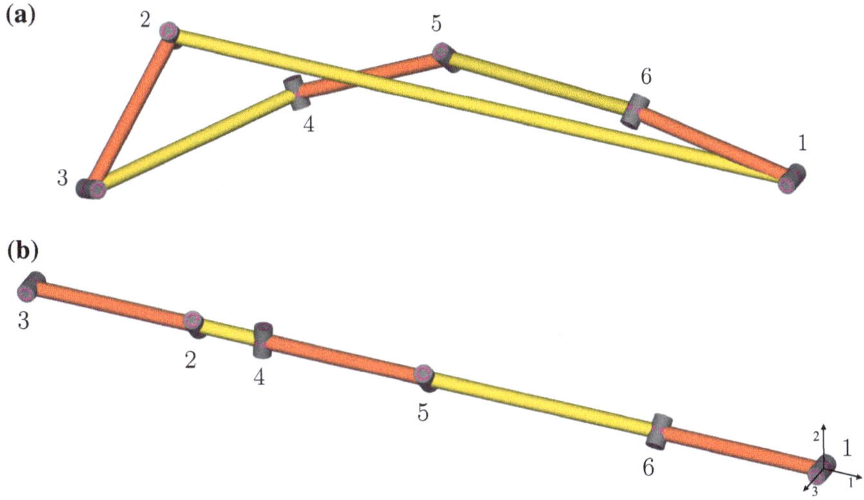

Fig. 9 Overconstrained 6R linkage in **a** regular configuration and **b** in a kinematic singularity

The 6R linkage in Fig. 9a is such linkages. It is a variant of the linkages presented by Song and Chen (2012), with Denavit–Hartenberg parameters $\alpha_{12} = \frac{3}{4}\pi, \alpha_{23} = \alpha_{61} = \frac{\pi}{4}, \alpha_{34} = \frac{\pi}{2}, \alpha_{45} = -\frac{\pi}{4}, \alpha_{56} = \frac{\pi}{2}, a_{12} = 2 + \frac{1}{\sqrt{2}}, a_{34} = a_{56} = 1, a_{23} = a_{61} = \frac{1}{\sqrt{2}}, a_{45} = \frac{1}{\sqrt{2}}, d_i = 0.3, i = 1, \ldots, 6$. Its c-space is a smooth one-dimensional manifold, so that its finite (global) DOF is 1. As one would expect, the constraint Jacobian is rank deficient with rank 5 (as for the Bricard linkage). However, in the configuration in Fig. 9b, where all joint axes intersect with a common line, the rank of the Jacobian further drops to 4. Yet, the c-space is smooth, and in contrast to common belief, the drop of rank, and thus increase of differential mobility, cannot be attributed to a singularity of the c-space, like the bifurcation in case of the four bars. That is, the singularity is not reflected in the differential geometric properties of the c-space but rather encoded in the defining equations.

Secondly, it is frequently said that a linkage is singular whenever the loop constraints it is subjected to are dependent, i.e., whenever the constraint Jacobian *is* rank deficient. This is indeed not in agreement with the notion of a 'singularity' as an exceptional point, which requires the Jacobian to *change* its rank. In case of the Bricard linkage (Fig. 2b), for example, the loop closure imposes six constraints on the six revolute joint angles. Obviously, the linkage can only be mobile if there is a submanifold of V where the constraints are permanently redundant. That is, the constraint Jacobian is permanently singular, while the c-space is a smooth curve. It is important to distinguish singularities of the constraint mapping (4) and the c-space V.

Definition 4.6 (Golubitsky and Guillemin 1973) Points $\mathbf{q} \in \mathbb{V}^n$ where $\mathbf{J}(\mathbf{q})$ is not full rank are called *critical points* of f.

Now, with the mobility as the defining kinematic quality of a linkage, singularities can be defined as follows. Denote with $d_{\max} := \max_{q \in \mathbb{V}^n} \operatorname{rank} \mathbf{J}(\mathbf{q})$ the maximal number of possibly independent geometric loop constraints.

Definition 4.7 A point $\mathbf{q} \in \mathbb{V}^n$ is called a *constraint singularity* if it is a critical point of the constraint mapping f, i.e., if rank $\mathbf{J}(\mathbf{q}) < d_{\max}$.

Definition 4.8 A configuration $\mathbf{q} \in V$ is a *kinematic singularity* if and only if δ_{diff} is not constant in any neighborhood of $\mathbf{q}$ in V. Otherwise, $\mathbf{q}$ is a *regular configuration*.

Definition 4.9 A configuration $\mathbf{q} \in V$ is a *c-space singularity* (topological singularity of V) if and only if V is not a smooth manifold at $\mathbf{q}$.

Notice that a c-space singularity is a kinematic singularity, while the reverse is not necessarily true (as for the above 6R linkage).

Definition 4.10 A linkage is called *overconstrained* (*redundantly constrained*) if and only if V consists of submanifolds of critical points of f.

A linkage is thus overconstrained if and only if there are regular configurations where the constraint Jacobian is singular.

Definition 4.11 A linkage is called *shaky* (*underconstrained*) if and only if there is a regular configuration $\mathbf{q} \in V$ where $\delta_{\text{diff}}(\mathbf{q}) > \delta(\mathbf{q})$.

Notice that the concept of shakiness is frequently restricted to structures, (i.e., immobile linkages), while Definition 4.11 also applies to mobile linkages. A shaky linkage is automatically overconstrained (with redundant constraints). The pathological four-bar linkage in Fig. 5 is an overconstrained and shaky structure, while the extended Watt linkage in Fig. 3 is a shaky mobile linkage. Vectors $\mathbf{x} \in \ker \mathbf{J}(\mathbf{q})$ are generally tangents to curves in the c-space and can be identified with the joint velocities $\dot{\mathbf{q}}$. This is not true at kinematic singularities and for shaky linkages.

These definitions relate the phenomena of interest to the corresponding mathematical objects. Clearly, analyzing kinematic singularities involves analysis of the c-space geometry as well of the actual motion. This will be addressed in the following by assessing local properties of smooth curves (the actual motions) as well as the local geometry of the c-space.

Before embarking into the analysis, a final example shall clarify the significance of separately considering the geometry of finite motions and of the c-space. The planar linkage in Fig. 10 was presented by Connelly and Servatius (1994), which is constructed by a combination of two Watt linkages whose coupler curves approximate straight lines. While this was possibly the first linkage that was reported to have a cusp singularity in c-space, other linkages with cusps were reported recently (López-Custodio et al. 2018a, b). A two-dimensional section of the local c-space geometry at the cusp is schematically depicted in Fig. 11.

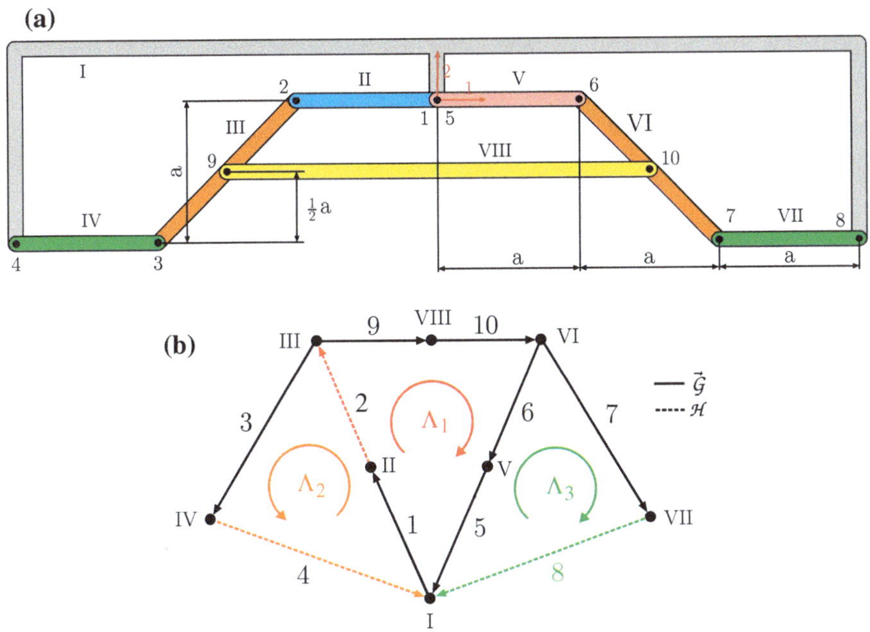

Fig. 10 **a** Double-Watt linkage in the reference configuration. **b** Topological graph and FCs

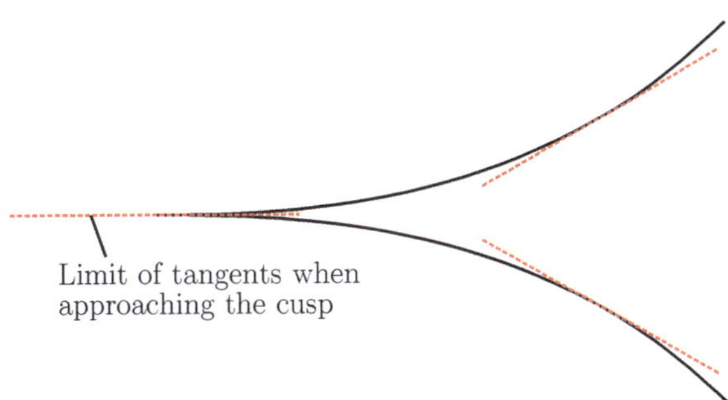

Fig. 11 Schematic representation of a cusp and its 'tangents'

The fundamental difference between a cusp and a bifurcation is that there is no continuous motion possible through the cusp so that the linkage must come to a halt before proceeding along one of the branches (which is why cusps are sometimes called stationary singularities). It is clear that care must be taken when speaking of tangents at a cusp. Furthermore, locally the two branches of the cusp become parallel (see Sect. 7). Consequently, the tangential aspects of finite motions through a cusp singularity do not reveal anything about the finite motion characteristic. Surprisingly, the c-space singularities of virtually all mechanisms that were reported in the literature are bifurcations. Even at the singularity of the kinematotropic Watt linkage in

Fig. 7, where a curve and a surface intersect, there are well-defined curves through the singularity. Not surprisingly all approaches reported in the literature make (implicit) use of the continuity assumption.

Definition 4.12 The *rank of the configuration* $\mathbf{q}$ is understood as the rank of $\mathbf{J}(\mathbf{q})$. The set of constraint singularities of rank less than k is denoted with

$$L_k := \{\mathbf{q} \in V | \mathrm{rank}\, \mathbf{J}(\mathbf{q}) < k\}. \tag{15}$$

The set of constraint singularities of rank k is thus $R_k := L_{k+1} \backslash L_k$. It is clear that $L_1 \subsetneqq L_2 \subsetneqq \ldots \subsetneqq L_{d_{\max}}$.

Definition 4.13 The set of constraint singularities is denoted with $\Sigma := L_{d_{\max}}$. The set of c-space singularities (singularities of V) is denoted with Σ_V and the set of kinematic singularities with Σ_K.

A constraint singularity may not be a c-space singularity (as for overconstrained and underconstrained mechanisms), and it may neither be a kinematic singularity. Also, a kinematic singularity is not necessarily a c-space singularity (e.g., in Fig. 9). This leads to the inclusion $\Sigma_V \subseteq \Sigma$ and $\Sigma_K \subset \Sigma$.

Definition 4.14 If Σ_V is not a smooth manifold at $\mathbf{q} \in V$, then $\mathbf{q}$ is a *higher-order c-space singularity.*

Definition 4.15 If δ_{diff} is not constant in any neighborhood of $\mathbf{q}$ in Σ_K, then $\mathbf{q}$ is a *higher-order kinematic singularity.*

Remark 4.16 Almost all points are regular points of V since Σ_V is a lower-dimensional closed subset of V. If Σ_V is not smooth, the variety of c-space singularities has itself singularities. The definition of higher-order singularities of smooth mappings (Golubitsky and Guillemin 1973) must not be confused with the above notion of higher-order singularities. Denote the critical points of f (i.e., where its Jacobian is not full rank) with Σ_f, and let f_Σ be the mapping f restricted to Σ_f. A critical point $\mathbf{q} \in \Sigma_f$ is a higher-order singularity of f if it is a critical point of f_Σ. The mapping f_Σ may also possess critical points and can be restricted to the set of its critical points. Proceeding in this way leads to the Thom-Boardman classification (Gibson 1979).

Thus, a kinematic singularity $\mathbf{q} \in V$ of an overconstrained mechanism is a higher-order singularity of f, since its c-space consists of manifolds of critical points of the constraint mapping, i.e., locally $V \cap U(\mathbf{q}) = \Sigma_f \cap U(\mathbf{q})$, and a kinematic singularity is then a critical point of f when restricted to Σ_f (see Examples 5.4 III and 5.4 IV).

Throughout this chapter, the rank k is used to label L_k and R_k in (15) rather than the corank, since it may occasionally be appropriate to restrict the image of f to a lower-dimensional subalgebra, as for planar linkages for instance.

Aiming at a computational procedure, the above definitions can be rephrased using the set of constraint singularities as follows:

1. $L_{d_{\max}}$ is the restriction of the set of critical points of f to V.
2. A point $\mathbf{q}$ of rank r belongs to L_k, for all $k > r$.
3. If $L_{d_{\max}} = \emptyset$, then there are no kinematic singularities.
4. A configuration $\mathbf{q}$ of rank r is regular if and only if V is smooth at $\mathbf{q}$ and $L_{r+1} \cap U(\mathbf{q}) = V \cap U(\mathbf{q})$. Otherwise, $\mathbf{q}$ is a kinematic singularity. If $\mathbf{q}$ is regular, then δ_{diff} is locally constant and equal to $n - r$.
5. Let $\mathbf{q}$ be a regular point with rank $r < d_{\max}$. If in addition $\delta_{\text{diff}}(\mathbf{q}) = \delta(\mathbf{q})$, then the mechanism is overconstrained, and if $\delta_{\text{diff}}(\mathbf{q}) > \delta(\mathbf{q})$, it is underconstrained.
6. At a point $\mathbf{q}$ of rank r, R_r and L_{r+1} are locally identical, that is $R_r \cap U(\mathbf{q}) = L_{r+1} \cap U(\mathbf{q})$ (because $L_r \cap U(\mathbf{q}) = \emptyset$).
7. At a point $\mathbf{q}$ of rank r, R_r and L_{r+1} are smooth manifolds.

Statement 4 is simply repeating that kinematic singularities are not necessarily c-space singularities, so that for a point being regular it is not sufficient to require that V is locally smooth. Statements 6 and 7 follow from the fact that points with rank less than k form a closed subvariety of L_k and thus of V, so that the rank cannot decrease in a sufficiently small neighborhood of any point.

The presented examples show that an exhaustive and conclusive local analysis involves investigation of possible smooth curves, the rank at these curves, and the analysis of the local c-space geometry. It will be seen that the latter is the most difficult.

With slight abuse of language, finite motions with constant rank k will be called 'motions with rank k.' Curves in V consisting of regular points will be called regular motions.

5 Local Analysis of Smooth Finite Motions

Finite motions of the linkage correspond to curves in its c-space. A smooth motion is represented by an analytic curve, which possesses a well-defined tangent vector. At any point $\mathbf{q} \in V$, the set of tangents to smooth curves passing through that point locally approximates the geometry of smooth finite motions. This set forms a cone in general rather than a vector space. It could be argued that this provides a faithful local approximation of the c-space. However, the crucial point is that it is defined by tangents to analytic curves. That is, situations where there are no smooth curves through the point in question, so that a tangent vector is not defined, are not captured. Moreover, it may appear inconsistent to consider smooth motions when aiming to investigate singularities. An intuitive justification is that almost all linkages exhibit singularities that are due to non-tangential intersection. A simple explanation might be that cusps are non-generic. An exhaustive local singularity and mobility analysis hence consists of an approximation of finite motions as well as of the actual c-space geometry. In this section, a formal framework for the local approximation of smooth motions is presented along with the necessary computational means.

5.1 Kinematic Tangent Cone to the C-Space

The tangency of the family of arcs passing through a point is formalized by the following.

Definition 5.1 The *kinematic tangent cone* of V at $\mathbf{q} \in V$ (i.e., of a linkage with c-space variety V), denoted with $C_{\mathbf{q}}^{\mathrm{K}} V$, is the set of tangent vectors to smooth arcs in V through $\mathbf{q}$.

A finite motion $\mathbf{q}(t)$ of the linkage satisfies the velocity constraints and all its time derivatives. For later use, the ith-order constraint is written as

$$H_l^{(i)}(\mathbf{q}, \dot{\mathbf{q}}, \ldots, \mathbf{q}^{(i)}) = \mathbf{0} \tag{16}$$

with the mappings $H_l^{(i)} : \mathbb{V}^n \times \mathbb{R}^n \times \ldots \times \mathbb{R}^n \to \mathbb{R}^6 \cong se(3)$ defined as

$$\begin{aligned} H_l^{(1)}(\mathbf{q}, \dot{\mathbf{q}}) &:= \mathbf{J}_l(\mathbf{q})\dot{\mathbf{q}} \\ H_l^{(2)}(\mathbf{q}, \dot{\mathbf{q}}, \ddot{\mathbf{q}}) &:= \frac{d}{dt} H_l^{(1)}(\mathbf{q}, \dot{\mathbf{q}}) \\ &\cdots \\ H_l^{(i)}(\mathbf{q}, \dot{\mathbf{q}}, \ldots, \mathbf{q}^{(i)}) &:= \frac{d^{i-1}}{dt^{i-1}} H_l^{(1)}(\mathbf{q}, \dot{\mathbf{q}}) \end{aligned} \tag{17}$$

where, for the sake of compactness, the notation $\mathrm{D}^{(k)} := \frac{d^k}{dt^k}$ and $\mathbf{q}^{(k)} := \frac{d^k}{dt^k}\mathbf{q}$ will be used whenever appropriate.

For a velocity vector $\dot{\mathbf{q}}$, i.e., the tangent to a curve, there are higher-order time derivatives of $\mathbf{q}(t)$ satisfying the constraints (16) for any i. The set of tangent vectors to smooth curves through $\mathbf{q}$ in V thus defines the kinematic tangent cone as

$$C_{\mathbf{q}}^{\mathrm{K}} V = K_{\mathbf{q}}^{\kappa} \subseteq \ldots \subseteq K_{\mathbf{q}}^{3} \subseteq K_{\mathbf{q}}^{2} \subseteq K_{\mathbf{q}}^{1} \tag{18}$$

where

$$\begin{aligned} K_{\mathbf{q}}^{i} := \{\mathbf{x} | \exists \mathbf{y}, \mathbf{z}, \ldots \in \mathbb{R}^n : \; & H_l^{(1)}(\mathbf{q}, \mathbf{x}) = \mathbf{0}, \\ & H_l^{(2)}(\mathbf{q}, \mathbf{x}, \mathbf{y}) = \mathbf{0}, \\ & H_l^{(3)}(\mathbf{q}, \mathbf{x}, \mathbf{y}, \mathbf{z}) = \mathbf{0}, \\ & \cdots \\ & H_l^{(i)}(\mathbf{q}, \mathbf{x}, \mathbf{y}, \mathbf{z}, \ldots) = \mathbf{0}, l = 1, \ldots, \gamma. \end{aligned} \tag{19}$$

There is a finite κ for which the series in (18) terminates. The inclusion is not strict. From its definition follows that $K_{\mathbf{q}}^{i}$, and thus $C_{\mathbf{q}}^{\mathrm{K}} V$, is a cone in $\mathbb{R}^n$, i.e., when $\mathbf{x} \in C_{\mathbf{q}}^{\mathrm{K}} V$ then also $a\mathbf{x} \in C_{\mathbf{q}}^{\mathrm{K}} V$, $a \in \mathbb{R}$. This is the set of tangents to smooth finite curves through $\mathbf{q} \in V$, which motivates the name *kinematic tangent cone* of V at $\mathbf{q}$. The adjective 'kinematic' refers to the fact that it characterizes the tangent aspects to finite curves through a configuration $\mathbf{q} \in V$. The ith-order cone $K^i_{\mathbf{q}}$ reveals possible differential motions of order i (which may not be tangent to a curve). Clearly, $K^1_{\mathbf{q}} = \ker\mathbf{J}(\mathbf{q})$ is the vector space of first-order differential motions.

The mappings $H^{(i)}_l(\mathbf{q}, \mathbf{x}, \mathbf{y}, \mathbf{z}, \ldots)$ consists of polynomials of order i in $\mathbf{x}$, of order $i-1$ in $\mathbf{y}$, etc. The kinematic tangent cone is an algebraic variety defined by these polynomials. It provides an algebraic model that locally approximates the set of possible smooth curves in V through $\mathbf{q}$. In contrast to an algebraic formulation of the constraints (Remark 2.2), this algebraic variety is of (low) order i.

It is important to emphasize that it does not necessarily approximate the local geometry of the c-space V (as for the linkage in Fig. 10). Such cases are rare but have just recently attracted attention (López-Custodio et al. 2018b; Müller 2018a; López-Custodio et al. 2018a). More on this will be discussed in Sect. 7. Nevertheless, the following statement is obvious.

Lemma 5.2 *If the c-space is a smooth manifold at* $\mathbf{q} \in V$*, then* $C^{\mathrm{K}}_{\mathbf{q}}V$ *is a vector space, i.e.,* $C^{\mathrm{K}}_{\mathbf{q}}V = \mathrm{span}\, C^{\mathrm{K}}_{\mathbf{q}}V$.

The converse statement is not always true, so that $C^{\mathrm{K}}_{\mathbf{q}}V$ being a vector space is not sufficient for $\mathbf{q}$ being a regular point of V.

Lemma 5.3 *If* $\mathbf{q} \in V$ *is a kinematic singularity, then* $K^i_{\mathbf{q}} \neq K^{i+1}_{\mathbf{q}}$ *for some* i.

This condition is only necessary but not sufficient. For immobile linkages, it is $\dim K^1_{\mathbf{q}} > 0$ and $\dim K^{\kappa}_{\mathbf{q}} = 0$. For shaky mobile linkage in regular configuration, it is always $\dim K^1_{\mathbf{q}} > \dim K^{\kappa}_{\mathbf{q}} > 0$.

Lemma 5.4 *The dimension of the kinematic tangent cone is a lower bound on the local finite DOF at* $\mathbf{q} \in V$*, i.e.,* $\delta(\mathbf{q}) \geq \dim C^{\mathrm{K}}_{\mathbf{q}}V$.

The example in Sect. 5.4.5 is a case where the dimension of $C^{\mathrm{K}}_{\mathbf{q}}V$ is strictly less than $\delta(\mathbf{q})$.

Definition 5.5 A linkage is called shaky of order k iff $\dim K^k_{\mathbf{q}} > \delta_{\mathrm{loc}}(\mathbf{q})$ at a regular configuration $\mathbf{q} \in V$.

This is a refinement of Definition 4.11. It is instructive to distinguish the shakiness of a certain higher order. At a (regular) configuration $\mathbf{q} \in V$, the linkage possesses kth-order motions of dimension $\dim K^k_{\mathbf{q}}$. For instance, the linkage in Fig. 2 of Wohlhart (1999) is immobile but exhibits one-dimensional first-, second-, and third-order motions (see Müller 2016b). Another example is the extended Watt linkage in Fig. 3 (see Sect. 5.4.3).

5.2 *Higher-Order Loop Closure Constraints*

5.2.1 Closed-Form Relations

Evaluating the defining equations in (19) requires the higher-order derivatives of the constraint Jacobian. The partial derivative of the instantaneous joint screw coordinates (6) can be expressed in terms of the Lie bracket as (Müller 2014)

$$\frac{\partial \mathbf{S}_i}{\partial q_j} = [\mathbf{S}_j, \mathbf{S}_i] \\ = \mathbf{ad}_{\mathbf{S}_j}\mathbf{S}_i, \ j < i \tag{20}$$

and vanish for $j \geq i$, where $\mathbf{ad}$ is the matrix form of the Lie bracket (Murray et al. 1994). This allows for a closed-form algebraic expression of the second-order (acceleration) constraints

$$\sum_{j\leq i} \mathbf{S}_j \ddot{q}_j + \sum_{k<j\leq i} \left[\mathbf{S}_k, \mathbf{S}_j\right] \dot{q}_j \dot{q}_k = \mathbf{0}. \tag{21}$$

Also for higher-order partial derivatives, there exists a closed form (Müller 2014) that can be written as

$$\frac{\partial^\nu \mathbf{S}_i}{\partial q_{\alpha_1} \partial q_{\alpha_2} \cdots \partial q_{\alpha_\nu}} = \left[\mathbf{S}_{\beta_\nu}, \left[\mathbf{S}_{\beta_{\nu-1}}, \left[\mathbf{S}_{\beta_{\nu-2}}, \ldots \left[\mathbf{S}_{\beta_1}, \mathbf{S}_i\right] \ldots\right]\right]\right] \\ = \mathbf{ad}_{\mathbf{S}_{\beta_\nu}} \mathbf{ad}_{\mathbf{S}_{\beta_{\nu-1}}} \mathbf{ad}_{\mathbf{S}_{\beta_{\nu-2}}} \cdots \mathbf{ad}_{\mathbf{S}_{\beta_1}} \mathbf{S}_i, \quad \text{if} \alpha_1, \ldots, a_\nu < i \tag{22}$$

where $\beta_\nu \leq \beta_{\nu-1} \leq \ldots \leq \beta_2 \leq \beta_1 < i$ is the ordered set of indexes $\{\alpha_1, \ldots, \alpha_\nu\}$. Therewith, all higher-order constraints can be formulated. For instance, the second-order (acceleration) and third-order (jerk) constraints are

$$\mathbf{0} = \sum_{j\leq n} \mathbf{S}_j \ddot{q}_j + \sum_{k<j\leq n} \left[\mathbf{S}_k, \mathbf{S}_j\right] \dot{q}_j \dot{q}_k \tag{23}$$

$$\begin{aligned} \mathbf{0} &= \sum_{j\leq n} \mathbf{S}_j \dddot{q}_j + 2 \sum_{k<j\leq n} \left[\mathbf{S}_k, \mathbf{S}_j\right] \dot{q}_k \ddot{q}_j + \sum_{k<j\leq n} \left[\mathbf{S}_k, \mathbf{S}_j\right] \ddot{q}_k \dot{q}_j \\ &+ \sum_{l<k<j\leq n} [[\mathbf{S}_l, \mathbf{S}_k], \mathbf{S}_j] \dot{q}_l \dot{q}_k \dot{q}_j + \sum_{l,k<j\leq n} [\mathbf{S}_k, [\mathbf{S}_l, \mathbf{S}_j]] \dot{q}_l \dot{q}_k \dot{q}_j \\ &= \sum_{j\leq n} \mathbf{S}_j \dddot{q}_j + 2 \sum_{l<k<j\leq n} \left[\mathbf{S}_l, \left[\mathbf{S}_k, \mathbf{S}_j\right]\right] \dot{q}_l \dot{q}_k \dot{q}_j \\ &+ \sum_{k<j\leq n} \left(\left[\mathbf{S}_k, \mathbf{S}_j\right] \left(\ddot{q}_k \dot{q}_j + 2\dot{q}_k \ddot{q}_j\right) + \left[\mathbf{S}_k, [\mathbf{S}_k, \mathbf{S}_j]\right] \dot{q}_k^2 \dot{q}_j \right). \end{aligned} \tag{24}$$

The last form in (24) is obtained using properties of the Lie bracket. The manipulation of these relations gets involved for higher orders.

5.2.2 Recursive Relation

The number of Lie brackets in the explicit closed form grows with order $O\ (n\ (\nu - 1))$, where ν is the order of the constraints. The computational burden of such an explicit construction can be alleviated by a recursive evaluation. To this end, introduce

$$\mathsf{S}_i(\mathbf{q},\dot{\mathbf{q}}) := \sum_{j\le i} \mathbf{S}_j\left(\mathbf{q}\right)\dot{q}_j \tag{25}$$

so that the velocity constraints (7) read

$$\mathsf{S}_n\left(\mathbf{q},\dot{\mathbf{q}}\right) = \mathbf{0}. \tag{26}$$

A straightforward derivation shows that (Müller 2016c)

$$\mathrm{D}^{(k)}\mathsf{S}_i = \sum_{j\le i}\sum_{l=0}^{k} \tbinom{k}{l}\mathrm{D}^{(l)}\mathbf{S}_j q_j^{(k-l+1)} \tag{27}$$

$$\mathrm{D}^{(k)}\mathbf{S}_i = \sum_{l=0}^{k-1} \tbinom{k-1}{l}[\mathrm{D}^{(l)}\mathsf{S}_i, \mathrm{D}^{(k-l-1)}\mathbf{S}_i]. \tag{28}$$

The kth time derivative of the loop constraints, i.e., of S_n, is thus given in terms of the time derivatives (28) of joint screws up to order k, which in turn are given in terms of time derivative of S_i, $i \le n$ up to order $k-1$.

Therewith, the ith-order constraint mapping in (16) is given as

$$H^{(i)}(\mathbf{q},\dot{\mathbf{q}},\ldots,\mathbf{q}^{(i)}) := \mathrm{D}^{(i-1)}\mathsf{S}_n\left(\mathbf{q},\dot{\mathbf{q}}\right). \tag{29}$$

Constraints of arbitrary order can thus be recursively constructed in symbolic form or evaluated for given arguments.

5.3 Computational Procedure

Without loss of generality, the configuration under consideration can always be chosen to be the reference $\mathbf{q} = \mathbf{0}$. Then, the instantaneous joint screws in (27) and (28) are simply $\mathbf{S}_i\left(\mathbf{0}\right) = \mathbf{Y}_i$. Thus, the computation involves only simple algebraic vector operations.

The determination of the kinematic tangent cone consists of the following three main steps:

1. Determination of the joint screw coordinates $\mathbf{Y}_i$ in the reference configuration $\mathbf{q} = \mathbf{0}$
2. Construction of the polynomials $H^{(i)}(\mathbf{q},\mathbf{x},\mathbf{y},\mathbf{z},\ldots)$, $i = 1,\ldots,\kappa$, $l = 1,\ldots,\gamma$
3. Explicitly or implicitly describe $K_{\mathbf{q}}^{i}$ by either
   (a) Reduction to a system of polynomials $p_1\left(\mathbf{x}\right),\ldots,p_N\left(\mathbf{x}\right)$ of degree i in $\mathbf{x}$ that implicitly defines $K_{\mathbf{q}}^{i} = \mathbf{V}\left(p_1,\ldots,p_N\right)$
   (b) Solving the polynomial system for $\mathbf{x}$ in order to obtain an explicit parameterization of $K_{\mathbf{q}}^{i}$

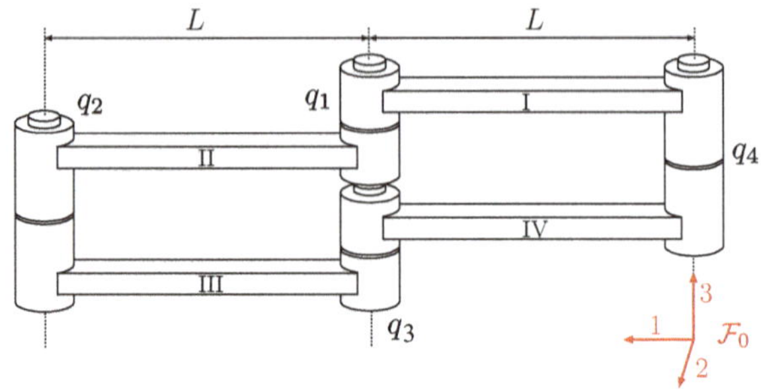

Fig. 12 Planar four-bar linkage in the reference configuration $\mathbf{q}_0$

In step 3, and in the following,

$$\mathbf{V}(p_1, \ldots, p_N) := \{\mathbf{x} \in \mathbb{R}^n | p_1(\mathbf{x}) = \ldots = p_N(\mathbf{x}) = 0\}$$

denotes the algebraic variety defined by the zero set of a system of N polynomials $p_i(\mathbf{x}), i = 1, \ldots, N$ in $\mathbf{x}$.

The first two steps are straightforward and can be easily implemented either symbolically or numerically using the above recursive relations. The third step deserves more attention. The symbolic elimination in order to obtain a system in $\mathbf{x}$ must take into account that only real solutions can be used. Most computer algebra systems treat the polynomials over the complex field, however.

5.4 Examples

All examples are analyzed in the reference configuration $\mathbf{q}_0 = \mathbf{0}$. That is, the instantaneous screw coordinates are $\mathbf{S}_i(\mathbf{0}) = \mathbf{Y}_i$.

5.4.1 Planar Four-Bar Linkage with Identical Link Lengths

The planar four-bar linkage with equal link lengths in its reference configuration is shown in Fig. 12. The joint screw coordinates deduced from the shown frame $\mathcal{F}_0$ are

$$\mathbf{Y}_1 = \mathbf{Y}_3 = (0, 0, 1, 0, -L, 0)^T, \mathbf{Y}_2 = (0, 0, 1, 0, -2L, 0)^T, \mathbf{Y}_4 = (0, 0, 1, 0, 0, 0)^T. \tag{30}$$

Since only two of these $\mathbf{Y}_i$ are linearly independent, i.e., rank $\mathbf{J}(\mathbf{q}_0) = 2$, the differential DOF of the linkage is $\delta_{\text{diff}}(\mathbf{q}_0) = 4 - \text{rank}\,\mathbf{J}(\mathbf{q}_0) = 2$. The mappings (17) up to third order are

$$H^{(1)}\left(\mathbf{q}_0,\mathbf{x}\right)=\begin{pmatrix}0\\0\\x_1+x_2+x_3+x_4\\0\\-L\left(x_1+2x_2+x_3\right)\\0\end{pmatrix}$$

$$H^{(2)}\left(\mathbf{q}_0,\mathbf{x},\mathbf{y}\right)=\begin{pmatrix}0\\0\\y_1+y_2+y_3+y_4\\-L\left(x_2(x_1-x_3)-(x_1+2x_2+x_3)x_4\right)\\-L\left(y_1+2y_2+y_3\right)\\0\end{pmatrix}$$

$$H^{(3)}\left(\mathbf{q}_0,\mathbf{x},\mathbf{y},\mathbf{z}\right)=\begin{pmatrix}0\\0\\z_1+z_2+z_3+z_4\\L\Big((2x_1-x_3)y_2-x_4(y_1+2y2+y_3)\\-2(x_1+x_3)y_4+x_2(y_1-2(y_3+2y_4))\Big)\\L\Big(x_2(x_1^2-(2x_1+x_2)x_3)-z_1-2z_2-z_3\\-(x_1^2+4x_1x_2+2x_2^2+2(x_1+x_2)x_3+x_3^2)x_4\Big)\\0\end{pmatrix}.$$

The first-order cone (19) is the vector space of dimension $\delta_{\text{diff}}\left(\mathbf{q}_0\right)=2$ defined by $H^{(1)}\left(\mathbf{q}_0,\mathbf{x}\right)=\mathbf{0}$

$$K_{\mathbf{q}_0}^1=\mathbf{V}(x_1+x_3+2x_2,x_2-x_4)\subset\mathbb{R}^4.$$

Eliminating the variables $\mathbf{y}$ from $H^{(1)}\left(\mathbf{q}_0,\mathbf{x}\right)=H^{(2)}\left(\mathbf{q}_0,\mathbf{x},\mathbf{y}\right)=\mathbf{0}$ yields the equations $x_1+x_2+x_3+x_4=0$, $L\left(x_2x_3+x_2^2\right)=0$, $L\left(x_4-x_2\right)=0$ that determine the second-order cone

$$K_{\mathbf{q}_0}^2=K_{\mathbf{q}_0}^{2(I)}\cup K_{\mathbf{q}_0}^{2(II)}\quad\text{with}\quad \begin{aligned}K_{\mathbf{q}_0}^{2(I)}&=\mathbf{V}(x_1+x_3,x_2,x_4)\\K_{\mathbf{q}_0}^{2(II)}&=\mathbf{V}(x_1+x_2,x_3+x_2,x_4-x_2).\end{aligned}\tag{31}$$

Proceeding for higher orders shows that $K_{\mathbf{q}_0}^i=\ldots=K_{\mathbf{q}_0}^3=K_{\mathbf{q}_0}^2$, which shows that $C_{\mathbf{q}_0}^{\mathrm{K}}V=K_{\mathbf{q}_0}^2$. The kinematic tangent cone is the union of the two vector spaces $K_{\mathbf{q}_0}^{2(I)}$ and $K_{\mathbf{q}_0}^{2(II)}$. The latter are the tangent spaces to the set of smooth motions. From $C_{\mathbf{q}_0}^{\mathrm{K}}V$ can be concluded that the linkage has the local DOF $\delta\left(\mathbf{q}_0\right)\geq\dim C_{\mathbf{q}_0}^{\mathrm{K}}V=1$, and since it is not a vector space, with Lemma 5.2, the configuration $\mathbf{q}_0$ is a c-space singularity and thus a kinematic singularity. Because $\dim K_{\mathbf{q}_0}^1=2$ and $\dim K_{\mathbf{q}_0}^2=\dim C_{\mathbf{q}_0}^{\mathrm{K}}V=1$, the linkage can perform two-dimensional first-order motion and one-dimensional second- and higher-order motions.

In principle, the kinematic tangent cone does not reveal the c-space geometry. For this example, it is well know that two manifolds intersect at $\mathbf{q}_0$, which correspond to the two motion modes of the four-bar linkage, and that $K_{\mathbf{q}_0}^{2(I)}$ and $K_{\mathbf{q}_0}^{2(II)}$ are in fact the respective tangent spaces. This corresponds to the situation shown in Fig. 4 (although the linkage dimensions and thus the c-spaces are different, the c-space is locally the same).

5.4.2 Planar Four-Bar Linkage with Different Link Lengths

Figure 1 shows a planar four-bar linkage where the links have pairwise equal length. Assuming the lengths of links 1 and 4 are $2L$, and the lengths of links 2 and 3 are L, the joint screw coordinates relative to $\mathcal{F}_0$ are

$$\begin{aligned} \mathbf{Y}_1 &= \mathbf{Y}_3 = (0,0,1,0,-2L,0)^T \\ \mathbf{Y}_2 &= (0,0,1,0,-3L,0)^T, \mathbf{Y}_4 = (0,0,1,0,0,0)^T. \end{aligned} \tag{32}$$

The first- and second-order cones are found to be

$$\begin{aligned} K_{\mathbf{q}_0}^1 &= \mathbf{V}(2x_1 + 2x_3 + 3x_2, x_2 - 2x_4) \subset \mathbb{R}^4 \\ K_{\mathbf{q}_0}^2 &= K_{\mathbf{q}_0}^{2(I)} \cup K_{\mathbf{q}_0}^{2(II)} \text{ with } K_{\mathbf{q}_0}^{2(I)} = \mathbf{V}(x_1 + x_3, x_2, x_4) \\ & \qquad\qquad\qquad\qquad\quad K_{\mathbf{q}_0}^{2(II)} = \mathbf{V}(4x_1 + 3x_2, x_1 + x_3, 2x_1 + 3x_4). \end{aligned} \tag{33}$$

All higher-order cones are identical to $K_{\mathbf{q}_0}^2$ so that $C_{\mathbf{q}_0}^{\mathrm{K}} V = K_{\mathbf{q}_0}^2$. The kinematic tangent cone is thus again the intersection of two vector spaces, which are the tangent spaces to the two manifolds intersecting at $\mathbf{q}_0$. This is the situation shown in Fig. 4.

5.4.3 Extended Watt Linkage—Underconstrained Kinematotropic Two-Loop Planar Linkage

The planar two-loop linkage in Fig. 3a comprises eight revolute joints. The $\gamma = 2$ fundamental cycles (FCs) Λ_1 and Λ_2 are chosen as shown in Fig. 3b. For the analysis, the geometry as in Fig. 13 is used. The corresponding constraint mappings are

$$\begin{aligned} f_1(\mathbf{q}) &= \exp(\mathbf{Y}_2 q_2) \exp(\mathbf{Y}_3 q_3) \exp(\mathbf{Y}_8 q_8) \exp(\mathbf{Y}_1 q_1) \\ f_2(\mathbf{q}) &= \exp(\mathbf{Y}_2 q_2) \exp(\mathbf{Y}_4 q s_4) \exp(\mathbf{Y}_5 q_5) \exp(\mathbf{Y}_6 q_6) \exp(\mathbf{Y}_7 q_7) \end{aligned} \tag{34}$$

and the constraint Jacobians corresponding the individual FCs are shown in (14). These constitute the overall 12×8 constraint Jacobian (11). For the planar linkage, the constraint Jacobian has maximal rank $d_{\max} = 6$. At $\mathbf{q}_0$, it is rank $\mathbf{J}(\mathbf{q}_0) = 5$ so that $\mathbf{q}_0 \in L_6$.

The joint screw coordinates in the reference configuration $\mathbf{q}_0$ are

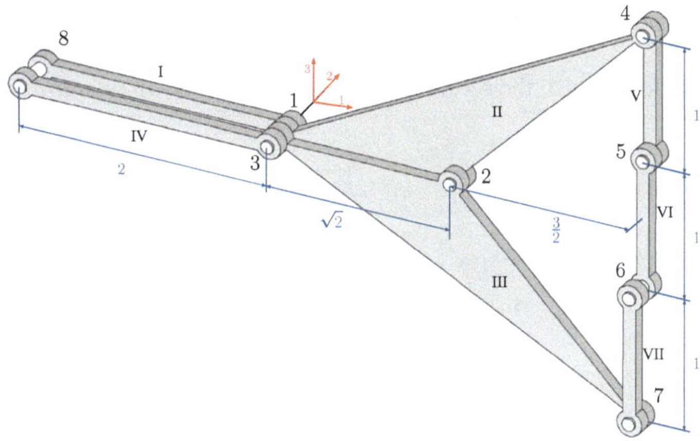

Fig. 13 Extended Watt linkage in singular configuration $\mathbf{q}_0$ with special geometry

$$
\begin{aligned}
&\mathbf{Y}_1 = \mathbf{Y}_3 = (0,0,1,0,0,0)^T, \mathbf{Y}_2 = (0,0,1,0,-\sqrt{2},0)^T \\
&\mathbf{Y}_4 = (0,0,1,\frac{3}{2},\frac{-3}{2}-\sqrt{2},0)^T, \mathbf{Y}_5 = (0,0,1,\frac{1}{2},\frac{-3}{2}-\sqrt{2},0)^T \\
&\mathbf{Y}_6 = (0,0,1,\frac{-1}{2},\frac{-3}{2}-\sqrt{2},0)^T, \mathbf{Y}_8 = (0,0,1,0,2,0)^T \\
&\mathbf{Y}_7 = (0,0,1,\frac{-3}{2},\frac{-3}{2}-\sqrt{2},0)^T.
\end{aligned} \tag{35}
$$

This gives rise to the cones

$$
\begin{aligned}
K^1_{\mathbf{q}_0} &= K^2_{\mathbf{q}_0} = \mathbf{V}\big(x_1+x_3, x_4-x_6-2x_7, x_5+2x_6+3x_7, x_2, x_8\big) \subset \mathbb{R}^8 \\
K^3_{\mathbf{q}_0} &= K^{3(I)}_{\mathbf{q}_0} \cup K^{3(II)}_{\mathbf{q}_0} \\
&\quad \text{with } \; K^{3(I)}_{\mathbf{q}_0} = \mathbf{V}\big(x_1+x_3, x_2, x_4, x_5, x_6, x_7, x_8\big) \\
&\qquad\quad\;\; K^{3(II)}_{\mathbf{q}_0} = \mathbf{V}\big(x_4-x_6-2x_7, x_5+2x_6+3x_7, x_1, x_2, x_3, x_8\big) \\
K^4_{\mathbf{q}_0} &= K^3_{\mathbf{q}_0}.
\end{aligned} \tag{36}
$$

The kinematic tangent cone is found to be $C^{\mathrm{K}}_{\mathbf{q}_0}V = K^4_{\mathbf{q}_0}$. This is the union of a one-dimensional and a two-dimensional vector space. It is not a vector space itself and thus $\mathbf{q}_0$ a c-space singularity. $K^{3(I)}_{\mathbf{q}_0}$ and $K^{3(II)}_{\mathbf{q}_0}$ are the tangent spaces to smooth motions through $\mathbf{q}_0$. It can be concluded that the local DOF of the linkage is $\delta(\mathbf{q}_0) \geq \dim C^{\mathrm{K}}_{\mathbf{q}_0}V = 2$. It is clear by inspection that these are the tangent spaces to the one-dimensional and two-dimensional manifolds, corresponding to the one-DOF and two-DOF motion modes, intersecting at $\mathbf{q}_0$. Consequently, the linkage is kinematotropic. This situation is depicted in Fig. 8. But it must be reminded that this cannot strictly be inferred from the analysis of regular curves.

5.4.4 Kinematic Singularity of a 6R Linkage with Smooth C-Space

The screw coordinates of the 6R linkage in Fig. 9b in its shown reference configuration are

$$\begin{aligned}
\mathbf{Y}_1 &= (0,0,1,0,0,0)^T\,, \mathbf{Y}_2 = (0, \frac{1}{\sqrt{2}}, \frac{-1}{\sqrt{2}}, 0, -\frac{1}{2}-\sqrt{2}, -\frac{1}{2}-\sqrt{2})^T \\
\mathbf{Y}_3 &= (0,0,-1,0,-2-\sqrt{2},0)^T, \mathbf{Y}_4 = (0,1,0,0,0,-1-\sqrt{2})^T \\
\mathbf{Y}_5 &= (0, \frac{1}{\sqrt{2}}, \frac{-1}{\sqrt{2}}, 0, -\frac{1}{2}-\frac{1}{\sqrt{2}}, -\frac{1}{2}-\frac{1}{\sqrt{2}})^T \\
\mathbf{Y}_6 &= (0, \frac{1}{\sqrt{2}}, \frac{1}{\sqrt{2}}, 0, \frac{1}{2}, -\frac{1}{2})^T.
\end{aligned} \tag{37}$$

At the reference configuration, it is rank $\mathbf{J}(\mathbf{q}_0) = 4$. The kinematic tangent cone is found as

$$\begin{aligned}
K^1_{\mathbf{q}_0} &= \mathbf{V}\big(2x_1 + x_5 + (\sqrt{2}-1)x_6, 2x_3 - x_5 + (\sqrt{2}-1)x_6), \\
&\qquad x_2 + (1+\sqrt{2})x_5 - x_6, x_4 - x_5 + \sqrt{2}x_6\big) \\
K^2_{\mathbf{q}_0} &= \mathbf{V}\big(x_1 + x_5, x_2, x_3, x_4 + (1+\sqrt{2})x_5, x_6 - (1+\sqrt{2})x_5\big)
\end{aligned}$$

and $K^2_{\mathbf{q}_0} = K^3_{\mathbf{q}_0} = \ldots = C^{\mathrm{K}}_{\mathbf{q}_0} V$. Since this is a vector space, the set of smooth curves through $\mathbf{q}_0$ should be locally a manifold. The linkage could be underconstrained and $\mathbf{q}_0$ a regular point. Yet, it might be a c-space singularity and/or a kinematic singularity. Checking the latter necessitates analysis of the rank along such curves (Sect. 6.3), while checking for c-space singularities demands analyzing the c-space geometry (Sect. 8.3).

The first-order cone has dimension $\dim K^1_{\mathbf{q}_0} = \delta_{\text{diff}}(\mathbf{q}_0) = 2$, so that there are two-dimensional first-order motions, but one-dimensional higher-order motions.

5.4.5 Double-Watt Linkage with Cusp Singularity in C-Space

Connelly and Servatius (1994) presented the planar one-DOF linkage in Fig. 10a as an example where the c-space exhibits a cusp singularity. It is formed by a mirror symmetric arrangement of two straight-line-generating Watt linkages that are connected by the rod VIII. It comprises $n = 10$ revolute joints with parallel axes and $V \subset \mathbb{T}^{10}$.

For the following analysis, the geometry parameter is set to $a = 1$. Then, in the reference configuration $\mathbf{q}_0$ in Fig. 10, the joint screws are

$$\begin{aligned}
\mathbf{Y}_1 &= \mathbf{Y}_5 = (0,0,1,0,0,0)^T, \mathbf{Y}_2 = (0,0,1,0,1,0)^T \\
\mathbf{Y}_3 &= (0,0,1,-1,2,0)^T, \mathbf{Y}_4 = (0,0,1,-1,3,0)^T \\
\mathbf{Y}_6 &= (0,0,1,0,-1,0)^T, \mathbf{Y}_7 = (0,0,1,-1,-2,0)^T \\
\mathbf{Y}_8 &= (0,0,1,-1,-3,0)^T, \mathbf{Y}_9 = (0,0,1,-1/2,3/2,0)^T \\
\mathbf{Y}_{10} &= (0,0,1,-1/2,-3/2,0)^T.
\end{aligned}$$

As shown in Fig. 10b, the linkage possesses $\gamma = 3$ FCs Λ_1, Λ_2, and Λ_3. The corresponding loop constraint mappings are

$$\begin{aligned}
f_1(\mathbf{q}) &= \exp(q_1\mathbf{Y}_1)\exp(q_2\mathbf{Y}_2)\exp(q_9\mathbf{Y}_9)\cdot \\
&\quad \exp(q_{10}\mathbf{Y}_{10})\exp(q_6\mathbf{Y}_6)\exp(q_5\mathbf{Y}_5) \\
f_2(\mathbf{q}) &= \exp(q_1\mathbf{Y}_1)\exp(q_2\mathbf{Y}_2)\exp(q_3\mathbf{Y}_3)\exp(q_4\mathbf{Y}_4) \\
f_3(\mathbf{q}) &= \exp(-q_5\mathbf{Y}_5)\exp(-q_6\mathbf{Y}_6)\exp(q_7\mathbf{Y}_7)\exp(q_8\mathbf{Y}_8).
\end{aligned} \tag{38}$$

The overall 18×10 constraint Jacobian (11) is constituted by the loop constraint Jacobians

$$\begin{aligned}
\mathbf{J}_1 &= \left(\mathbf{S}_1, \mathbf{S}_2, \mathbf{S}_3, \mathbf{S}_4, \mathbf{0}, \mathbf{0}, \mathbf{0}, \mathbf{0}, \mathbf{0}, \mathbf{0}\right) \\
\mathbf{J}_2 &= \left(\mathbf{S}_1, \mathbf{S}_2, \mathbf{0}, \mathbf{0}, \mathbf{S}_5, \mathbf{S}_6, \mathbf{0}, \mathbf{0}, \mathbf{S}_9, \mathbf{S}_{10}\right) \\
\mathbf{J}_3 &= \left(\mathbf{0}, \mathbf{0}, \mathbf{0}, \mathbf{0}, -\mathbf{S}_5, -\mathbf{S}_6, \mathbf{S}_7, \mathbf{S}_8, \mathbf{0}, \mathbf{0}\right).
\end{aligned}$$

For this planar linkage, the maximal rank of $\mathbf{J}$ is $d_{\max} = 9$.

At $\mathbf{q}_0$, it is $r = \operatorname{rank} \mathbf{J}(\mathbf{q}_0) = 8$, and the differential DOF is $\delta_{\text{diff}}(\mathbf{q}_0) = 2$.

The first- and second-order cones are, respectively,

$$\begin{aligned}
K^1_{\mathbf{q}_0} &= \mathbf{V}\big(x_1 + x_2, x_1 + x_3, x_1 - x_4, x_5 + x_6, x_5 - x_7, \\
&\qquad x_5 + x_8, x_1 - x_5 - 3x_9, x_1 + 3x_{10} - x_5\big) \\
K^2_{\mathbf{q}_0} &= \mathbf{V}\big(x_1 + x_2, x_1 + x_3, x_4 - x_1, x_5 - x_1, x_1 + x_6, \\
&\qquad x_7 - x_1, x_1 + x_8, x_9, x_{10}\big).
\end{aligned} \tag{39}$$

The third-order cone is $K^3_{\mathbf{q}_0} = \{\mathbf{0}\}$, and the kinematic tangent cone is thus $C^{\mathrm{K}}_{\mathbf{q}_0} V = \{\mathbf{0}\}$. Since the latter is zero-dimensional, there is no smooth curve through $\mathbf{q}_0$. This suggests that the finite DOF of the linkage is zero and that the linkage is finitely immobile but shaky (since $\delta_{\text{diff}}(\mathbf{q}_0) = 2$) structure. It will be shown in Sect. 8.3.5 that it does possess a finite mobility, so that $\delta(\mathbf{q}_0) > \dim C^{\mathrm{K}}_{\mathbf{q}_0} V = 0$ is in fact a strictly lower estimate.

Remark 5.6 The fact that the third-order cone is zero-dimensional, i.e., there are no third-order motions, while the linkage is mobile with finite mobility $\delta_{\text{loc}} = 1$, has led Connelly and Servatius (1994) to question the notion of higher-order rigidity. According to this notion, a linkage is called k-order rigid if $\dim K^k_{\mathbf{q}} = 0$ and $\dim K^i_{\mathbf{q}} > 0, i < k$, for $k > 0$. That is, a linkage is called rigid of order $\kappa > 0$ if $\dim K^{\kappa}_{\mathbf{q}} = 0$ in (18). The double-Watt linkage was thus called third-order rigid. This definition is problematic, however, since it does not reflect whether a linkage is rigid, but rather whether there are higher-order motions corresponding to smooth finite motions.

The examples in Sects. 5.4.4 and 5.4.5 show that the tangents to smooth curves do not necessarily indicate that the considered point is a kinematic singularity or a c-space (and thus kinematic) singularity. The first requires investigating whether the rank of the Jacobian is locally constant and the latter requires analyzing the local geometry of V.

6 Local Analysis of Smooth Finite Motions with Certain Rank

Analyzing smooth curves through a given point $\mathbf{q} \in V$ reveals possible finite motions through that point. For identification of kinematic singularities, it is necessary to check whether the rank of $\mathbf{J}$ changes in any neighborhood of $\mathbf{q}$ in V. A first step toward such an analysis is to investigate the rank for possible smooth motions through $\mathbf{q}$. This may, however, not provide a complete picture since, as already pointed out, there are situations where there are no smooth curves through a point (e.g., cusp).

6.1 *Kinematic Tangent Cone*

Definition 6.1 The kinematic tangent cone to $L_k \subseteq V$ at $\mathbf{q} \in V$, denoted with $C^{\mathrm{K}}_{\mathbf{q}} L_k$, is the set of tangent vectors to smooth arcs in V passing through $\mathbf{q}$ at which all points have rank less than k.

This is the restriction of the kinematic tangent cone to V to the subvariety L_k. The rank condition in the definition (15) will be expressed in terms of the k-minors of the constraint Jacobian, since a matrix has rank k if and only if at least one of its k-minors is nonzero.

The Jacobian (11) of a multi-loop linkage is a $6\gamma \times n$ matrix. In order to construct the minors, the index sets $\boldsymbol{\alpha} = \{\alpha_1, \ldots, \alpha_k\}, \alpha_{i-1} < \alpha_i$ and $\boldsymbol{\beta} = \{\beta_1, \ldots, \beta_k\}, \beta_{j-1} < \beta_j$ are used to select the rows and columns of $\mathbf{J}$. Denote with $\mathbf{J}_{\alpha\beta}$ the $k \times k$ submatrix of $\mathbf{J}$, consisting of the elements of the rows $\alpha_i \in \{1, \ldots, 6\gamma\}$ and columns $\beta_j \in \{1, \ldots, n\}$ of $\mathbf{J}$. The determinant of this matrix, denoted with $m_{\alpha\beta}(\mathbf{q}) := \det \mathbf{J}_{\alpha\beta}(\mathbf{q})$, is referred to as the $\boldsymbol{\alpha\beta}$-minor of $\mathbf{J}$ of order k. Notice that this differs from the conventional definition of minors where the index set would indicate the rows and columns that are eliminated from $\mathbf{J}$.

Since the rank of $\mathbf{J}(\mathbf{q})$ is less than k if and only if all its minors of order k vanish, the set (15) of configurations of rank less than k is given as

$$L_k = \{\mathbf{q} \in \mathbb{V}^n \,|\, f_l(\mathbf{q}) = \mathbf{I},\, m_{\boldsymbol{\alpha\beta}}(\mathbf{q}) = 0 \\ \forall \boldsymbol{\alpha} \subseteq \{1,\ldots,6\gamma\}, \boldsymbol{\beta} \subseteq \{1,\ldots,n\}, |\boldsymbol{\alpha}|=|\boldsymbol{\beta}|=k, l = 1, ..., \gamma\}. \quad (40)$$

The number of minors of order k in (40) is $\binom{6\gamma}{k} \cdot \binom{n}{k}$. Since this number increases dramatically with the order of the minor, a systematic and efficient way for its evaluation is crucial.

The time derivatives of the minors are now written as

$$\begin{aligned} M_{\boldsymbol{\alpha\beta}}^{(1)}(\mathbf{q}, \dot{\mathbf{q}}) &:= \frac{d}{dt} m_{\boldsymbol{\alpha\beta}}(\mathbf{q}) \\ M_{\boldsymbol{\alpha\beta}}^{(2)}(\mathbf{q}, \dot{\mathbf{q}}, \ddot{\mathbf{q}}) &:= \frac{d^2}{dt^2} m_{\boldsymbol{\alpha\beta}}(\mathbf{q}) \\ M_{\boldsymbol{\alpha\beta}}^{(i)}(\mathbf{q}, \dot{\mathbf{q}}, \ldots, \mathbf{q}^{(i)}) &:= \frac{d^i}{dt^i} m_{\boldsymbol{\alpha\beta}}(\mathbf{q}). \end{aligned} \quad (41)$$

A vector $\mathbf{x} \in \mathbb{R}^n$ is a feasible first-order motion in L_k if $H_l^{(1)}(\mathbf{q}, \mathbf{x}) = \mathbf{0}$, $M_{\boldsymbol{\alpha\beta}}^{(1)}(\mathbf{q}, \mathbf{x}) = 0$, $|\boldsymbol{\alpha}|=|\boldsymbol{\beta}|=k$ at $\mathbf{q} \in L_k, l = 1, \ldots, \gamma$. The set of tangent vectors to ith-order motions through $\mathbf{q} \in L_k$ is

$$K_{\mathbf{q}}^{k,i} := \{\mathbf{x} | \exists \mathbf{y}, \mathbf{z}, \ldots \in \mathbb{R}^n : \quad (42)$$

$$\begin{aligned} &H_l^{(1)}(\mathbf{q}, \mathbf{x}) = \mathbf{0}, M_{\alpha\beta}^{(1)}(\mathbf{q}, \mathbf{x}) = 0, \\ &H_l^{(2)}(\mathbf{q}, \mathbf{x}, \mathbf{y}) = \mathbf{0}, M_{\alpha\beta}^{(2)}(\mathbf{q}, \mathbf{x}, \mathbf{y}) = 0, \\ &H_l^{(3)}(\mathbf{q}, \mathbf{x}, \mathbf{y}, \mathbf{z}) = \mathbf{0}, M_{\alpha\beta}^{(3)}(\mathbf{q}, \mathbf{x}, \mathbf{y}, \mathbf{z}) = 0, \\ &\quad \ldots \\ &H_l^{(i)}(\mathbf{q}, \mathbf{x}, \mathbf{y}, \mathbf{z}, \ldots) = \mathbf{0}, M_{\alpha\beta}^{(i)}(\mathbf{q}, \mathbf{x}, \mathbf{y}, \mathbf{z}, \ldots) = 0, \\ &\forall \boldsymbol{\alpha} \subseteq \{1,\ldots,6\}, \boldsymbol{\beta} \subseteq \{1,\ldots,n\}, |\boldsymbol{\alpha}|=|\boldsymbol{\beta}|= k, l = 1, ..., \gamma\}. \end{aligned}$$

The kinematic tangent cone to L_k can now be defined via the sequence

$$C_{\mathbf{q}}^{\mathrm{K}} L_k = K_{\mathbf{q}}^{k,\kappa} \subseteq \ldots \subseteq K_{\mathbf{q}}^{k,3} \subseteq K_{\mathbf{q}}^{k,2} \subseteq K_{\mathbf{q}}^{k,1} \quad (43)$$

which terminates with finite κ. As for (19), $K_{\mathbf{q}}^{k,i}$ is a cone in $\mathbb{R}^n$ so that $\mathbf{x} \in K_{\mathbf{q}}^{k,i}$ implies that $a\mathbf{x} \in K_{\mathbf{q}}^{k,i}$ for $a \in \mathbb{R}$.

Lemma 6.2 *Let* $\mathbf{q}$ *be a point of rank* r. *If* $C_{\mathbf{q}}^{\mathrm{K}} L_{r+1} \neq C_{\mathbf{q}}^{\mathrm{K}} V$, *then* $\mathbf{q} \in V$ *is a kinematic singularity.*

The converse is not always true, e.g., when V possesses cusp singularities (see Sect. 6.3.5).

Remark 6.3 For particular linkages, such as planar or spherical, the number of independent constraints per loop is a priori less than 6, and thus the maximal rank is $d_{\max} \leq 6\kappa$. Moreover, the loop closure constraints are formulated so that some components are identically zero. This is in particular so for planar linkages where only three components of the constraints (7) are not identically zero. The Jacobian can be reduced accordingly. For a planar linkage with γ FC, the number of minors of order k in (40) is then reduced to $\binom{3\gamma}{k} \cdot \binom{n}{k}$.

Remark 6.4 The kinematic tangent cone $C^{\mathrm{K}}_{\mathbf{q}} L_k$ is the set of tangents to smooth finite motions through $\mathbf{q}$ with rank less than k. It does in general not consist of tangents to smooth curves with constant rank except for $L_{r+1}, r = \operatorname{rank} \mathbf{J}(\mathbf{q})$. $C^{\mathrm{K}}_{\mathbf{q}} L_{r+1}$ is in fact the set of tangents to smooth curves of rank $r = \operatorname{rank} \mathbf{J}(\mathbf{q})$ through $\mathbf{q}$.

6.2 Higher-Order Time Derivatives of Jacobian Minors

First, consider the $6 \times n$ Jacobian of a single-loop linkage in (8). The jth column of $\mathbf{J}$ is the instantaneous joint screw coordinate vector with components $\mathbf{S}_j = \left(S_{1j}, \ldots, S_{6j}\right)^T$. From this, the vector $\mathbf{S}_{\boldsymbol{\alpha} j} = \left(S_{a_1 j}, \ldots, S_{a_k j}\right)^T$ is constructed with elements according to the row indexes in $\boldsymbol{\alpha}$. The $\boldsymbol{\alpha\beta}$-minor is thus

$$m_{\boldsymbol{\alpha\beta}}(\mathbf{q}) := \det \mathbf{J}_{\boldsymbol{\alpha\beta}}(\mathbf{q}) = \left|\mathbf{S}_{\boldsymbol{\alpha} b_1}(\mathbf{q}) \cdots \mathbf{S}_{\boldsymbol{\alpha} b_k}(\mathbf{q})\right|. \tag{44}$$

The first time derivative of $m_{\boldsymbol{\alpha\beta}}$ is easily found, using basic properties of determinants (Steeb 1991)

$$\frac{d}{dt} m_{\boldsymbol{\alpha\beta}} = \sum_{\mu \in \boldsymbol{\beta}} \left|\mathbf{S}_{\boldsymbol{\alpha}\beta_1} \cdots \dot{\mathbf{S}}_{\boldsymbol{\alpha}\mu} \cdots \mathbf{S}_{\boldsymbol{\alpha}\beta_k}\right|.$$

It is instructive to inspect higher time derivatives. Evaluating derivative of order five, for example, yields

$$\begin{aligned}
\frac{d^5}{dt^5} m_{\boldsymbol{\alpha\beta}} &= \sum_{\beta_i \in \boldsymbol{\beta}} \left|\mathbf{S}_{\boldsymbol{\alpha}\beta_1} \cdots \mathbf{S}^{(5)}_{\boldsymbol{\alpha}\beta_i} \cdots \mathbf{S}_{\boldsymbol{\alpha}\beta_k}\right| \\
&+ 5 \sum_{\beta_i \neq \beta_j \in \boldsymbol{\beta}} \left|\mathbf{S}_{\boldsymbol{\alpha}\beta_1} \cdots \mathbf{S}^{(4)}_{\boldsymbol{\alpha}\beta_i} \cdots \dot{\mathbf{S}}_{\boldsymbol{\alpha}\beta_j} \cdots \mathbf{S}_{\boldsymbol{\alpha}\beta_k}\right| \\
&+ 10 \sum_{\beta_i \neq \beta_i \in \boldsymbol{\beta}} \left|\mathbf{S}_{\boldsymbol{\alpha}\beta_1} \cdots \dddot{\mathbf{S}}_{\boldsymbol{\alpha}\beta_i} \cdots \ddot{\mathbf{S}}_{\boldsymbol{\alpha}\beta_j} \cdots \mathbf{S}_{\boldsymbol{\alpha}\beta_k}\right| \\
&+ 10 \sum_{\beta_i \neq \beta_i \neq \beta_l \in \boldsymbol{\beta}} \left|\mathbf{S}_{\boldsymbol{\alpha}\beta_1} \cdots \dddot{\mathbf{S}}_{\boldsymbol{\alpha}\beta_i} \cdots \dot{\mathbf{S}}_{\boldsymbol{\alpha}\beta_j} \cdots \dot{\mathbf{S}}_{\boldsymbol{\alpha}\beta_l} \cdots \mathbf{S}_{\boldsymbol{\alpha}\beta_k}\right| \\
&+ 10 \sum_{\beta_i \neq \beta_i \neq \beta_l \neq \beta_r \in \boldsymbol{\beta}} \left|\mathbf{S}_{\boldsymbol{\alpha}\beta_1} \cdots \ddot{\mathbf{S}}_{\boldsymbol{\alpha}\beta_i} \cdots \dot{\mathbf{S}}_{\boldsymbol{\alpha}\beta_j} \cdots \dot{\mathbf{S}}_{\boldsymbol{\alpha}\beta_l} \cdots \dot{\mathbf{S}}_{\boldsymbol{\alpha}\beta_r} \cdots \mathbf{S}_{\boldsymbol{\alpha}\beta_k}\right| \quad (*)
\end{aligned}$$

$$+ 15 \sum_{\beta_i \neq \beta_i \neq \beta_l \in \boldsymbol{\beta}} \left| \mathbf{S}_{\alpha\beta_1} \cdots \ddot{\mathbf{S}}_{\alpha\beta_i} \cdots \ddot{\mathbf{S}}_{\alpha\beta_j} \cdots \dot{\mathbf{S}}_{\alpha\beta_l} \cdots \mathbf{S}_{\alpha\beta_k} \right| \quad (**)$$

$$+ \sum_{\beta_i \neq \beta_i \neq \beta_l \neq \beta_r \neq \beta_s \in \boldsymbol{\beta}} \left| \mathbf{S}_{\alpha\beta_1} \cdots \dot{\mathbf{S}}_{\alpha\beta_i} \cdots \dot{\mathbf{S}}_{\alpha\beta_j} \cdots \dot{\mathbf{S}}_{\alpha\beta_l} \cdots \dot{\mathbf{S}}_{\alpha\beta_r} \cdots \dot{\mathbf{S}}_{\alpha\beta_s} \cdots \mathbf{S}_{\alpha\beta_k} \right| .$$

In order to derive a compact expression for general order, a multi-index, denoted $\mathbf{a}_k = (a_1, \ldots, a_k) \in \mathbb{N}^k$, is used to indicate the degrees of time derivatives of $\mathbf{S}_{\alpha\beta_j}$, $1 \leq j \leq k$. For instance, the minor $\left|\mathbf{S}_{\alpha\beta_1} \dddot{\mathbf{S}}_{\alpha\beta_2} \dot{\mathbf{S}}_{\alpha\beta_3} \mathbf{S}_{\alpha\beta_4} \dot{\mathbf{S}}_{\alpha\beta_5}\right|$ of order $k = 5$ is written as $\left|\mathbf{S}^{(a_1)}_{\alpha\beta_1} \mathbf{S}^{(a_2)}_{\alpha\beta_2} \mathbf{S}^{(a_3)}_{\alpha\beta_3} \mathbf{S}^{(a_4)}_{\alpha\beta_4} \mathbf{S}^{(a_5)}_{\alpha\beta_5}\right|$ with $\mathbf{a}_k = (0, 3, 1, 0, 1)$. Apparently, in the expressions for $\frac{d^\nu}{dt^\nu} m_{\alpha\beta}$ it is $|\mathbf{a}_k| = a_1 + \ldots + a_k = \nu$. The νth time derivative can now be written in the compact form

$$\frac{d^\nu m_{\alpha\beta}}{dt^\nu} = \sum_{|\mathbf{a}_k| = \nu} \left| \mathbf{S}^{(a_1)}_{\alpha\beta_1} \, \mathbf{S}^{(a_2)}_{\alpha\beta_2} \cdots \mathbf{S}^{(a_k)}_{\alpha\beta_k} \right| \frac{\nu!}{\mathbf{a}_k! n_1! \cdots n_\nu!} \qquad (45)$$

with $\mathbf{a}_k! = a_1! \cdot \ldots \cdot a_k!$, where $n_i := |\{a_j | a_j = i\}|$ is the number of times the ith derivative occurs in the multi-index. The term $\nu!/(a_1! \ldots a_k!)$ accounts for the number of times the index β_i is repeated, whereas the division by $n_i!$ accounts for the repetition of the ith time derivative. For instance, the sum $(*)$ contains 3! times the same term since $\dot{\mathbf{S}}$ is repeated, and $(**)$ contains 2! times the same summand due to the repetition of $\ddot{\mathbf{S}}$.

The relation (45) can be easily evaluated using the recursive form of the time derivatives $\mathbf{S}^{(i)}_{\alpha j} \equiv \mathrm{D}^{(i)} \mathbf{S}_{\alpha j}$ in (28). The latter are already evaluated for the higher-order constraints. The derivatives of the minor could also be constructed using the explicit form (22) of the partial derivatives, which leads to rather complex, however.

6.3 Examples

6.3.1 Planar Four-Bar Linkage with Identical Link Lengths

The joint screw coordinate vectors are given in (30). The rank of the constraint Jacobian in the reference configuration is rank $\mathbf{J}(\mathbf{q}_0) = 2$ so that $\mathbf{q}_0 \in L_k, k \geq 3$. Since the maximal rank for this planar linkage is 3, only L_3 needs to be investigated.

There are $\binom{6}{3} \cdot \binom{4}{3} = 20 \cdot 4 = 80$ minors of order 3 of the 6×4 constraint Jacobian. The only non-trivial derivatives $M^{(1)}_{\alpha\beta}(\mathbf{q}, \mathbf{x}) = 0$, $|\boldsymbol{\alpha}| = |\boldsymbol{\beta}| = 3$ in (42) are

$$0 = -L^2 x_2, \quad 0 = L^2 x_3, \quad 0 = -L^2 (2x_2 + x_3),$$

the non-trivial second derivatives $M^{(2)}_{\alpha\beta}(\mathbf{q}, \mathbf{x}, \mathbf{y}) = 0$, $|\boldsymbol{\alpha}| = |\boldsymbol{\beta}| = 3$ are

$$0 = -L^2 y_2, \quad 0 = L^2 y_3, \quad 0 = -L^2 (2y_2 + y_3),$$

and the non-trivial third derivatives $M^{(3)}_{\alpha\beta}(\mathbf{q},\mathbf{x},\mathbf{y},\mathbf{z})=0$, $|\boldsymbol{\alpha}|=|\boldsymbol{\beta}|=3$ are

$$0=L^2(x_2^3-z_2),\quad 0=L^2(x_2^3+3x_2^2x_3+3x_2x_3^2-z_2)$$
$$0=L^2(z_3-x_3^3),\quad 0=L^2(2x_2^3+3x_2^2x_3+3x_2x_3^2+x_3^3-2z_2-z_3).$$

The first-order cone $K^{3,1}_{\mathbf{q}_0}=\{\mathbf{0}\}$ is sufficient to show that $C^{\mathrm{K}}_{\mathbf{q}_0}L_3=K^{3,1}_{\mathbf{q}_0}=K^{3,2}_{\mathbf{q}_0}=K^{3,3}_{\mathbf{q}_0}=\{\mathbf{0}\}$.

It should be remarked that with the chosen reference frame $\mathcal{F}_0$ the constraint Jacobian can be reduced to the 3×4 matrix comprising rows 3, 4, and 5. Then, there are only $\binom{3}{3}\cdot\binom{4}{3}=3$ minors.

The configuration $\mathbf{q}_0$ is a kinematic singularity (Lemma 6.2) since $C^{\mathrm{K}}_{\mathbf{q}_0}L_3\neq C^{\mathrm{K}}_{\mathbf{q}_0}V$. This means that the rank is not constant for arbitrarily small finite motions through $\mathbf{q}_0$. This is clear since $\mathbf{q}_0$ is a c-space singularity as shown in Sect. 5.4.1. The differential DOF is locally $\delta_{\mathrm{diff}}=1$ except at $\mathbf{q}_0$ where $\delta_{\mathrm{diff}}(\mathbf{q}_0)=2$. Because $C^{\mathrm{K}}_{\mathbf{q}_0}L_3\subset C^{\mathrm{K}}_{\mathbf{q}_0}L_4=C^{\mathrm{K}}_{\mathbf{q}_0}V$, the manifolds intersecting at $\mathbf{q}_0$ consist of points of rank 3.

6.3.2 Planar Four-Bar Linkage with Different Link Lengths

The joint screw coordinate vectors are given in (32). As before, the rank of the constraint Jacobian in the reference configuration is rank $\mathbf{J}(\mathbf{q}_0)=2$ so that $\mathbf{q}_0\in L_k,k\geq 3$. The non-trivial derivatives $M^{(1)}_{\alpha\beta}(\mathbf{q},\mathbf{x})=0$, $|\boldsymbol{\alpha}|=|\boldsymbol{\beta}|=3$ in (42) are

$$0=-2L^2x_2,\quad 0=-L^2x_2,\quad 0=2L^2x_3,\quad 0=-L^2(3x_2+2x_3).$$

The first-order cone is thus $K^{3,1}_{\mathbf{q}_0}=\{\mathbf{0}\}$ and so $C^{\mathrm{K}}_{\mathbf{q}_0}L_3=\{\mathbf{0}\}$.

6.3.3 Extended Watt Linkage—Underconstrained Kinematotropic Two-Loop Planar Linkage

With rank $\mathbf{J}(\mathbf{q}_0)=5$, it is $\mathbf{q}_0\in L_k,k\geq 6$. Since the maximal rank of the constraint Jacobian (of a planar linkage with 2 FCs) is 6, only L_6 will be investigated. The nonvanishing first derivatives of the $\binom{12}{6}\cdot\binom{8}{6}=924\cdot 28$ minors $M^{(1)}_{\alpha\beta}(\mathbf{q},\mathbf{x})$ of order $|\boldsymbol{\alpha}|=|\boldsymbol{\beta}|=6$ are

$$3\sqrt{2}x_2,6\sqrt{2}x_2,9\sqrt{2}x_2.$$

They define the first-order cone

$$K^{6,1}_{\mathbf{q}_0}=\mathbf{V}\big(x_1+x_3,x_2,3x_4+2x_5+x_6,2x_4+x_5-x_7,x_8\big).\tag{46}$$

Higher derivatives of the minors yield $C^{\mathrm{K}}_{\mathbf{q}_0}L_6=K^{6,2}_{\mathbf{q}_0}=K^{6,3}_{\mathbf{q}_0}=\ldots=\mathbf{V}\big(x_1+x_3,x_2,x_4,x_5,x_6,x_7,x_8\big)=K^{3(I)}_{\mathbf{q}_0}$, with $K^{3(I)}_{\mathbf{q}_0}$ in (36). Hence, for the one-dimensional finite motions through $\mathbf{q}_0$ the differential DOF is 2, whereas for the two-dimensional finite motions the differential DOF is 1. In conclusion, the linkage can perform smooth motions through $\mathbf{q}_0$ with $\delta = 1$, $\delta_{\text{diff}} = 3$ as well as motions with $\delta = \delta_{\text{diff}} = 2$. These two motion modes of different dimensions could not be concluded from the analysis of finite motions (Sect. 5.4.3) nor from a c-space analysis as they do not reflect the differential DOF.

A peculiar feature of this linkage is that when the linkage approaches the singularity from the two-dimensional motion mode the rank drops from 6 to 5 and *remains* 5 when entering the one-dimensional motion mode. This is a remarkable example where, at a singularity, the rank drops but is equal to the rank at regular points in the neighborhood. Yet, the configuration is singular since the rank changes.

6.3.4 Kinematic Singularity of a 6R Linkage with Smooth C-Space

The only sets of critical points to analyze are L_6 and L_5 since rank $\mathbf{J}(\mathbf{q}_0) = 4$. The only minor of order 6 is $\det \mathbf{J}(\mathbf{q}_0)$. Its first-order derivative vanishes so that $K^{6,1}_{\mathbf{q}_0} = K^1_{\mathbf{q}_0}$. Evaluating the higher-order cones yields $K^{6,2}_{\mathbf{q}_0} = K^{6,3}_{\mathbf{q}_0} = \ldots = C^{\mathrm{K}}_{\mathbf{q}_0} L_6 = C^{\mathrm{K}}_{\mathbf{q}_0} V$. Thus, all curves passing through $\mathbf{q}_0$ are of rank less than 6, which is expected for a one-DOF 6R linkage.

Evaluating the derivatives of the $6 \cdot 6 = 36$ minors of order 5 yields the kinematic tangent cone $K^{5,1}_{\mathbf{q}_0} = K^{5,2}_{\mathbf{q}_0} = K^{5,3}_{\mathbf{q}_0} = \ldots = C^{\mathrm{K}}_{\mathbf{q}_0} L_5 = \{\mathbf{0}\}$. The rank can locally not drop below 5 so that the set of curves with rank 4 is locally the zero-dimensional manifold $L_5 \cap U(\mathbf{q}_0) = \{\mathbf{q}_0\}$.

The subvariety formed by the set of smooth curves is a smooth one-dimensional manifold at $\mathbf{q}_0$. At the same time, $\mathbf{q}_0$ is a kinematic singularity since $C^{\mathrm{K}}_{\mathbf{q}_0} L_5 \neq C^{\mathrm{K}}_{\mathbf{q}_0} V$ (Lemma 6.2).

That is, since $\{\mathbf{0}\} = C^{\mathrm{K}}_{\mathbf{q}_0} L_5 \subset C^{\mathrm{K}}_{\mathbf{q}_0} L_6 = C^{\mathrm{K}}_{\mathbf{q}_0} V$, any configuration $\mathbf{q} \in V \cap U(\mathbf{q}_0)$ in a sufficiently small neighborhood $U(\mathbf{q}_0)$ of $\mathbf{q}_0$ has rank 5, and so $\delta_{\text{diff}}(\mathbf{q}) = 1$, except at $\mathbf{q}_0$.

The linkage is overconstrained since $C^{\mathrm{K}}_{\mathbf{q}_0} L_6 = C^{\mathrm{K}}_{\mathbf{q}_0} V$ and $d_{\max} = 6$. This linkage is remarkable in that the smooth curves through $\mathbf{q}_0$ form a one-dimensional manifold but yet $\mathbf{q}_0$ is a kinematic singularity where the rank changes. Whether $\mathbf{q}_0$ is regular or singular points of the c-space is to be checked by means of a local c-space analysis (Sect. 8.3.4).

6.3.5 Double-Watt Linkage with Cusp Singularity in C-Space

Because $r = \text{rank}\,\mathbf{J}(\mathbf{q}_0) = 8$, the reference configuration $\mathbf{q}_0$ belongs to $L_k, k \geq 9$. Since the maximal rank of $\mathbf{J}$ is 9, the analysis of L_9 is sufficient.

The first derivatives of the minors of order 9 yield

$$\begin{aligned} K^{9,1}_{\mathbf{q}_0} =& \mathbf{V}\big(x_1 + x_2, x_{10} + x_9, 6x_2 + x_6 + 6x_9 - x_7, x_3 + x_4, x_2 + x_4, \\ & 2x_5 + 3x_9, x_5 + x_6, x_7 + x_8, x_5 + x_8\big). \end{aligned} \tag{47}$$

This is the one-dimensional vector subspace of $K^1_{\mathbf{q}_0}$ defined by the condition $x_1 = -x_5$. The second-order cone is $K^{9,2}_{\mathbf{q}_0} = \{\mathbf{0}\}$, and thus $C^{\mathrm{K}}_{\mathbf{q}_0} L_9 = \{\mathbf{0}\}$. This should be expected since there are no finite smooth motions through $\mathbf{q}_0$. It cannot be concluded that $\mathbf{q}_0$ is a kinematic singularity.

7 Interlude: The Tangent Cone to a Variety

The analysis so far was concerned with possible smooth curves through a given point for which the basis was provided by the constraint equations. Although the latter define the c-space V, their tangents reveal the geometry of smooth motions rather than the tangential aspects of V. Moreover, it is crucial to understand the relation of a particular set of equations and the variety they generate. Therefore, the general situation is briefly discussed in this section.

It is first important to distinguish between a variety and the corresponding topological set. Let V be a real point set in $\mathbb{V}^n$. V is an analytic variety if and only if it is closed, and for each $\mathbf{q} \in V$ there is a neighborhood $U(\mathbf{q})$ in $\mathbb{V}^n$ and analytic functions $f_1, \ldots, f_m$ such that $V \cap U(\mathbf{q})$ is the set of common zeros of f_i in $U(\mathbf{q})$.

A central concept is that of the ideal of analytic germs vanishing over V at $\mathbf{q}$, denoted with $I(V, \mathbf{q})$. This ideal takes care of the fact the point set V can be defined as the set of zeros of different functions. The kinematic constraints, for instance, are just a special (but well justified) choice of functions. But it is important to notice that a particular set of generating functions may not yield a proper description of the tangential aspects of the set V.

As a simple example, consider the system of two equations

$$\begin{aligned} f_1(x, y, z) &= z + x^2 \\ f_2(x, y, z) &= z \end{aligned} \tag{48}$$

defined on $\mathbb{R}^3$. The set of zeros is the y-axis

$$V = \mathbf{V}(f_1, f_2) = \{(0, y, 0) \in \mathbb{R}^3 | y \in \mathbb{R}\}$$

i.e., $\dim V = 1$. This is indeed a smooth one-dimensional manifold. The Jacobian of this system of equations is

$$\mathbf{J}(x, y, z) = \begin{pmatrix} 2x & 0 & 1 \\ 0 & 0 & 1 \end{pmatrix}.$$

At any point $(x, y, z) \in V$, the Jacobian is singular, since $x = 0$, and the two-dimensional null-space

$$\ker \mathbf{J}(x, y, z) = \{(u, v, 0) \in \mathbb{R}^3 | u, v \in \mathbb{R}\}$$

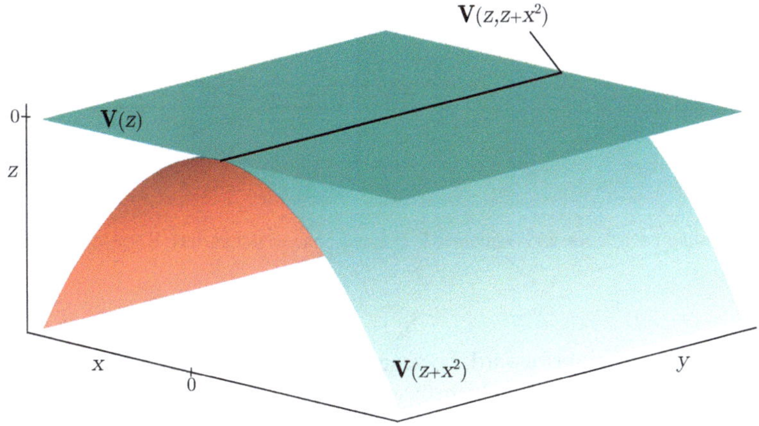

Fig. 14 Solution set of $f_1 = f_2 = 0$ in (48)

is the x-y-plane. The null-space does clearly not describe the tangent space to the smooth manifold V (see Fig. 14). Indeed, other functions can be used to define V so that the Jacobian has full rank and its null-space properly defines its tangent space (e.g., $f_1 = x$, $f_2 = z$). The tangent space to the c-space V can in general not be defined as the null-space of the Jacobian. While this is true in general, there is a good reason for choosing the kinematic constraints as generating functions. The extended Watt linkage in Sect. 5.4.3 shows the situation just described. In the one-dimensional motion mode (Fig. 6), where V is a smooth one-dimensional manifold, the constraint Jacobian is rank deficient so that $\delta_{\text{diff}} = 3$. This rank deficiency does have a significance as it is related to the kinematics. Yet, the tangent space must be defined independently of the equations defining V. To this end, V is regarded as the analytic variety so that at any $\mathbf{q} \in V$ it is generated by the ideal $I(V, \mathbf{q})$, denoted as $V = \mathbf{V}(I(V, \mathbf{q}))$. That is, V is locally defined as the set of common zeros of all analytic germs at $\mathbf{q}$.

The *tangent space* $T_{\mathbf{q}}V$ to the analytic variety V at $\mathbf{q}$ is the vector space defined by vectors $\mathbf{v} \in \mathbb{R}^n$ such that $df(\mathbf{q})\mathbf{v} = 0, \forall f \in I(V, \mathbf{q})$. Then, $\mathbf{q} \in V$ is a regular point of the variety V if and only if there is a neighborhood $U(\mathbf{q}) \subset \mathbb{V}^n$ and m analytic functions such that the differentials $df_1, \ldots, df_m$ are linearly independent and $V \cap U(\mathbf{q})$ is the set of common zeros of f_i in $U(\mathbf{q})$. In other words, $\mathbf{q} \in V$ is a regular point of the variety if and only if the dimension of the tangent space is constant in some neighborhood $U(\mathbf{q}) \cap V$. The set V is a manifold at $\mathbf{q} \in V$ if $\mathbf{q}$ is a regular point of the variety V. At regular points, the tangent space to the manifold at $\mathbf{q}$ is given by the equivalence classes of curves through $\mathbf{q}$.

At regular points of V, the tangent space well describes the local geometry. But at c-space singularities, it fails to do so. Therefore, Whitney (1965) introduced the tangent cone to a (complex) analytic variety. This was already indicated in Fig. 4. Since it is also defined in terms of the ideal, it has become an established concept for algebraic varieties (which to some extent allows for ideal computations). Its

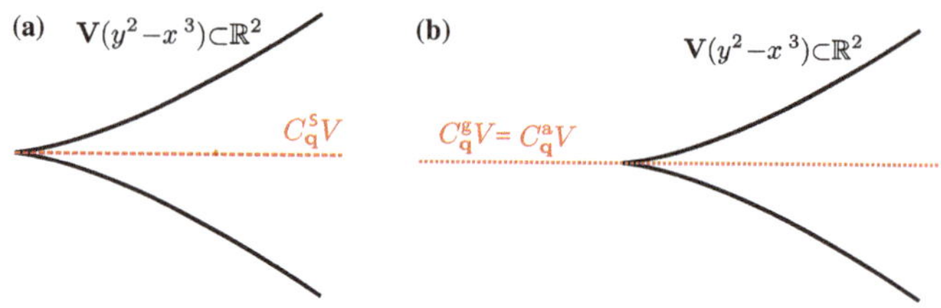

Fig. 15 **a** Tangent semicone and **b** geometric tangent cone to the variety $\mathbf{V}\left(y^2 - x^3\right) \in \mathbb{R}^2$ at the origin

application is not straightforward, however, and there are even different definitions of tangent cones. This was already discussed by Whitney (1965). A recent discussion can be found in O'Shea and Wilson (2004). The following three definitions are mostly used.

The *tangent semicone*, denoted $C^{\mathrm{s}}_{\mathbf{q}}V$, of a variety V at the point $\mathbf{q} \in V$ consists of the limits of all secant *rays* originating from $\mathbf{q}$ that pass through a sequence of points that converges to $\mathbf{q}$. The *geometric tangent* cone $C^{\mathrm{g}}_{\mathbf{q}}V$ of V at $\mathbf{q} \in V$ consists of the limits of all secant *lines* originating from $\mathbf{q}$ that pass through a sequence of points that converges to $\mathbf{q}$ (this is denoted with $C_3\,(V, \mathbf{q})$ in Whitney 1965 p. 211). That is, it consists of tangent vectors to differentiable arcs in V from $\mathbf{q}$. The *algebraic tangent* cone $C^{\mathrm{a}}_{\mathbf{q}}V$ of an algebraic variety V at $\mathbf{q} \in V$ is defined to be the set of common zeros of all initial forms (the homogeneous part of lowest degree) of all polynomials that vanish identically on V. This is adopted to analytic varieties by replacing the polynomial ideal of the algebraic variety by the ideal of analytic germs that vanish over V at $\mathbf{q}$.

These definitions are clearly different. Figure 15 shows this for the (algebraic) variety $\mathbf{V}\left(y^2 - x^3\right) \in \mathbb{R}^2$. The tangent semicone is the positive x-axis, whereas the geometric tangent cone is the complete x-axis. While these are constructed geometrically, the algebraic tangent cone is defined by the leading polynomial of $y^2 - x^3$, i.e., by $y^2 = 0$, which is again the x-axis. The origin is a double point with a repeated tangent, also called 'cuspidal tangent' by Gibson (1998). The geometric and algebraic tangent cones are symmetric in the sense that for $\mathbf{x} \in C^{\mathrm{g}}_{\mathbf{q}}V$ also $-\mathbf{x} \in C^{\mathrm{g}}_{\mathbf{q}}V$. The tangent semicone may not be symmetric, as for the above example. The tangent semicone may not be the limit of tangent spaces (O'Shea and Wilson 2004), as it does not consist of tangent lines to V. For example, the tangent semicone of $V = \mathbf{V}(z^2 - x^3y^3) \subset \mathbb{R}^3$ at the origin is the first and third quadrant of the xy -plane (Fig. 16), which is a sensible local approximation of V. The geometric and algebraic tangent cones, on the other hand, are the entire xy-plane: $C^{\mathrm{g}}_{\mathbf{q}}V = C^{\mathrm{a}}_{\mathbf{q}}V = \mathbf{V}\left(z^2\right) = \{(u, v, w)\,|w = 0\}$. All three versions fail to locally approximate V and thus cannot be applied to describe the local mobility. Even more, they also fail to at least reveal the local (finite) DOF since their dimension may be different from the local dimension of V. As an example, consider the variety $\mathbf{V}\left(x^2 + y^2 - z^3\right) \in \mathbb{R}^3$ (Fig. 17). The tangent semicone is the positive z-axis, and the geometric and (real) algebraic tangent cones are the entire z-axis. Thus, all tangent cones are one-dimensional, while the variety is two-dimensional.

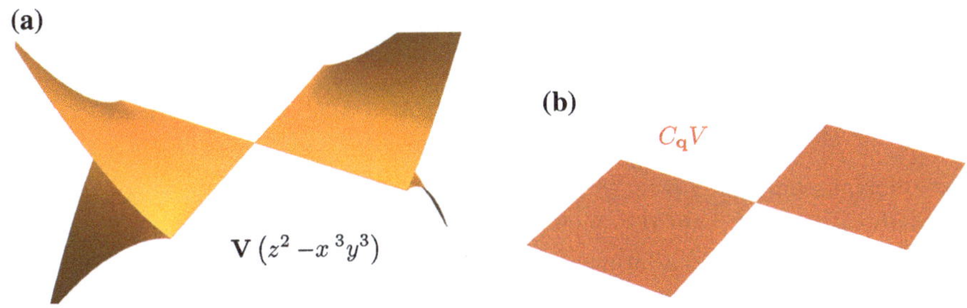

Fig. 16 **a** Variety $\mathbf{V}(z^2 - x^3y^3) \subset \mathbb{R}^3$ and **b** its tangent semicone

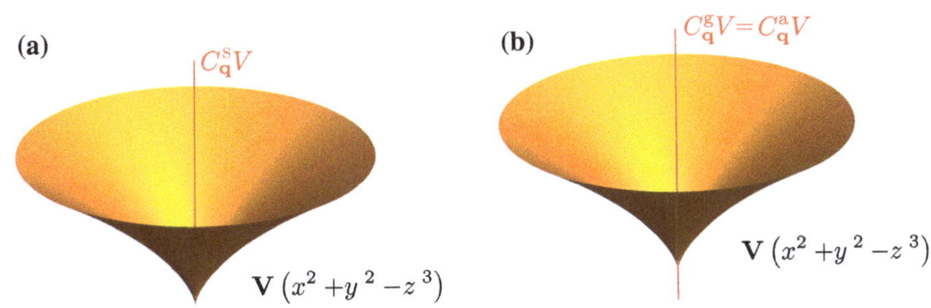

Fig. 17 **a** Tangent semicone and **b** geometric tangent cone of the variety $\mathbf{V}(x^2 + y^2 - z^3) \subset \mathbb{R}^3$

The algebraic tangent cone (resting on the Zariski topology) cannot capture the 'unilateral' nature of V. However, it was proven by Whitney (1965) that for complex algebraic varieties it does provide a faithful local approximation and is identical to the geometric tangent cone, and in particular $\dim C^{\mathrm{g}}_{\mathbf{q}}V = \dim_{\mathbf{q}} V$. The important point is that this strictly applies to complex varieties only.

The semicone and geometric cone at $\mathbf{q}$ refer to V as a real topological set and are defined by analytic arcs from $\mathbf{q}$. A real analytic arc starting at any (regular or singular) point has a well-defined tangent *ray* Whitney (1965) so that the tangent semicone (consisting of tangent rays to arcs from $\mathbf{q} \in V$) is well defined. But, as for the algebraic tangent cone, the tangent cone to a real variety may not represent a local approximation, as Fig. 17 clearly shows.

An example where the algebraic and geometric tangent cones are different is $\mathbf{V}\left(y^3 + yx^2 - x^4\right)$ considered as real variety (Bochnak et al. 1998 Sect. 3.3). The geometric tangent cone is the x-axis, whereas the algebraic tangent cone, defined by $y\left(y^2 + 2x^2\right) = 0$, is the union of the origin and the x-axis, hence the x-axis. If considered as complex variety, the algebraic tangent cone is indeed the union of two one-dimensional varieties.

It should be clear from the discussion so far that the concept of tangent cone cannot be applied straightforwardly. Therefore, in the remaining part of this chapter, a rather direct approach to the local approximation of the c-space is pursued by means of a series expansion of the constraints.

8 Local Analysis of the Configuration Space

8.1 Higher-Order local Approximation

The kinematic tangent cone describes the local geometry of smooth curves through a (possibly singular) point $\mathbf{q} \in V$, but not necessarily the geometry of the c-space V (as in case of cusp singularities of V). A local νth-order approximation of V is defined by the νth-order truncation of the series expansion of $f(\mathbf{q}) - \mathbf{I}$ at $\mathbf{q} \in V$ as

$$V_{\mathbf{q}}^{\nu} := \{\mathbf{x} \in \mathbb{R}^n | \mathrm{d} f_{l,\mathbf{q}}(\mathbf{x}) + \frac{1}{2}\mathrm{d}^2 f_{l,\mathbf{q}}(\mathbf{x}) + \ldots + \frac{1}{\nu!}\mathrm{d}^{\nu} f_{l,\mathbf{q}}(\mathbf{x}) = \mathbf{0}\}. \tag{49}$$

This is an algebraic variety of order ν. There is a neighborhood $U(\mathbf{q})$ and finite order κ so that

$$V_{\mathbf{q}}^{\kappa} \cap U(\mathbf{q}) = V \cap U(\mathbf{q}). \tag{50}$$

Therewith, the analytic variety V is locally approximated by an algebraic variety. This allows for approximation of arbitrary order, but there are two principal issues. One is that the sufficient order κ is unknown. The second, more critical issue, is that analyzing an algebraic variety of higher order is computationally very challenging.

Lemma 8.1 *The point $\mathbf{q} \in V$ is a c-space singularity if and only if $V_{\mathbf{q}}^{\kappa}$ is not a smooth manifold.*

Since checking whether the approximation $V_{\mathbf{q}}^{\kappa}$ is smooth becomes rather involved, checking this condition can be difficult to use. Chen (2011) proposed a second-order approximation using a second-order Taylor expansion of the constraints.

8.2 Higher-Order Series Expansion of the Constraint Mapping

The Taylor series expansion of the constraint mapping (4) is

$$f(\mathbf{q} + \mathbf{x}) = f(\mathbf{q}) + \sum_{k \geq 1} \frac{1}{k!}\mathrm{d}^k f_{\mathbf{q}}(\mathbf{x}) \tag{51}$$

with the kth differential of f

$$\begin{aligned}
\mathrm{d}^k f_{\mathbf{q}}(\mathbf{x}) &= \sum_{|\mathbf{a}|=k} \frac{k!}{a_1!a_2!\cdots a_n!} \frac{\partial^k f}{\partial q_1^{a_1}\partial q_2^{a_2}\ldots\partial q_n^{a_n}} x_1^{a_1}x_2^{a_2}\cdots x_n^{a_n} \\
&= \sum_{|\mathbf{a}|=k} \frac{k!}{\mathbf{a}!}\mathbf{x}^{\mathbf{a}}\partial^{\mathbf{a}} f \\
&= \sum_{\alpha_1,\alpha_2,\ldots,\alpha_k} x_{\alpha_1}x_{\alpha_2}\cdots x_{\alpha_n} \frac{\partial^k f}{\partial q_{\alpha_1}\partial q_{\alpha_2}\ldots\partial q_{\alpha_k}} \\
&= (\mathrm{D}^k_{\mathbf{q}^T} f)(\mathbf{x}^{\otimes k}\otimes \mathbf{I}_4)
\end{aligned} \tag{52}$$

evaluated at $\mathbf{q}$ (which is not indicated for sake simplicity), with the multi-index $\mathbf{a} = (a_1, \ldots, a_n) \in \mathbb{N}^n$ and $\mathbf{a}! = a_1!a_2!\cdots a_n!$.

The differentials of higher degree at $\mathbf{q} \in V$ admit the recursive relation[2]

$$\mathrm{d}^k f_{\mathbf{q}}(\mathbf{x}) = h^{(k)}_{\mathbf{q}}(\mathbf{x}) - \sum_{i=1}^{k-1}\binom{k-1}{i-1}\mathrm{d}^i f_{\mathbf{q}}(\mathbf{x})\, \mathrm{d}^{k-i} f^{-1}_{\mathbf{q}}(\mathbf{x}),\ \text{for}\ k>1 \tag{53}$$

where $\mathrm{d} f_{\mathbf{q}}(\mathbf{x}) = h^{(1)}_{\mathbf{q}}(\mathbf{x})$ for $k = 1$, and

$$h^{(k)}_{\mathbf{q}}(\mathbf{x}) := \sum_{i\leq n} x_i \mathrm{d}^{k-1}\widehat{\mathbf{S}}_{i,\mathbf{q}}(\mathbf{x}) \tag{54}$$

with the matrix $\widehat{\mathbf{S}} \in se(3)$ associated with the screw vector $\mathbf{S}$ (Murray et al. 1994). The differentials of the inverse $f^{-1}(\mathbf{q})$ admit the recursive relation

$$\mathrm{d}^k f^{-1}_{\mathbf{q}}(\mathbf{x}) = -\sum_{i=1}^{k}\binom{k}{i}\mathrm{d}^i f_{\mathbf{q}}(\mathbf{x})\, \mathrm{d}^{k-i} f^{-1}_{\mathbf{q}}(\mathbf{x}). \tag{55}$$

Also, the differentials of the joint screws $\mathbf{S}_i$ are given recursively by

$$\mathrm{d}^k\mathbf{S}_{i,\mathbf{q}}(\mathbf{x}) = \sum_{j\leq i}\sum_{r=0}^{k-1}\binom{k-1}{r}[\mathrm{d}^r\mathbf{S}_{j,\mathbf{q}}(\mathbf{x}), \mathrm{d}^{k-r-1}\mathbf{S}_{i,\mathbf{q}}(\mathbf{x})]x_j,\ k\geq 1. \tag{56}$$

In the above relations, it is $\mathrm{d}^k\mathbf{S}_{j,\mathbf{q}} = \mathbf{S}_j(\mathbf{q})$ and $\mathrm{d}^k f^{-1}_{\mathbf{q}} = f^{-1}(\mathbf{q})$, for $k = 0$. If the point of expansion is the reference configuration $\mathbf{q} = \mathbf{0}$, then $f^{-1}(\mathbf{0}) = \mathbf{I}$ and $\mathbf{S}_j(\mathbf{0}) = \mathbf{Y}_j$, and (51) is the Maclaurin series. Clearly, $\mathrm{d}^k f_{\mathbf{q}}(\mathbf{x})$, $\mathrm{d}^k f^{-1}_{\mathbf{q}}(\mathbf{x})$, $h^{(k)}_{\mathbf{q}}(\mathbf{x})$, $\mathrm{d}^k\mathbf{S}_{i,\mathbf{q}}(\mathbf{x})$ are homogenous polynomials of degree k.

[2] With slight abuse of notation, $\mathrm{d} f^{-1}_{\mathbf{q}}(\mathbf{x})$ denotes the differential of $f(\mathbf{q})^{-1} \in SE(3)$.

8.3 Examples

8.3.1 Planar Four-Bar Linkage with Identical Link Lengths

The Taylor series in (49) up to order 3 are

$$
\mathrm{d} f_{\mathbf{q}_0}(\mathbf{x}) = \begin{pmatrix} 0 & -x_1 - x_2 - x_3 - x_4 & 0 & 0 \\ x_1 + x_2 + x_3 + x_4 & 0 & 0 & -L(x_1 + 2x_2 + x_3) \\ 0 & 0 & 0 & 0 \\ 0 & 0 & 0 & 0 \end{pmatrix}
$$

$$
\mathrm{d} f_{\mathbf{q}_0}(\mathbf{x}) + \frac{1}{2}\mathrm{d}^2 f_{\mathbf{q}_0}(\mathbf{x}) = \begin{pmatrix} -\frac{1}{2}(x_1 + x_2 + x_3 + x_4)^2 & -x_1 - x_2 - x_3 - x_4 & 0 & \begin{matrix}\frac{1}{2}L(x_1^2 + 4x_2x_1 + 2x_2^2 \\ +x_3^2 + 2(x_1 + x_2)x_3)\end{matrix} \\ x_1 + x_2 + x_3 + x_4 & -\frac{1}{2}(x_1 + x_2 + x_3 + x_4)^2 & 0 & -L(x_1 + 2x_2 + x_3) \\ 0 & 0 & 0 & 0 \\ 0 & 0 & 0 & 0 \end{pmatrix}
$$

$$
\mathrm{d} f_{\mathbf{q}_0}(\mathbf{x}) + \frac{1}{2}\mathrm{d}^2 f_{\mathbf{q}_0}(\mathbf{x}) + \frac{1}{6}\mathrm{d}^3 f_{\mathbf{q}_0}(\mathbf{x}) =
$$

$$
\begin{pmatrix} -\frac{1}{2}(x_1 + x_2 + x_3 + x_4)^2 & \begin{matrix}\frac{1}{6}(x_1 + x_2 + x_3 + x_4)^3 \\ -x_1 - x_2 - x_3 - x_4\end{matrix} & 0 & \begin{matrix}\frac{1}{2}L(x_1^2 + 4x_2x_1 + 2x_2^2 \\ +x_3^2 + 2(x_1 + x_2)x_3)\end{matrix} \\ \begin{matrix}-\frac{1}{6}(x_1 + x_2 + x_3 + x_4)^3 \\ +x_1 + x_2 + x_3 + x_4\end{matrix} & -\frac{1}{2}(x_1 + x_2 + x_3 + x_4)^2 & 0 & \begin{matrix}\frac{1}{6}L(x_1^3 + 6x_2x_1^2 + 6x_2^2x_1 \\ +2x_2^3 + x_3^3 + 3(x_1 + x_2)x_3^2 \\ +3(x_1 + x_2)^2x_3 \\ -6(x_1 + 2x_2 + x_3))\end{matrix} \\ 0 & 0 & 0 & 0 \\ 0 & 0 & 0 & 0 \end{pmatrix}. \quad (57)
$$

This leads to $V_{\mathbf{q}_0}^1 = \ker \mathbf{J}(\mathbf{q}_0) = K_{\mathbf{q}_0}^1$, $V_{\mathbf{q}_0}^2 = V_{\mathbf{q}_0}^3 = C_{\mathbf{q}_0}^{\mathrm{K}} V = K_{\mathbf{q}_0}^2$, with $K_{\mathbf{q}_0}^1$ and $K_{\mathbf{q}_0}^2$ in (31). The c-space is thus locally the union of the two one-dimensional manifolds $\mathbf{V}(q_1 + q_3, q_2, q_4)$ and $\mathbf{V}(q_1 + q_2, q_3 + q_2, q_4 - q_2)$.

8.3.2 Planar Four-Bar Linkage with Identical Link Lengths

The analysis yields $V_{\mathbf{q}_0}^1 = \ker \mathbf{J}(\mathbf{q}_0) = K_{\mathbf{q}_0}^1$, $V_{\mathbf{q}_0}^2 = C_{\mathbf{q}_0}^{\mathrm{K}} V = K_{\mathbf{q}_0}^2$, with $K_{\mathbf{q}_0}^1$ and $K_{\mathbf{q}_0}^2$ in (33). The third-order approximation yields $V_{\mathbf{q}_0}^3 = V_{\mathbf{q}_0}^{3(I)} \cup V_{\mathbf{q}_0}^{3(II)}$ with $V_{\mathbf{q}_0}^{3(I)} = K_{\mathbf{q}_0}^{2(I)}$ and

$$
\begin{aligned}
V_{\mathbf{q}_0}^{3(II)} = \mathbf{V}(&-324x_1 + 3x_1^5 + 324x_3 - 3x_1^4x_3 + 36x_1x_3^2 - 48x_3^3 + 2x_1^2x_3^3 - 2x_1x_3^4 + 2x_3^5, \\
&288x_1 + 216x_2 + 24x_1^2x_2 + 18x_1^4x_2 + 24x_1x_2^2 + 44x_1^3x_2^2 + 42x_1^2x_2^3 + 18x_1x_2^4 + 3x_2^5, \\
&72x_1 - 12x_1^3 + 108x_4 - 24x_1^2x_4 + 3x_1^4x_4 - 12x_1x_4^2 + 2x_1^3x_4^2 - 6x_1^2x_4^3 + 6x_4^5)
\end{aligned}
$$

implicitly parameterized in terms of x_1, for instance.

8.3.3 Extended Watt Linkage—Underconstrained Kinematotropic Two-Loop Planar Linkage

The first-order approximation of the c-space is by definition $V^1_{\mathbf{q}_0} = \ker \mathbf{J}(\mathbf{q}_0) = K^1_{\mathbf{q}_0}$, with $K^1_{\mathbf{q}_0}$ in (36). The second-order differentials (52) of the constraint mappings (34) evaluated at $\mathbf{q}_0 = \mathbf{0}$ in terms of the screw coordinates (35) yield the second-order approximation

$$\begin{aligned} V^2_{\mathbf{q}_0} &= \{\mathbf{x} \in \mathbb{R}^8 |\ \mathrm{d} f_{1,\mathbf{q}_0}(\mathbf{x}) + \tfrac{1}{2}\mathrm{d}^2 f_{1,\mathbf{q}_0}(\mathbf{x}) = \mathbf{0}, \\ &\qquad \mathrm{d} f_{7,\mathbf{q}_0}(\mathbf{x}) + \tfrac{1}{2}\mathrm{d}^2 f_{7,\mathbf{q}_0}(\mathbf{x}) = \mathbf{0}\} \\ &= V^{2(I)}_{\mathbf{q}_0} \cup V^{2(II)}_{\mathbf{q}_0} \end{aligned} \tag{58}$$

where

$$V^{2(I)}_{\mathbf{q}_0} = \mathbf{V}(x_1 + x_3, x_2, x_4, x_5, x_6, x_7, x_8)$$

$$\begin{aligned} V^{2(II)}_{\mathbf{q}_0} = \{\mathbf{x} \in \mathbb{R}^8 | x_1 &= -\tfrac{1}{4}\left(\sqrt{2}+2\right) x_2, x_3 = -\tfrac{1}{4}\left(\sqrt{2}+2\right) x_2, \\ x_5 &= \tfrac{1}{8}(3(x_2-2)x_2 - 12x_4) \\ &\quad -\sqrt{24(x_2-2)x_2x_4 - 3x_2(x_2(x_2+2)(3x_2-2)+32) - 48{x_4}^2}, \\ x_6 &= \pm\tfrac{1}{4}\sqrt{24(x_2-2)x_2x_4 - 3x_2(x_2(x_2+2)(3x_2-2)+32) - 48{x_4}^2}, \\ x_7 &= \tfrac{1}{8}(4x_4 - x_2(3x_2+2) \\ &\quad -\sqrt{24(x_2-2)x_2x_4 - 3x_2(x_2(x_2+2)(3x_2-2)+32) - 48{x_4}^2}), \\ x_8 &= \tfrac{x_2}{\sqrt{2}}\}. \end{aligned}$$

$V^{2(I)}_{\mathbf{q}_0}$ is the second-order approximation of the manifold corresponding to the one-DOF motion mode, while $V^{2(II)}_{\mathbf{q}_0}$ approximates the two-DOF mode. The latter is parameterized in terms of the joint variables of joints 2 and 4. The term in the square root appearing for x_5, x_6, and x_7 must be positive. This is the condition that the four-bar chain formed by the joints 4, 5, 6, 7 cannot be stretched further. This relation can be seen by eliminating x_1, x_3, x_5, x_6 from the equations in (58), which gives the constraints in (58) in terms of x_2, x_4, and x_7 as

$$\begin{aligned} 0 &= \sqrt{2}x_2 - 2x_8 \\ 0 &= x_4\left(-24x_2^2 + 16x_2 - 32x_7\right) + 32x_4^2 + 9x_2^4 + 12x_2^3 \\ &\quad + 24x_2^2x_7 - 4x_2^2 + 16x_2x_7 + 48x_2 + 32x_7^2. \end{aligned}$$

These define the two-dimensional variety shown in Fig. 18. Clearly, the joint angle of joint 2 can only be negative, which means that the four-bar chain can only be folded.

The third-order analysis yields a local approximation $V^3_{\mathbf{q}_0} = V^{3(I)}_{\mathbf{q}_0} \cup V^{3(II)}_{\mathbf{q}_0}$. Here, $V^{3(I)}_{\mathbf{q}_0} = V^{2(I)}_{\mathbf{q}_0}$ since the one-DOF motion mode is globally defined by the linear rela-

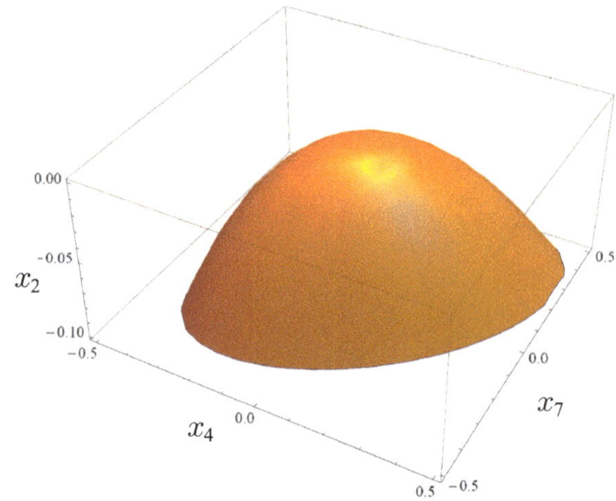

Fig. 18 x_2-x_4-x_7 section of the local approximation $V_{\mathbf{q}_0}^{2(II)}$ of the c-space of the extended Watt linkage

tions $x_1 = -x_3, x_2 = x_4 = x_5 = x_6 = x_7 = x_8 = 0$. The expression for the component $V_{\mathbf{q}_0}^{3(II)}$ is rather involved, and the higher-order analyses become very complicated. While all defining equations in (49) can be determined explicitly, the explicit resolution of the higher-order approximation becomes almost impossible.

8.3.4 Kinematic Singularity of a 6R Linkage with Smooth C-Space

With the joint screw coordinates (37), the first- and second-order local approximations of V are $V_{\mathbf{q}_0}^1 = K_{\mathbf{q}_0}^1$ and $V_{\mathbf{q}_0}^2 = K_{\mathbf{q}_0}^2 = C_{\mathbf{q}_0}^{\mathrm{K}}$. The c-space is thus locally up to second-order a vector space. Since $C_{\mathbf{q}_0}^{\mathrm{K}}$ is the tangent space to smooth motions, the c-space is locally a one-dimensional manifold.

In conclusion, this linkage exhibits a kinematic singularity at $\mathbf{q}_0$, while the c-space is a smooth manifold. This is an example where a constraint singularity is not a c-space singularity as conjectured by Zlatanov et al. (2002).

The local DOF of the linkage is $\delta(\mathbf{q}_0) = 1$, and the differential DOF is $\delta_{\mathrm{diff}}(\mathbf{q}_0) = 2$.

8.3.5 Double-Watt Linkage with Cusp Singularity in C-Space

The differentials of the constraint mappings (38) give rise to the first-order approximation $V_{\mathbf{q}_0}^1 = K_{\mathbf{q}_0}^1$, which is the two-dimensional vector space in (39) and the second-order approximation

$$
\begin{aligned}
V^2_{\mathbf{q}_0} = \mathbf{V}\big(\, & x_5 + x_6 - x_7 - x_8, -\tfrac{1}{2}(x_5 + x_6 - x_7 - x_8)^2, -x_5 - x_6 + x_7 + x_8, \\
& \tfrac{1}{2}(x_7 + x_8)(2x_5 + 2x_6 - x_7 - x_8) + x_6 - 2x_7 - 3x_8, \\
& x_5(x_6 - 2x_7 - 3x_8) + \tfrac{1}{2}(x_6^2 - 4x_6x_7 - 6x_6x_8 + 2x_7^2 + 6x_7x_8 \\
& -2x_7 + 3x_8^2 - 2x_8), \tfrac{1}{4}(-x_{10} - 4x_3 - 4x_4 - x_9), \\
& \tfrac{1}{2}(-3x_1 - x_{10} - 3x_2 - 2x_3 - 2x_4 - x_5 - x_6 - x_9), \\
& x_1 + x_{10} + x_2 + x_5 + x_6 + x_9, -\tfrac{1}{2}(x_1 + x_{10} + x_2 + x_5 + x_6 + x_9)^2, \\
& \tfrac{1}{4}(-3x_{10} + 6x_2 + 8x_3 + 12x_4 - 2x_6 + 3x_9), \\
& x_2 + \tfrac{3x_9}{2} - \tfrac{1}{4}(x_{10} + x_9)(2x_1 + x_{10} + 2x_2 + x_9) - \tfrac{3x_{10}}{2} - x_6, \\
& \tfrac{1}{4}(x_{10}(6x_1 + 6x_2 + 4x_6 + 6x_9 - 2) + x_1(-4x_2 + 4x_6 - 6x_9), \\
& +3x_{10}^2 - 2x_2^2 + 4x_2x_6 - 6x_2x_9 + 2x_6^2 + 4x_6x_9 - 3x_9^2 - 2x_9) \\
& \tfrac{1}{2}(x_1 + x_{10} + x_2 + x_5 + x_6 + x_9) + x_1 + x_2 + x_3 + x_4, \\
& -x_1 - x_{10} - x_2 - x_5 - x_6 - x_9\big).
\end{aligned}
\tag{59}
$$

This is a one-dimensional real algebraic variety. The analysis shows that the third- and higher-order approximations are all one-dimensional. Therewith, the local DOF is $\delta(\mathbf{q}_0) = \dim V^2_{\mathbf{q}_0} = 1$. The configuration $\mathbf{q}_0$ is a c-space singularity since $V^2_{\mathbf{q}_0}$ is not a smooth manifold (Lemma 8.1).

It must be remarked that the actual analysis of the algebraic variety $V^2_{\mathbf{q}_0}$ is difficult. In order to describe the one-dimensional variety, it is desirable to have at least an implicit parameterization. The reduction to a bivariate polynomial in terms of x_1 and x_5, for instance, required highly complex Gröbner basis computations.[3] This polynomial is of order 32.

It is instructive to relate the local c-space geometry to the possible motion through $\mathbf{q}_0$. The local analysis of V shows that the one-dimensional c-space has a cusp singularity at $\mathbf{q}_0$. Figure 19 shows the x_1-x_5 and x_1-x_2 sections of $V^2_{\mathbf{q}_0}$, and Fig. 20 the x_1-x_3 and x_1-x_4 sections. As described by Connelly and Servatius (1994), the

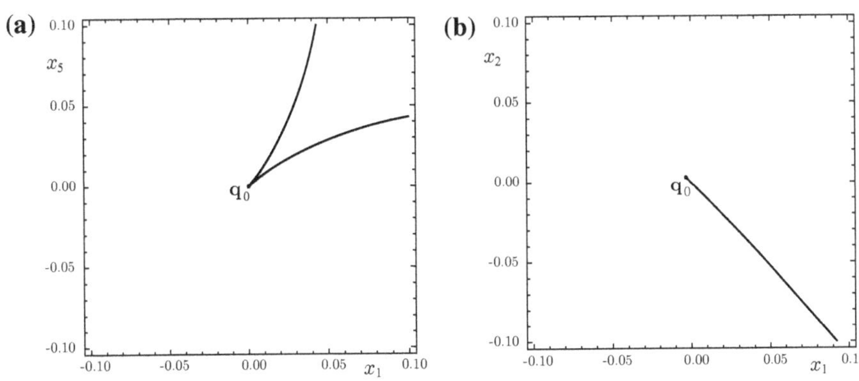

Fig. 19 Second-order approximation $V^2_{\mathbf{q}_0}$ of the c-space of the double-Watt linkage: **a** x_1-x_5 section and **b** x_1-x_2 section

[3]The author thanks Dr. Zijia Li for helping with the computation using the facilities of the Johann Radon Institute for Computational and Applied Mathematics (RICAM), Linz, Austria.

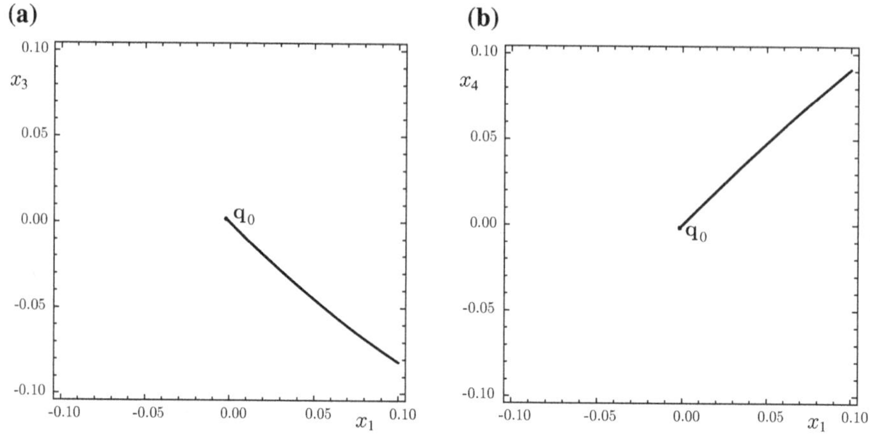

Fig. 20 Second-order approximation $V^2_{\mathbf{q}_0}$ of the c-space of the double-Watt linkage: **a** x_1-x_3 section and **b** x_1-x_4 section

finite (global) DOF is 1, and $\mathbf{q}_0$ is a cusp singularity of V. A cusp is a dead point so that there is no smooth curve passing through and there is hence no definable tangent. The linkage must stop in order to pass between the two branches of the cusp, hence $C^{\mathrm{K}}_{\mathbf{q}_0}V = \{\mathbf{0}\}$ (Sect. 5.4.5). As discussed by Connelly and Servatius (1994), in rigidity theory this linkage is called third-order rigid but not rigid since $K^3_{\mathbf{q}_0} = \{\mathbf{0}\}$. As apparent from this example, kth-order rigidity does not imply rigidity (the linkage may have to come to rest but is finitely mobile). On the other hand, in general, a linkage may be rigid but not kth-order rigid for some k (e.g., underconstrained and shaky mechanisms).

This example shows that the notion of a tangent space or tangent cone to V is not appropriate at cusps. The geometric tangent cone would simply be the line along the geometric union of the two branches of the curve, which would not indicate the fact that the linkage is at a dead point, nor that the motion curve through $\mathbf{q}_0$ has a cusp.

9 Local Analysis of Subvarieties of the C-Space of Points with Certain Rank

The methodology developed so far allows for investigating the rank at smooth curves through a given point. This does not yield a stratification of the c-space itself according to the rank since there may be points where there are no smooth curves passing (like cusps), and the kinematic tangent cone $C^{\mathrm{K}}_{\mathbf{q}}L_k$ does not give an exhaustive characterization of the subvarieties $L_k \cap V$ at such singularities.

9.1 Higher-Order Local Approximation

A νth-order local approximation of the set L_k at a point $\mathbf{q} \in L_k$, where the k-minor is $m_{\boldsymbol{\alpha\beta}}(\mathbf{q}) = 0$, is

$$L^{\nu}_{k,\mathbf{q}} = \{\mathbf{x} \in \mathbb{R}^n | \mathrm{d} f_{l,\mathbf{q}}(\mathbf{x}) + \frac{1}{2}\mathrm{d}^2 f_{l,\mathbf{q}}(\mathbf{x}) + \ldots + \frac{1}{\nu!}\mathrm{d}^{\nu} f_{l,\mathbf{q}}(\mathbf{x}) = \mathbf{0},$$
$$\mathrm{d}m_{\boldsymbol{\alpha\beta},\mathbf{q}}(\mathbf{x}) + \frac{1}{2}\mathrm{d}^2 m_{\boldsymbol{\alpha\beta},\mathbf{q}}(\mathbf{x}) + \ldots + \frac{1}{\nu!}\mathrm{d}^{\nu} m_{\boldsymbol{\alpha\beta},\mathbf{q}}(\mathbf{x}) = \mathbf{0},$$
$$|\boldsymbol{\alpha}| = |\boldsymbol{\beta}| = k, l = 1, \ldots, \gamma\} \tag{60}$$

with the differentials of f and $m_{\boldsymbol{\alpha\beta}}$, respectively. There is a neighborhood $U(\mathbf{q})$ and finite order κ so that

$$L^{\kappa}_{k,\mathbf{q}} \cap U(\mathbf{q}) = L_k \cap U(\mathbf{q}). \tag{61}$$

Definition 4.8 of kinematic singularities along with the local approximation of $V_{\mathbf{q}}$ and $L_{k,\mathbf{q}}$ gives rise to the following obvious necessary and sufficient condition.

Lemma 9.1 *A point* $\mathbf{q} \in V$ *is a kinematic singularity if and only if* $\dim V^{\kappa_1}_{\mathbf{q}} \neq \dim L^{\kappa_2}_{r+1,\mathbf{q}}$, *with* κ_1 *and* κ_2 *in* (50) *and* (61), *respectively, where* $r = \operatorname{rank} \mathbf{J}(\mathbf{q})$.

9.2 Higher-Order Series Expansion of Jacobian Minors

The expression for the time derivative (45) can be carried over directly giving rise to the differentials

$$\mathrm{d}^{\nu} m_{\boldsymbol{\alpha\beta}} = \sum_{|\mathbf{a}_k| = \nu} \left| \mathrm{d}^{a_1}\mathbf{S}_{\boldsymbol{\alpha}\beta_1,\mathbf{q}}(\mathbf{x})\ \mathrm{d}^{a_2}\mathbf{S}_{\boldsymbol{\alpha}\beta_2,\mathbf{q}}(\mathbf{x})\ \cdots\ \mathrm{d}^{a_k}\mathbf{S}_{\boldsymbol{\alpha}\beta_k,\mathbf{q}}(\mathbf{x}) \right| \frac{\nu!}{\mathbf{a}_k! n_1! \cdots n_{\nu}!}. \tag{62}$$

The differentials of $\mathbf{S}_i$ are determined by (56).

9.3 Examples

9.3.1 Extended Watt Linkage—Underconstrained Kinematotropic Two-Loop Planar Linkage

It turns out that $L^1_{6,\mathbf{q}_0} = K^{6,1}_{\mathbf{q}_0}$ in (46), and $L^k_{6,\mathbf{q}_0} = C^{\mathrm{K}}_{\mathbf{q}_0} L_6$ for $k \geq 2$. This confirms that points of rank 6 form a two-dimensional and points of rank 5 a one-dimensional manifold.

9.3.2 Kinematic Singularity of a 6R Linkage with Smooth C-Space

It has already been shown in Sect. 6.3.4 that $\mathbf{q}_0$ is a kinematic singularity, and in Sect. 8.3.4 that it is a regular point of V. The approximation of L_6 yields $L^2_{6,\mathbf{q}_0} = \ldots = C^{\mathrm{K}}_{\mathbf{q}_0} L_6 = C^{\mathrm{K}}_{\mathbf{q}_0} V$ and that of L_5 shows that $K^1_{5,\mathbf{q}_0} = \ldots = C^{\mathrm{K}}_{\mathbf{q}_0} L_5 = \{\mathbf{0}\}$. This reconfirms that $\mathbf{q}_0$ is a kinematic singularity.

9.3.3 Double-Watt Linkage with Cusp Singularity in C-Space

The first-order approximation is $L^1_{9,\mathbf{q}_0} = K^{9,1}_{\mathbf{q}_0}$ in (47), and the second-order approximation is $L^2_{9,\mathbf{q}_0} = \{\mathbf{0}\}$. Thus, $L^k_{9,\mathbf{q}_0} = \{\mathbf{0}\}, k \geq 2$, and the points of rank 8 form locally the zero-dimensional manifold $\{\mathbf{q}_0\}$, while points in the neighborhood have rank 9. Consequently, with Lemma 9.1, $\mathbf{q}_0$ is a kinematic singularity since $\dim L^2_{9,\mathbf{q}_0} \neq \dim V^2_{\mathbf{q}_0}$.

10 Discussion

This chapter was concerned with the basic notion of kinematic singularities of linkages and with computational approaches to investigate them symbolically and numerically. A configuration is a kinematic singularity if and only if the differential DOF, i.e., the rank of the constraint Jacobian $\mathbf{J}$, changes in any neighborhood of that point. Singularities of the c-space V (as topological set) are hence automatically kinematic singularities. The converse is not always true, however, since the c-space can be a smooth manifold even at a kinematic singularity. Possible configurations of a mechanism are represented by points in its c-space V. Finite motions of a mechanism are represented by curves in V, and smooth motions are represented by analytic curves in V. These rather trivial, but important, facts must be observed when investigating kinematic singularities.

Investigating kinematic singularities thus involves the analysis of

1. The local geometry of the c-space V
2. The local geometry of the subset in V of points of a certain rank
3. Possible smooth curves in c-space
4. Possible smooth curves in c-space with a certain rank.

The 6R linkage in Sect. 5.4.4 and the double-Watt linkage in Sect. 5.4.5 are two key examples of this chapter. The 6R linkage exhibits a kinematic singularity of rank 4, although its c-space is a smooth manifold. This singularity was identified using the kinematic tangent cone of V and of L_5. The kinematic tangent cone allows for detecting kinematic singularities if the c-space is the union of smooth manifolds (which applies to almost all linkages). The double-Watt linkage is exceptional in that it does not allow for smooth motions, as it is in a dead point at the considered singularity. The latter was identified as a c-space singularity (cusp) by means of a local approximation of the c-space.

It was the aim of this chapter to show that an exhaustive singularity analysis must address all of the above four points. The main statements that support such analysis can be summarized as follows:

- For V being a smooth manifold at $\mathbf{q}$, it is necessary that $C_{\mathbf{q}}^{\mathrm{K}}V$ is a vector space (Lemma 5.2).
- For $\mathbf{q} \in V$ being a kinematic singularity, it is sufficient that $C_{\mathbf{q}}^{\mathrm{K}}L_{r+1} \neq C_{\mathbf{q}}^{\mathrm{K}}V$, where $r = \mathrm{rank}\,\mathbf{J}\,(\mathbf{q})$ (Lemma 6.2).
- Necessary and sufficient for $\mathbf{q}$ being a kinematic singularity is that V and L_{r+1} have the same dimensions at $\mathbf{q}$, i.e., $\dim V_{\mathbf{q}}^{\kappa_1} \neq \dim L_{r+1,\mathbf{q}}^{\kappa_2}$, with κ_1 and κ_2 the sufficient approximation orders (Lemma 9.1) and $r = \mathrm{rank}\,\mathbf{J}\,(\mathbf{q})$.
- Necessary and sufficient for $\mathbf{q}$ being a c-space (and thus kinematic) singularity is that $V_{\mathbf{q}}^{\kappa}$ is not a smooth manifold (Lemma 8.1).

The only available necessary and sufficient condition involves analysis of the c-space geometry, which should be clear from the investigation of the double-Watt linkage. The somewhat unpleasant conclusion drawn from this example is that the mobility and singularity analysis may in fact require investigating the actual solution set V. This is alleviated using the approximating algebraic variety $V_{\mathbf{q}_0}^{\kappa}$. However, this too gets complicated very easily. An algebraic formulation of the constraints might be helpful to this end so that the c-space is an algebraic variety, which can be analyzed with algorithms from computation algebra. The latter have NP complexity, however, restricting their actual application. How to derive tailored formulations that may allow using these methods has been discussed by Manfred Husty in the previous chapter.

References

T. Arponen, A. Müller, S. Piipponen, J. Tuomela, Computational algebraic geometry and global analysis of regional manipulators. Appl. Math Comput. **232**, 820–835 (2004)

S. Bandyopadhyay, A. Ghosal, Analysis of configuration space singularities of closed-loop mechanisms and parallel manipulators. Mech. Mach. Theory **39**, 519–544 (2004)

J. Bochnak, M. Coste, M. Roy. *Real Algebraic Geometry*. (Springer, Heidelberg, 1998)

I. Bonev, C. Gosselin, Singularity analysis of 3-DOF planar parallel mechanisms via screw theory. ASME J. Mech. Des. **125**, 573–581 (2003)

O. Bottema, B. Roth. *Theoretical Kinematics*. (North Holland Press, Amsterdam, 1979)

R.W. Brockett, Robotic manipulators and the product of exponentials formula, in *Mathematical Theory of Networks and Systems*. Lecture Notes in Control and Information Sciences, vol. 58 (1984), pp. 120–129

W.H. Chai, Y. Chen, The line-symmetric octahedral bricard linkage and its structural closure. Mech. Mach. Theory **45**, 772–779 (2010)

C. Chen, The order of local mobility of mechanisms. Mech. Mach. Theory **46**, 1251–1264 (2011)

R. Connelly, H. Servatius, Higher-order rigidity-what is the proper definition? Discrete. Comput. Geom. **11**, 193–200 (1994)

S.K. Agrawal, Rate kinematics of in-parallel manipulator systems, in *IEEE International Conference on Robotics and Automation* (1990), pp. 104–109

I.F. de Bustos, J. Aguirrebeitia, R. Aviles, R. Ansola, Second order mobility analysis of mechanisms using closure equations. Meccanica **47**(7), 1695–1704 (2012)

J. Gallardo-Alvarado, J.M. Rico-Martinez, Jerk influence coefficients, via screw theory, of closed chains. Meccanica **36**, 213–228 (2001)

C.G. Gibson, *Singular Points of Smooth Mappings* (Pitman, London, 1979)

C.G. Gibson, *Elementary Geometry of Algebraic Curves* (Cambridge University Press, New York, 1998)

C.G. Gibson, K.H. Hunt, Geometry of screw systems-1: screws: genesis and geometry. Mech. Mach. Theory **25**(1), 1–10 (1990a)

C.G. Gibson, K.H. Hunt, Geometry of screw systems-2: classification of screw systems. Mech. Mach. Theory **25**(1), 11–27 (1990b)

M. Golubitsky, V. Guillemin, *Stable Mappings and Their Singularities* (Springer, New York, 1973)

C.M. Gosselin, J. Angeles, Singular analysis of closed-loop kinematic chains. IEEE Trans. Robot. Autom. **6**(3), 281–290 (1990)

J.M. Herve, Intrinsic formulation of problems of geometry and kinematics of mechanisms. Mech. Mach. Theory **17**(3), 179–184 (1982)

M. Husty, M. Pfurner, H.-P. Schröcker, K. Brunnthaler, Algebraic methods in mechanism analysis and synthesis. Robotica **25**(6), 661–675 (2007)

J. Kieffer, Differential analysis of bifurcations and isolated singularities of robots and mechanisms. IEEE Trans. Robot. Autom. **10**(1), 1–10 (1994)

J. Lerbet, Analytic geometry and singularities of mechanisms. ZAMM Z. Angew Math. Mech. **78**(10b), 687–694 (1999)

P.C.López-Custodio, A. Müller, J.M. Ricoand J.S. Dai, A synthesis method for 1-DOF mechanisms with a cusp in the configuration space. Mech. Mach. Theory (2018a). https://doi.org/10.1016/j.mechmachtheory.2018.09.008

P.C.López-Custodio, A. Müller, J.S. Dai, The double-koenigs mechanism–a spatial linkage with cusp singularities and multiple branches in the configuration space, in *4th IEEE/IFToMM International Conference on Reconfigurable Mechanisms and Robots (ReMAR)*, Delft, Netherlands, 20-22 June (2018b)

J.M. Rico Martinez, J. Duffy, Classification of screw systems-I. One- and two-systems. Mech. Mach. Theory **27**(4), 459–470 (1992a)

J.M. Rico Martinez, J. Duffy, Classification of screw systems-II. Three-systems. Mech. Mach. Theory **27**(4), 471–490 (1992b)

J.M. Rico Martinez, J. Gallardo, J. Duffy, Screw theory and higher order kinematic analysis of open serial and closed chains. Mech. Mach. Theory **34**(4), 559–586 (1999)

J.P. Merlet, Singularity configurations of parallel manipulators and Grassman geometry. Int. J. Robot. Res. **10**(2), 123–134 (1991)

A. Müller, Higher order local analysis of singularities in parallel mechanisms, in *27th Mechanisms and Robotics Conference, ASME Design Engineering Technical Conferences (DETC), Montreal, CA* (2002), pp. 515–522

A. Müller, Higher derivatives of the kinematic mapping and some applications. Mech. Mach. Theory **76**, 70–85 (2014)

A. Müller, Coordinate mappings for rigid body motions. ASME J. Comput. Nonlinear Dyn. **12**(2) (2016a)

A. Müller, Local kinematic analysis of closed-loop linkages-mobility, singularities, and shakiness. ASME J. Mech. Rob. **8** (2016b)

A. Müller, Higher-order constraints for linkages with lower kinematic pairs. Mech. Mach. Theory **100**, 33–43 (2016c)

A. Müller. A screw approach to the approximation of the local geometry of the configuration space and of the set of configurations of certain rank of lower pair linkages, in *42nd Mechanisms and Robotics Conference (MR)/ASME International Design Engineering Technical Conferences (IDETC), Quebec City, Canada, 26–29 August* (2018a)

A. Müller, Higher-order analysis of kinematic singularities of lower pair linkages and serial manipulators. ASME J. Mech. Rob. **10**(1) (2018b)

A. Müller, Topology, kinematics, and constraints of multi-loop linkages. Robotica **36**(11), 1641–1663 (2018c)

A. Müller, J.M. Rico, Mobility and higher order local analysis of the configuration space of single-loop mechanisms, in *11th International Symposium Advances in Robot Kinematics, Batz-sur-Mer, France* (Springer, Dordrecht 2008)

A. Müller, D. Zlatanov, Kinematic singularities of mechanisms revisited, in *IMA Conference on Mathematics of Robotics (IMAR), Oxford, UK, 9-11 September* (Kluwer Academic Publishers, 2015)

R.M. Murray, Z. Li, S.S. Sastry, *A Mathematical Introduction to Robotic Manipulation* (CRC Press, Boca Raton, 1994)

D.B. O'Shea, L.C. Wilson, Limits of tangent spaces to real surfaces. Am. J. Math. **126**(5), 951–980 (2004)

J. Selig, *Geometric Fundamentals of Robotics* (Springer, New York, 2005)

C.Y. Song, Y. Chen, A family of mixed double-goldberg 6R linkages. Proc. R. Soc. A **468**, 871–890 (2012)

W.H. Steeb, *Kronecker Products of Matrices and Applications* (BI-Wissenschaftsvlg, 1991)

Y. Takeda, Kinematic analysis of parallel mechanisms at singular points where a connecting chain has local mobility. Mech. Mach. Theory **41**, 945–957 (2006)

H. Whitney, Local properties of analytic varieties, in *Differential and Combinatorial Topology (A Symposium in Honor of M. Morse)* (Princeton University Press, Princeton, 1965)

K. Wohlhart, Degrees of shakiness. Mech. Mach. Theory **34**, 1103–1126 (1999)

K. Wohlhart, From higher degrees of shakiness to mobility. Mech. Mach. Theory **45**, 467–476 (2010)

D. Zlatanov, R.G. Fenton, B. Benhabib, A unifying framework for classification and interpretation of mechanism singularities. ASME J. Mech. Des. **117**(4), 566–572 (1995)

D. Zlatanov, R.G. Fenton, B. Benhabib, Identification and classification of the singular configurations of mechanisms. Mech. Mach. Theory **36**(6), 743–760 (1998)

D. Zlatanov, I.A. Bonev, C.M. Gosselin, Constraint singularity as c-space singularities, in *Advances, Robot Kinematics: Theory and Application, Caldes de Malavella, Spain* (Kluwer Academic Publishers, 2002), pp. 183–192

K. Wohlhart, Kinematotropic Linkages, in *Recent Advances in Robot Kinematics* (Kluwer, 1996), pp. 359–368

FSC
www.fsc.org
MIX
Papier | Fördert gute Waldnutzung
FSC® C083411

Kolibri 360 GmbH
Ferdinand-Jühlke-Straße 7
99095 Erfurt, Deutschland
produktsicherheit@kolibri360.de